Mathematical Cognition and Learning

Development of Mathematical Cognition

Neural Substrates and Genetic Influences

Mathematical Cognition and Learning

Series Editors

Daniel B. Berch
David C. Geary
Kathleen Mann Koepke

VOLUME 2

Development of Mathematical Cognition: Neural Substrates and Genetic Influences

Volume Editors

Daniel B. Berch
David C. Geary
Kathleen Mann Koepke

Mathematical Cognition and Learning

Development of Mathematical Cognition

Neural Substrates and Genetic Influences

Edited by

Daniel B. Berch
Curry School of Education
University of Virginia
Charlottesville, VA, USA

David C. Geary
Psychological Sciences
University of Missouri
Columbia, MO, USA

Kathleen Mann Koepke
Eunice Kennedy Shriver National Institute of Child
Health and Human Development (NICHD)
Bethesda, MD, USA

ELSEVIER

AMSTERDAM • BOSTON • HEIDELBERG • LONDON
NEW YORK • OXFORD • PARIS • SAN DIEGO
SAN FRANCISCO • SINGAPORE • SYDNEY • TOKYO
Academic Press is an imprint of Elsevier

Academic Press is an imprint of Elsevier
125 London Wall, London, EC2Y 5AS, UK
525 B Street, Suite 1800, San Diego, CA 92101–4495, USA
225 Wyman Street, Waltham, MA 02451, USA
The Boulevard, Langford Lane, Kidlington, Oxford OX5 1GB, UK

British Library Cataloguing in Publication Data
A catalogue record for this book is available from the British Library

Library of Congress Cataloging-in-Publication Data
A catalog record for this book is available from the Library of Congress

ISBN: 978-0-12-801871-2

For information on all Academic Press publications
visit our website at http://store.elsevier.com/

Printed and bound in the United States of America

Publisher: Nikki Levy
Acquisition Editor: Emily Ekle
Editorial Project Manager: Barbara Makinster
Production Project Manager: Caroline Johnson
Designer: Greg Harris

Contents

Contributors

Numbers in parenthesis indicate the pages on which the authors' contributions begin.

Daniel Ansari (27), Department of Psychology and Brain and Mind Institute, The University of Western Ontario, London, Ontario, Canada

Daniel B. Berch (1), Curry School of Education, UVA Cognitive Science Program, University of Virginia, Charlottesville, Virginia, USA

Ilaria Berteletti (109), Department of Psychology, University of Illinois, Champaign, Illinois, USA

James R. Booth (109), Department of Communication Sciences and Disorders, The University of Texas at Austin, Austin, Texas, USA

Bert De Smedt (219), Faculty of Psychology and Educational Sciences, University of Leuven, Leuven, Belgium

Wim Fias (195), Department of Experimental Psychology, Ghent University, Ghent, Belgium

David C. Geary (1), Department of Psychological Sciences, Interdisciplinary Neuroscience Program, University of Missouri, Columbia, Missouri, USA

Edward M. Hubbard (141), Department of Educational Psychology, University of Wisconsin, Madison, Wisconsin, USA

Daniel C. Hyde (51), Department of Psychology, University of Illinois Urbana-Champaign, Champaign, Illinois, USA

Roi Cohen Kadosh (245), Department of Experimental Psychology, University of Oxford, Oxford, UK

Kathleen Mann Koepke (1), NICHD/CDBB, Rockville, Maryland, USA

Yulia Kovas (299), Tomsk State University, Tomsk, Russia; Department of Psychology, Goldsmiths, University of London, and Social, Genetic and Developmental Psychiatry Centre, Institute of Psychiatry, King's College London, London, UK

Karin Kucian (165), Center for MR-Research and Children's Research Center, University Children's Hospital Zurich, Zurich, Switzerland

Mark Rose Lewis (141), Department of Educational Psychology, University of Wisconsin, Madison, Wisconsin, USA

Percival G. Matthews (141), Department of Educational Psychology, University of Wisconsin, Madison, Wisconsin, USA

Michèle M.M. Mazzocco (325), Institute of Child Development, and Center for Early Education and Development, University of Minnesota, Minneapolis, Minnesota, USA

Michael McCloskey (325), Department of Cognitive Science, Johns Hopkins University, Baltimore, Maryland, USA

Vinod Menon (79), Department of Psychiatry and Behavioral Sciences, Stanford University, Stanford, California, USA

Yi Mou (51), Department of Psychology, University of Illinois Urbana-Champaign, Champaign, Illinois, USA

Melissa M. Murphy (325), School of Education, Notre Dame of Maryland University, Baltimore, Maryland, USA

Stephen A. Petrill (299), Department of Psychology, The Ohio State University, Columbus, Ohio, USA

Andrea I. Quintero (325), MIND Institute, UC Davis Medical Center, and Department of Psychiatry and Behavioral Sciences, University of California at Davis, Davis, California, USA

Amar Sarkar (245), Department of Experimental Psychology, University of Oxford, Oxford, UK

Foreword

Poor numeracy is a serious handicap for individuals, more serious indeed than poor literacy, in determining life chances in a numerate society (Bynner & Parsons, 1997; Parsons & Bynner, 2005). Poor numerical skills in the population undermine the prosperity of society as a whole (OECD, 2010).

We need to understand not only what society provides to foster good numeracy, but also what the child brings to learning arithmetic, which we can think of as the learner's "starter kit." Therefore, understanding more precisely and more comprehensively the nature of the learner's starter kit in terms of its cognitive components, its neural basis, and its genetic origins is of great practical importance. This informs how we design learning contexts that optimize each learner's opportunity to become an effective citizen. This volume addresses these questions directly with new and exciting research focused on neural and genetic bases of arithmetical development.

One of the great virtues of Volume 1 in this series is that it draws attention to a domain-specific heritable component that could be part of the starter kit. Gelman and Gallistel, in their Foreword to Volume 1, introduce useful distinctions. They distinguish "our intuitive identification of number with countable quantity" from the conventional symbols we use to refer to these countable quantities, including counting words—one, two, three, and so on in English—and the familiar digits—1, 2, 3, ….

Now, of course, the human learner does not inherit conventional symbols. These are supplied by the culture, or rather most cultures. Some languages still exist that do not possess words for countable quantities, such as the Amazonian languages, Piraha (Gordon, 2004) and Munduruku (Pica, Lemer, Izard, & Dehaene, 2004), and most Australian languages (Butterworth, Reeve, Reynolds, & Lloyd, 2008; Dixon, 1980). Pinker, summarizing much research, concludes that conventional systems for counting "tend to be invented when a society develops agriculture, generates large quantities of indistinguishable objects, and needs to keep track of their exact magnitudes, particularly when they are taxed or traded." (Pinker, 2008, p. 138). That is, counting words probably date back to no longer than 10,000 years, and our familiar digits are very recent inventions, indeed, no more than 1500 years old (Menninger, 1969). This does not, of course, mean that without these words or other conventional symbols, humans lacked the capacity to represent and mentally manipulate countable quantities. The English philosopher, John Locke, observed that it is possible to conceive numbers for which one had no names. "The Tououpinambos [a Brazilian tribe] had no names for numbers above 5; any number beyond that, they made out by showing their fingers, and the others who were present." (Locke, 1690/1961, Book II Chapter XVI; based on explorer Jean de Léry's 1578 history of his voyage to Brazil).

Which cognitive tools are in the starter kit for arithmetical development? One does not have to go back far in the history of approaches to mathematical development to notice a radical difference from the content of these two volumes. There is Piagetian stress on the roles of perception and action in stages of development, and on sociocultural factors, especially in the nature of classroom interactions. Insofar as systematic individual differences are mentioned at all, it is often in terms of trait and knowledge differences, for example in discussions of "Aptitude-Treatment Interaction" [see Resnick and Ford (1981) for a brief discussion].

Differences in intelligence—as measured by a standardized IQ test—are occasionally invoked in research studies, though, for the layperson, including learners and their teachers, differences in mathematical ability are seen as a proxy for intelligence. This can be particularly harmful for the learner (Dweck & Leggett, 1988). Now, the relationship between intelligence and attainment is complex, as Petrill and Kovas (Chapter 11) point out. Studies quantify the genetic correlation of intelligence (*g*) to mathematical attainment at about 0.67. However, they found a similar correlation of mathematics achievement in measures of reading achievement, with a third of the genetic influence being specific to mathematics (Kovas, Harlaar, Petrill, & Plomin, 2005). More theoretically, Piaget and his colleagues (Piaget, 1952) have pointed to particular reasoning tools as necessary, and apparently sufficient, to acquire arithmetical competence: the ability to make valid transitive inferences, to understand classes and class inclusion, and to understand the reversibility of operations. The last is particularly important in solving number "conservation" tasks, which are a measure of the understanding of the concept of number. For Piagetians, individual differences could be defined in terms of the developmental stage reached by the individual.

Another way to look at this is to ask whether *inadequate* intelligence, at least as measured by IQ, prevents a child from acquiring arithmetical skills. Mazzocco et al. (Chapter 12) summarize the evidence as follows: "neurodevelopmental syndromes help demonstrate that IQ alone is neither the cause of MLD nor the sole source or variability in math performance – in syndrome groups or in the general population."

Adequate working memory capacity is frequently cited as an important component in the development of arithmetic, and indeed these capacities appear to underpin the mental manipulations required for calculation. Geary (1993), in an influential paper, proposes that Working Memory (WM) rehearsal is a necessary stage to forming long-term memories. The ability to retrieve simple number facts from long-term memory is an important stage in the development of arithmetical fluency (Chapter 4), and is the goal of educational policies. For example, the UK's national curriculum stresses "mathematical fluency," by which it means the ability to retrieve answers to simple sums and products from memory; for example, "By the end of year 2, pupils should know the number bonds to 20." Certainly, many studies have related individual differences in WM capacity and arithmetical attainment (e.g., De Smedt et al., 2009; Gathercole & Pickering, 2000; Gathercole, Pickering, Knight, & Stegmann, 2004). Kucian (Chapter 7) summarizes this position in the following way: "domain general, non-numerical skills ... are important for number processing, like working memory, attentional processes, or executive functions (planning, control, inhibition, monitoring, shifting) ... visuospatial working memory seems to be impaired in children with DD in particular." Deficits in these capacities, Kucian argues, can lead to dyscalculia—a topic to which I will return. Now Kucian distinguishes WM from attentional processes and executive functions, though many authors would include them as part of WM. In any case, these capacities are held to be domain-general rather than specific to the domain of number.

De Smedt (Chapter 9) reflects on the role of phonological processing. Although phonology is specific to language, language can be thought of as a domain-general resource for all kinds of thinking, reasoning, and memorization. Coding numerical information as words is important for acquiring and maintaining simple facts of addition and multiplication. The latter is frequently taught formally in schools through recitation. Dehaene and Cohen (1995) had already stressed phonological coding in the storage and retrieval of arithmetical facts in his "triple code" model, in which the parietal lobe, especially the angular gyrus, turns out to be engaged in both language processing and, in fact, retrieval tasks. De Smedt's own work in this area (De Smedt & Boets, 2010) has shown that "individual differences in phonological processing were uniquely related to problems that, in children, are likely to be retrieved from long-term memory, i.e., small additions, small subtractions and multiplications, but not to those that are likely to be solved by means of procedural strategies." Menon (Chapter 4) links the ability to retrieve arithmetic facts in relation to the hippocampus and the medial temporal lobe, the loci of long-term memory processing. Menon notes that, in the development of arithmetical competence, the role of these regions may change, with the role of the hippocampus becoming less important as other brain regions—the parietal lobes particularly—store facts and schemas as semantic memories.

Of course, the role of culture is crucial. The symbols we use to carry out numerical tasks, and which have taken several thousand years to develop, make mental life easier and more efficient. As the philosopher and mathematician, Alfred North Whitehead, noted, "By relieving the brain of all unnecessary work, a good notation sets it free to concentrate on more advanced problems, and in effect increases the mental power of the race." (Whitehead, 1948, p. 39).

Positional number notation is the example par excellence of a good notation. Ansari (Chapter 2) raises the critical question. "How do numerical symbols become representations of numerical quantity? It has frequently been hypothesized ... that symbols acquire numerical meaning by being connected to the approximate, non-symbolic representations of numerical quantity that humans share with other species and can be measured in very young infants." However, recent research, he observes, has cast doubt on a simple neural association in the parietal lobe between number symbols.

Many of the chapters in both this volume and the previous one, have sought to clarify the kind of domain-specific, non-verbal mechanisms we have inherited. Hyde and Mou (Chapter 3) point to two "core systems." One is concerned with the approximate numerosity of collections of objects, termed the Approximate Number System (ANS; sometimes as the Analogue Magnitude System), which can represent approximately the number of objects in a collection without an obvious limit. The other is the Object Tracking System (OTS) (or Parallel Individuation), which is limited to around four objects of attention (Feigenson, Dehaene, & Spelke, 2004). They review the evidence suggesting that these two systems are neurally distinct. These are mechanisms that are observable in human infants (Antell & Keating, 1983; Izard, Sann, Spelke, & Streri, 2009; Xu & Spelke, 2000), and which we share with other species (Agrillo & Beran, 2013; Agrillo, Piffer, Bisazza, & Butterworth, 2012). The OTS is held to underlie the process of "subitizing"—making an accurate estimate of one to four objects without serial enumeration, and it is proposed that the objects to be enumerated are held in visuo-spatial working memory so that the numerosities of small collections can be accurately estimated and compared (Carey, 2009; Le Corre & Carey, 2007; see also Chapter 3).

The question raised by Gelman and Gallistel in their Foreword to Volume 1 is whether either or both of these mechanisms are adequate to explain the development of numerical and arithmetical skills in humans.

Now it will come as no surprise to teachers of the first 3 years of school that children's numerical competence begins with whole numbers. However, these two foundational core systems do not involve whole numbers. The ANS comprises internal representations of different numerical magnitudes that can be thought of as Gaussian distributions of activation on a "mental number line." This is very different from the successor function used to characterize the natural (whole) numbers. It is typically tested by tasks involving clouds of dots (or other objects) typically too numerous to enumerate exactly in the time available. One common task is to compare two clouds of dots. (Addition and subtraction tasks for which the solution is compared with a third cloud of dots are also used.) Individual differences are described in terms of a psychophysical function, such as the Weber fraction, the smallest proportional difference between two clouds that can be reliably distinguished by the individual (Feigenson et al., 2004).

There has been considerable interest, indeed excitement, in many studies that show the performance on tasks designed to measure competence in the ANS—for example as a Weber fraction—correlates significantly with arithmetical performance in both children and adults (Barth et al., 2006; Gilmore, McCarthy, & Spelke, 2010; Halberda, Ly, Wilmer, Naiman, & Germine, 2012; Halberda, Mazzocco, & Feigenson, 2008).

But as we all know, correlation is not cause, and plausible mechanisms that have been proposed for the relationship are controversial. The problem is that the ANS deals only in approximate quantities, whereas ordinary school arithmetic deals with exact quantities, and the transition from approximations to whole number arithmetic still requires clarification.

The difficulty with the OTS is this: it has an upper limit of about 4. Now one key property of the number system, as Gelman and Gallistel point out in their Foreword to Volume 1, is that a valid operation on its elements always yields another element in the same system. If one such operation is addition, and if 3 is an element, then $3+3$ should yield an element in the system, but it cannot, since the limit is 4. To get round this, it has been proposed that noticing the number of objects being tracked can be linked to the number words children hear, and that they will be able to generalize—"bootstrap"—from these experiences to numbers above the limit (Carey, 2009; Le Corre & Carey, 2007). The key to bootstrapping is language, and in particular the counting words to get beyond the limits of the OTS. This idea is similar to one proposed by John Stuart Mill. He argued that our idea of "threeness" comes from generalizing from practical activities, in particular, from counting sets of objects using the counting words. That is we come to an understanding of threeness from many instances of counting or manipulating sets of three. In fact, he explicitly rejects the proposal that threeness or any other number is abstract. He writes, "All numbers must be numbers of something; there are no such things as numbers in the abstract. *Ten* must mean ten bodies, or ten sounds, or ten beatings of the pulse." This raises the problem of how the learner knows that the set of three coins has the same number as three wishes, except that they both involve counting to the word *three*.

While we do not doubt that these two core systems exist in the brain of human infants and other species, and one can imagine how making an approximate estimate can be helpful in some tasks, and Fuson has suggested that the ability to subitize can help the early learner check the accuracy of counting small sets (Fuson, 1988, 1992); nevertheless, it is hard to see how either can be foundational.

It has been argued that a quite different core system underlies the development of arithmetic, a mechanism that can represent the "numerosity" of a collection of objects; that is, the number of objects exactly, not approximately, up to a limit imposed by the developing brain. In a pioneering exploration, Gelman and Gallistel called these representations "numerons," and argued that learning to count is a process of learning how to map number words consistently onto numerons (Gelman & Gallistel, 1978).

A mechanism for the construction of numerons, and an equivalent system in other species, is the "accumulator" proposed by Meck and Church (Church & Meck, 1984; Meck & Church, 1983). The critical feature of the accumulator is a process of normalization such that each object to be counted is assigned the same representation or mental symbol. Meck and Church proposed a "gate" that opened to allow a fixed quantity of energy through to the accumulator for each object to be enumerated. Thus the number of objects is directly reflected in the quantity of energy in the accumulator, no matter what those objects are. A similar proposal can be found in the simulation by Dehaene and Changeux (1993) in which objects are normalized onto a "location map," such that each object has essentially the same representation, and is then summed by a "summation cluster." The models are in essence the same up to this point. Dehaene and Changeux map the state of the summation cluster onto the "numerosity cluster" which represents numbers as overlapping Gaussians on a vector (the ANS), while Meck and Church do not elaborate on the relationship of the accumulator to the identity of the numerosity. However, taking the accumulator a step further, Leslie, Gelman, and Gallistel (2008) show how a mapping can be done from the state of the accumulator to counting words and representations of numerosity. The virtue of the accumulator is that arithmetical operations are straightforward—accumulator contents can be added together, subtracted, and, as Leslie et al. show, multiplied.

A similar simulation can be found in work by Zorzi and his colleagues (Zorzi & Butterworth, 1997, 1999; Zorzi, Stoianov, & Umilta, 2005). Essentially their neural network starts with accumulator representations of numerical magnitude, but without the scalar noise in Meck and Church's model, and shows how it can model human addition performance better than a model with ANS-type representations.

One of the very interesting and important issues raised in this volume is why some children have such great trouble in learning arithmetic. This issue has been explored by the volume's editors previously in important and influential publications (Berch, 2005; Berch & Mazzocco, 2007; Geary et al., 2009; Mann Koepke & Miller, 2013). The developmental trajectory of individual arithmetical attainment is affected by external factors even before the learner encounters formal education. These include socioeconomic status, parental education, home learning environment and school effectiveness (Melhuish et al., 2008), all of which may impact both domain-general and domain-specific cognitive components.

There is also evidence that individual differences in various measures of numerosity processing, at least in older children, are key to understanding differences in arithmetical development (Geary et al., 2009; Piazza et al., 2010; Reeve, Reynolds, Humberstone, & Butterworth, 2012). Twin studies suggest that these cognitive differences appear to be at least partly genetic. Petrill and Kovas (Chapter 11) provide a comprehensive overview of a wide range of twin studies, showing a specific genetic factor in individual differences in mathematical ability, but also genetic overlap between mathematics and other cognitive abilities, for example between mathematics attainment and *g*. However, "molecular genetic studies to date have had difficulty elucidating specific replicable markers involved in mathematical variation."

Kucian (Chapter 7) seeks to characterize differences found in the brain of learners with developmental dyscalculia, a congenital but not always inherited disorder. She notes that DD is a serious handicap that can have "a negative impact on the employment prospects and mental and physical health of individuals, and on the economic status of countries." She stresses both the heterogeneity of the condition, and of the observed differences in the brains of those diagnosed with DD. Like many neurodevelopmental disorders, DD is often co-morbid with other neurodevelopmental disorders such as dyslexia, and this, Kucian says, is the rule rather than the exception, and studies should not be confined to cases of "pure" DD.

The genetic factor is reinforced by the finding that certain kinds of genetic anomaly, such as Turner Syndrome, affects numerical abilities, including very basic abilities such as selecting the larger of two numbers, or giving the number of dots in an array, even when general cognitive ability is normal, or even superior (Bruandet, Molko, Cohen, & Dehaene, 2004; Butterworth et al., 1999; Temple & Marriott, 1998). Findings like these suggest that it is the inherited domain-specific mechanism that is affected in Turner Syndrome, and is supported by findings of anomalies in the parietal lobes (Molko et al., 2003).

Mazzocco, Quintero, Murphy, and McCloskey (Chapter 12) show that careful analysis of the cognitive deficits, and assets of individuals with genetic anomalies, can be informative about the structure of normal arithmetical development. They ask

> *"how can we systematically study variation in a latent variable (i.e., MLD [mathematics learning disability]) for which there is no widely accepted diagnostic marker or working definition? One approach is to study known neurodevelopmental disorders associated with poor mathematics achievement. In this chapter we exemplify this approach by summarizing research on the observable intellectual characteristics (or cognitive phenotypes) of individuals with fragile X, Turner, or chromosome 22q11.2 deletion syndrome."*

Because each of these conditions is characterized by a specific mix of cognitive strengths and weaknesses, it is possible to use systematic analyses of associations and dissociations to evaluate their causal roles in MLD. To take one striking example: to what extent is verbal memory sufficient for typical numerical skills? "Girls with fragile X syndrome are far more accurate than their same-age MLD peers at oral number tasks such as reading numbers, counting aloud from 1, counting backwards, or skip counting (e.g., counting by tens; Murphy, Mazzocco, Gerner, & Henry, 2006). In fact, girls with fragile X perform nearly as well if not better than their peers *without MLD*, on all of these tasks. They are, however, as impaired as their MLD peers on more conceptual numerical tasks such as verbal magnitude comparison (reporting which of two numbers is more), identifying a specific ordinal position (e.g., identifying the 4th person in line), and using one-to-one correspondence when counting despite facility at counting aloud forwards or backwards, and skip counting." Turner Syndrome females can elucidate the causal link between intelligence and arithmetical competence: these individuals are in the normal and often superior range on FSIQ, yet they are invariably MLD, and, when tested, show deficits on very simple numerical tasks such as number comparison and dot enumeration. These dissociations point to deficiencies in core numerical competence at the root of their MLD.

No account of the development of number processing and arithmetic can avoid a discussion of education. Berteletti and Booth (Chapter 5) emphasize the role of fingers in the development of numerical competence and the way in which culture, including formal education, shapes the way in which fingers are used. Fingers have multiple neural representations, but the parietal lobe is a region that supports both finger and number processes. In Gerstmann's Syndrome, damage to the left parietal lobe affects both the representation of fingers and also calculation ability, and there are now neuroimaging studies of calculation tasks showing the activation of finger-relevant regions outside the parietal, for example, in motor cortex. Rusconi, Walsh, and Butterworth (2005) were the first to demonstrate that transcranial magnetic stimulation of this region impaired both finger responses and arithmetical fact retrieval. Berteletti and Booth review the close connection between early arithmetical counting strategies and finger use by children. What is lacking in our understanding of the connection between fingers and numbers is a longitudinal study tracking the onset of finger use, and its offset. This could shed light on a strong embodied cognition account in which finger use is a necessary stage in typical arithmetical development. What about children who do not use their fingers? Are they better or worse at arithmetic? Do culturally-shaped styles of finger counting affect arithmetical efficiency? Should we teach children how to use their fingers? Should we prevent them using their fingers to encourage more sophisticated arithmetical strategies? These are questions that teachers ask, and we should try to answer.

Sarkar and Cohen Kadosh (Chapter 10) describe some of their own pioneering work in the use of transcranial electrical stimulation (tES) to improve learning numerical abilities in typical learners and in dyscalculics, and also those with mathematics anxiety. Electrical stimulation affects neuronal excitability via critical neurotransmitters, such as GABA, that are involved in learning and memory. The idea here is that the more excitable the neurons are the more readily the brain can be modified and learn.

Sarkar and Cohen Kadosh document many studies now showing long-term improvements in numerical and arithmetical skills using combined training regimes with various kinds of electrical stimulation. They also show how this technique can modulate one crucial test of the ANS, a candidate core capacity for learning arithmetic. This study trained participants on the standard ANS task that required selecting the greater of two arrays of dots. Training improved discrimination (Weber fraction), but transcranial random noise stimulation dramatically improved discrimination up to 16 weeks post-training. Moreover, training transferred to proficiency in discriminating continuous time and space stimuli. Thus, there was both improvement and transfer. As noted above, many studies have shown a correlation between Weber fraction and arithmetic. Therefore, one might expect the improvement in the Weber fraction to transfer to an improvement in arithmetic accuracy or speed. However, this was not observed (Cappelletti et al., 2013).

The studies of significant improvements in arithmetic, not just in mathematics, but also in planning, could clearly have important consequences for education. Sarkar and Cohen Kadosh are well aware of the issues surrounding the transfer of a lab-validated technique into the classroom. They also note that improvements in one domain may have costs in another. Thus in one study, they found "a double-dissociation between two key aspects of learning: the memorization and recognition of stimuli, and the automaticity with which knowledge of those stimuli are accessed .... Therefore, there were costs in both cases: enhanced learning and recognition came at the cost of impaired automaticity, and enhanced automaticity came at the cost of impaired learning and recognition" (Iuculano & Cohen Kadosh, 2013). Even though tES involves delivery of a weak electrical current, usually in the range of 1-2 mA, there will be safety concerns when the technique falls into poorly trained hands.

Volume 2 develops themes propounded so powerfully in Volume 1. Both volumes have stressed the importance of an innate domain-specific capacity as a key tool in the learner's "starter kit," but this volume also raises important challenges for the future (for Volume 3?).

One future focus, touched on in this volume, concerns individual differences in development. First, we need to have a clearer picture of the way in which distinct cognitive tools, both domain-general and domain-specific, interact with each other. For example, can good intelligence and good working memory compensate for poor or defective domain-specific capacity? Are there reliable neural differences between the good arithmetic learners and the poor? If so, are the differences in grey matter, white matter, patterns of activity, resting state, functional connectivity, or indeed some new way of analyzing brain structure and function? To what extent are individual differences shaped by genetic factors, and, if so, which genes and their alleles are primarily responsible?

These questions inevitably lead to issues in education. How can we make learning contexts more effective for the unique mix of capacities and experience of each individual learner? How can we make best use of digital technologies in dealing with individual differences? Can good teaching compensate for the congenital handicap of developmental dyscalculia? If so, what should be the teaching targets and what should be the pedagogical approach?

I hope these introductory remarks will help the reader appreciate the theoretical and practical importance of work described in the following chapters.

Brian Butterworth

REFERENCES

Agrillo, C., & Beran, M. J. (2013). Number without language: Comparative psychology and the evolution of numerical cognition. *Frontiers in Psychology*, *4*, 295.

Agrillo, C., Piffer, L., Bisazza, A., & Butterworth, B. (2012). Evidence for two numerical systems that are similar in humans and guppies. *PLoS ONE*, *7*(2), e31923. http://dx.doi.org/10.1371/journal.pone.0031923.

Antell, S. E., & Keating, D. P. (1983). Perception of numerical invariance in neonates. *Child Development*, *54*, 695–701.

Barth, H., La Mont, K., Lipton, J., Dehaene, S., Kanwisher, N., & Spelke, E. (2006). Non-symbolic arithmetic in adults and young children. *Cognition*, *98*(3), 199–222.

Berch, D. B. (2005). Making sense of number sense: Implications for children with mathematical disabilities. *Journal of Learning Disabilities*, *38*(4), 333–339.

Berch, D. B., & Mazzocco, M. M. M. (Eds.), (2007). *Why is math so hard for some children? The nature and origins of mathematical learning difficulties and disabilities*. Baltimore, MD: Paul H. Brookes Publishing Co.

Bruandet, M., Molko, N., Cohen, L., & Dehaene, S. (2004). A cognitive characterization of dyscalculia in Turner syndrome. *Neuropsychologia*, *42*, 288–298.

Butterworth, B., Granà, A., Piazza, M., Girelli, L., Price, C., & Skuse, D. (1999). Language and the origins of number skills: Karyotypic differences in Turner's syndrome. *Brain & Language*, *69*, 486–488.

Butterworth, B., Reeve, R., Reynolds, F., & Lloyd, D. (2008). Numerical thought with and without words: Evidence from indigenous Australian children. *Proceedings of the National Academy of Sciences of the United States of America*, *105*, 13179–13184. http://dx.doi.org/10.1073/pnas.0806045105.

Bynner, J., & Parsons, S. (1997). *Does numeracy matter?*. London: The Basic Skills Agency.

Cappelletti, M., Gessaroli, E., Hithersay, R., Mitolo, M., Didino, D., Kanai, R., et al. (2013). Transfer of cognitive training across magnitude dimensions achieved with concurrent brain stimulation of the parietal lobe. *The Journal of Neuroscience*, *33*(37), 14899–14907. http://dx.doi.org/10.1523/jneurosci.1692-13.2013.

Carey, S. (2009). Where our number concepts come from. *Journal of Philosophy*, *106*(4), 220–254.

Church, R., & Meck, W. (1984). The numerical attribute of stimuli. In H. L. Roitblatt, T. Bever, & H. Terrace (Eds.), *Animal cognition*. Hillsdale, NJ: Erlbaum.

De Smedt, B., & Boets, B. (2010). Phonological processing and fact retrieval: Evidence from developmental dyslexia. *Neuropsychologia*, *48*, 3973–3981.

De Smedt, B., Janssen, R., Bouwens, K., Verschaffel, L., Boets, B., & Ghesquière, P. (2009). Working memory and individual differences in mathematics achievement: A longitudinal study from first grade to second grade. *Journal of Experimental Child Psychology*, *103*(2), 186–201.

Dehaene, S., & Changeux, J.-P. (1993). Development of elementary numerical abilities: A neuronal model. *Journal of Cognitive Neuroscience*, *5*, 390–407.

Dehaene, S., & Cohen, L. (1995). Towards and anatomical and functional model of number processing. *Mathematical Cognition*, *1*, 83–120.

Dixon, R. M. W. (1980). *The languages of Australia*. Cambridge: Cambridge University Press.

Dweck, C. S., & Leggett, E. L. (1988). A social-cognitive approach to motivation and personality. *Psychological Review*, *95*(2), 256–273. http://dx.doi.org/10.1037/0033-295X.95.2.256.

Feigenson, L., Dehaene, S., & Spelke, E. (2004). Core systems of number. *Trends in Cognitive Sciences*, *8*(7), 307–314.

Fuson, K. C. (1988). *Children's counting and concepts of number*. New York: Springer Verlag.

Fuson, K. C. (1992). Relationships between counting and cardinality from age 2 to 8. In J. Bideaud, C. Meljac, & J. P. Fisher (Eds.), *Pathways to number, children's developing numerical abilities*. Hillsdale, NJ: LEA.

Gathercole, S. E., & Pickering, S. J. (2000). Working memory deficits in children with low achievements in the national curriculum at 7 years of age. *British Journal of Educational Psychology*, *70*(2), 177–194.

Gathercole, S. E., Pickering, S. J., Knight, C., & Stegmann, Z. (2004). Working memory skills and educational attainment: Evidence from national curriculum assessments at 7 and 14 years of age. *Applied Cognitive Psychology*, *18*(1), 1–16. http://dx.doi.org/10.1002/acp.934.

Geary, D. C. (1993). Mathematical disabilities: Cognition, neuropsychological and genetic components. *Psychological Bulletin*, *114*, 345–362.

Geary, D. C., Bailey, D. H., Littlefield, A., Wood, P., Hoard, M. K., & Nugent, L. (2009). First-grade predictors of mathematical learning disability: A latent class trajectory analysis. *Cognitive Development*, *24*, 411–429.

Gelman, R., & Gallistel, C. R. (1978). *The child's understanding of number* (1986 ed.). Cambridge, MA: Harvard University Press.

Gilmore, C. K., McCarthy, S. E., & Spelke, E. S. (2010). Non-symbolic arithmetic abilities and mathematics achievement in the first year of formal schooling. *Cognition*, *115*(3), 394–406.

Gordon, P. (2004). Numerical cognition without words: Evidence from Amazonia. *Science*, *306*, 496–499.

Halberda, J., Ly, R., Wilmer, J. B., Naiman, D. Q., & Germine, L. (2012). Number sense across the lifespan as revealed by a massive Internet-based sample. *Proceedings of the National Academy of Sciences of the United States of America*, *109*(28), 11116–11120. http://dx.doi.org/10.1073/pnas.1200196109.

Halberda, J., Mazzocco, M. M. M., & Feigenson, L. (2008). Individual differences in non-verbal number acuity correlate with maths achievement. *Nature*, *455*, 665–668. http://dx.doi.org/10.1038/nature07246.

Iuculano, T., & Cohen Kadosh, R. (2013). The mental cost of cognitive enhancement. *The Journal of Neuroscience*, *33*(10), 4482–4486. http://dx.doi.org/10.1523/jneurosci.4927-12.2013.

Izard, V., Sann, C., Spelke, E. S., & Streri, A. (2009). Newborn infants perceive abstract numbers. *Proceedings of the National Academy of Sciences of the United States of America*, *106*(25), 10382–10385.

Kovas, Y., Harlaar, N., Petrill, S. A., & Plomin, R. (2005). 'Generalist genes' and mathematics in 7-year-old twins. *Intelligence*, *33*(5), 473–489.

Le Corre, M., & Carey, S. (2007). One, two, three, four, nothing more: An investigation of the conceptual sources of the verbal counting principles. *Cognition*, *105*(2), 395–438. http://dx.doi.org/10.1016/j.cognition.2006.10.005.

Leslie, A. M., Gelman, R., & Gallistel, C. R. (2008). The generative basis of natural number concepts. *Trends in Cognitive Sciences*, *12*(6), 213–218. http://dx.doi.org/10.1016/j.tics.2008.03.004.

Locke, J. (1690/1961). *An essay concerning human understanding* (Based on Fifth Edition, edited by J. W. Yolton). London: J. M. Dent.

Mann Koepke, K., & Miller, B. (2013). At the intersection of math and reading disabilities: Introduction to the special issue. *Journal of Learning Disabilities*, *46*(6), 483–489. http://dx.doi.org/10.1177/0022219413498200.

Meck, W. H., & Church, R. M. (1983). A mode control model of counting and timing processes. *Journal of Experimental Psychology: Animal Behavior Processes*, *9*(4), 320–334.

Melhuish, E. C., Sylva, K., Sammons, P., Siraj-Blatchford, I., Taggart, B., Phan, M. B., et al. (2008). Preschool influences on mathematics achievement. *Science*, *321*, 1161–1162.

Menninger, K. (1969). *Number words and number symbols: A cultural history of numbers* (P. Broneer, Trans.). Cambridge, MA: MIT Press.
Molko, N., Cachia, A., Rivière, D., Mangin, J.-F., Bruandet, M., Le Bihan, D., et al. (2003). Functional and structural alterations of the intraparietal sulcus in a developmental dyscalculia of genetic origin. *Neuron*, *40*, 847–858.
Murphy, M. M., Mazzocco, M. M. M., Gerner, G., & Henry, A. E. (2006). Mathematics learning disability in girls with Turner syndrome or fragile X syndrome. *Brain and Cognition*, *61*, 195–210.
OECD. (2010). *The high cost of low educational performance. The long-run economic impact of improving educational outcomes.*
Parsons, S., & Bynner, J. (2005). *Does numeracy matter more?*. London: National Research and Development Centre for Adult Literacy and Numeracy, Institute of Education.
Piaget, J. (1952). *The child's conception of number*. London: Routledge & Kegan Paul.
Piazza, M., Facoetti, A., Trussardi, A. N., Berteletti, I., Conte, S., Lucangeli, D., et al. (2010). Developmental trajectory of number acuity reveals a severe impairment in developmental dyscalculia. *Cognition*, *116*(1), 33–41.
Pica, P., Lemer, C., Izard, V., & Dehaene, S. (2004). Exact and approximate calculation in an Amazonian indigene group with a reduced number lexicon. *Science*, *306*, 499–503.
Pinker, S. (2008). *The stuff of thought. Language as a window into human nature*. London: Penguin Books.
Reeve, R., Reynolds, F., Humberstone, J., & Butterworth, B. (2012). Stability and change in markers of core numerical competencies. *Journal of Experimental Psychology: General*, *141*(4), 649–666. http://dx.doi.org/10.1037/a0027520.
Resnick, L. B., & Ford, W. W. (1981). *The psychology of mathematics for instruction*. Hillsdale, NJ: Lawrence Erlbaum Associates.
Rusconi, E., Walsh, V., & Butterworth, B. (2005). Dexterity with numbers: rTMS over left angular gyrus disrupts finger gnosis and number processing. *Neuropsychologia*, *43*(11), 1609–1624.
Temple, C. M., & Marriott, A. J. (1998). Arithmetical ability and disability in Turner's syndrome: A cognitive neuropsychological analysis. *Developmental Neuropsychology*, *14*, 47–67.
Whitehead, A. N. (1948). *An introduction to mathematics* (Originally published in 1911 ed.). London: Oxford University Press.
Xu, F., & Spelke, E. (2000). Large number discrimination in 6-month-old infants. *Cognition*, *74*, B1–B11.
Zorzi, M., & Butterworth, B. (1997). On the representation of number concepts. In M. G. Shafto & P. Langley (Eds.), *Proceedings of the nineteenth annual conference of the cognitive science society*. Mahwah, NJ: LEA.
Zorzi, M., & Butterworth, B. (1999). A computational model of number comparison. In *Paper presented at the proceedings of the twenty first annual meeting of the cognitive science society.*
Zorzi, M., Stoianov, I., & Umilta, C. (2005). Computational modelling of numerical cognition. In J. I. D. Campbell (Ed.), *Handbook of mathematical cognition* (pp. 67–84). Hove: Psychology Press.

Preface

Not a week goes by where there isn't at least one story in the news about mathematics learning, whether it's about the latest National Assessment of Educational Progress (NAEP) scores in the U.S., comparisons of performance by different countries on international tests such as the Trends in International Mathematics and Science Study (TIMSS) or the Program for International Student Assessment (PISA), or debates in op-ed pieces concerning the merits or shortcomings of the U.S. Common Core State Standards in Mathematics. Clearly, the myriad of opinions proffered by educators, administrators, teachers, journalists, and bloggers concerning the continuing difficulties experienced by many children in learning mathematics indicates there is still much to be learned. The science of the typical and atypical development of mathematical cognition and precisely what kinds of instructional interventions to improve mathematical fluency and achievement are best suited for each type of learner have yet to achieve a level of clarity that can settle the debates.

Although admirable progress has certainly been made with respect to our understanding of the basic cognitive operations, mental representations, and solution strategies used by children in various kinds of numerical and mathematical tasks, it is only fairly recently that the use of brain imaging tools have begun to reveal the neural underpinnings of developmental changes in mathematical thinking and learning. Furthermore, findings emerging from this research help to constrain cognitive theories of mathematical development, ensuring they remain neurobiologically plausible. Likewise, advances in molecular genetic methods when combined with more traditional quantitative genetic approaches are gradually providing us with a unique window on the relative contributions of heredity, shared environments, and nonshared environments to individual differences in mathematical ability and disability. The current volume, the second in the Mathematical Cognition and Learning Series, encompasses a unique blend of chapters that cover the landscape of these two domains. Taken together, they offer the reader not only comprehensive and insightful reviews of contemporary research in each of these domains, but also several nascent, neurocognitive models of numerical and arithmetic processing, along with the implications of this work for improving mathematics instruction. As such, this volume should be of particular interest to researchers and students in cognitive neuroscience, behavioral genetics, developmental psychology, educational psychology, special education, and mathematics education.

Although selected special issues of scholarly journals have occasionally covered brain imaging studies of children's mathematical development, no other edited books to date have focused exclusively on this topic nor included relevant work from behavioral and molecular genetics. Our volume aims to present the major biological evidences (genetic and neurobiologic) of mathematical development by including strong contributions from a group of leading North American and European researchers and scholars, as well as rising stars in the field who have already made important contributions to the literature. The resulting collection of chapters takes the reader on an interesting, unique, and informative tour of exciting ways in which contemporary as well as innovative brain imaging techniques and cutting-edge findings from modern genetic and genomic methods are being used to inform and illuminate our understanding of and ability to shape mathematical cognitive development through environmental and biologic means.

At issue in these pages are several theoretical debates, including how numerical symbols are linked to nonsymbolic representations of numerical quantity and how tightly these representations are connected in the brain (Chapter 2); how the malleable brains of children recruit different regions that will ultimately establish the mature neural system of arithmetic fact retrieval found in most adults (Chapter 4); what neurocognitive architecture is best suited to learning fractions (Chapter 6); whether mathematical proficiency is determined by a single unique, underlying cognitive factor or by multiple cognitive components such as memory, spatial processing, or executive function (Chapter 8); to what extent, if any, exogenous electrical stimulation of the brain can enhance numerical cognition and learning (Chapter 10); and the degree to which children presenting with different neurodevelopmental disorders differ in the specificity of their numerical and mathematical impairments (Chapter 12).

Beyond these kinds of issues, the other chapters deal with such matters as distinct ways in which two core, cognitive and brain systems contribute to symbolic numerical and mathematical development (Chapter 3); the use of fingers in the development of numerical representations and arithmetic processing (Chapter 5); evidence suggesting that at least some forms of developmental dyscalculia are associated with alterations in brain structure and function as well as metabolism and fiber connections in the parietal cortex (Chapter 7); the possible role of phonological processing and the recruitment of a widespread brain network in children's developing, arithmetic fact-retrieval skills (Chapter 9); and how both genes and environments influence the origins and development of mathematical ability and disability (Chapter 11). In the introductory chapter, we discuss some of the fundamental issues concerning the interpretation of findings from brain imaging and behavioral genetics research, and how these play out in the study of mathematical cognitive development.

Finally, we thank the Child Development and Behavior Branch of the *Eunice Kennedy Shriver* National Institute of Child Health and Human Development

(NICHD), NIH for the principal funding to support the conference on which this volume is based, and we are likewise grateful to the Institute of Education Sciences (IES) of the U.S. Department of Education for their continuing monetary and substantive support of this conference series.

Daniel B. Berch
Davis C. Geary
Kathleen Mann Koepke

Chapter 1

Introduction: How the Study of Neurobiological and Genetic Factors Can Enhance Our Understanding of Mathematical Cognitive Development

Daniel B. Berch[1], David C. Geary[2], Kathleen Mann Koepke[3]
[1]Curry School of Education, UVA Cognitive Science Program, University of Virginia, Charlottesville, Virginia, USA, [2]Department of Psychological Sciences, Interdisciplinary Neuroscience Program, University of Missouri, Columbia, Missouri, USA, [3]NICHD/CDBB, Rockville, Maryland, USA

INTRODUCTION

Our understanding of developmental changes in mathematical cognition has advanced significantly over the past 40 plus years. A variety of innovative tasks have been used to study the intuitive number sense of preverbal human infants, toddlers, and preschoolers (see Geary, Berch, & Mann Koepke, 2015 for reviews), as well as the formal mathematical learning of school-age children and adolescents. The findings that have emerged from the use of these paradigms and the consistent effects they have generated are so numerous that it would be challenging to list them all here, much less to explain their significance with respect to achieving a comprehensive understanding of both the typical and atypical development of mathematical thinking and learning. That being said, it is worth providing the reader with a representative set of topics that have been studied to date. These include subitizing (i.e., the rapid and accurate apprehension of the quantity of small collections of items); magnitude comparison of nonsymbolic quantities (e.g., random dot arrays) and symbolic quantities (e.g., Arabic numerals); transcoding (i.e., translating from one

Development of Mathematical Cognition. http://dx.doi.org/10.1016/B978-0-12-801871-2.00001-0

numerical format to another); other relations between nonsymbolic and symbolic numerical skills; the development of counting skills, single-digit arithmetic, and multidigit arithmetic processing; the acquisition of place value; conceptual and procedural knowledge of fractions; proportional reasoning; mathematical equivalence; and the role of domain general processes (e.g., working memory, processing speed) in mathematical cognitive development. Furthermore, some findings have been so robust that they have attained the status of "effects," such as the numerical distance effect (Moyer & Landauer, 1967); the problem-size effect (Ashcraft & Guillaume, 2009; Zbrodoff & Logan, 2005), the Spatial-Numerical Association of Response Codes (SNARC) effect (see Fischer & Shaki, 2014 for a review), and the operational momentum effect (also reviewed by Fischer & Shaki, 2014).

NEUROBIOLOGICAL PERSPECTIVES ON MATHEMATICAL COGNITIVE DEVELOPMENT

Few contemporary researchers would question that developmental changes in mathematical thinking and learning are a reflection of changes in neurobiological functions, whether induced from maturation per se, environmental triggers (e.g., informal home activities, formal education), or some combination of both. But precisely what can the study of neural substrates and genetic influences bring to this field and how can we best characterize these potential contributions, outline their value-added to cognitive accounts alone, as well as delineate their limitations? In this chapter, we explore these issues to provide a foundation for the more specific treatments of circumscribed research topics that the reader will encounter in the ensuing chapters. Among other themes, we address some longstanding as well as more recent controversies concerning the uses of neuroimaging methods in general and functional magnetic resonance imaging (fMRI) in particular—both because the latter has been used for studying the neural correlates of mathematical development far more extensively than other techniques and because some of the issues are relevant to other imaging procedures. Our analysis of these concerns is designed to acquaint the reader with some of the critical assumptions underlying the use of such techniques and how they can inform cognitive theory, thereby prospectively advancing our understanding of mathematical cognitive development.

USING NEUROIMAGING METHODS TO STUDY CHILDREN'S MATHEMATICAL DEVELOPMENT

Ten to fifteen years ago, we knew comparatively little about the neural underpinnings of developmental changes in mathematical cognition. As such, it would have been difficult to even conceive of putting together an entire edited volume focusing on this topic. However, the technical, methodological, and statistical advances achieved in neuroimaging in recent years, along with the increasing availability of scanners for use by cognitive neuroscientists have contributed to an increasing growth in the number of empirical studies of mathematical cognition and development, meriting the broad as well as in-depth treatment of this topic that the present volume affords. That being said, as it is generally acknowledged that work in this area is still in its early stages, stepping back to take a larger perspective on this entire enterprise is warranted.

We begin by reviewing the comparatively brief history of this kind of research, after which we discuss the importance of taking into consideration differences between the developing and mature brain. Then we explore some of the assumptions made when interpreting the meaning of fMRI data with respect to both behavioral outcomes and cognitive theories, examine criticisms concerning the meaning and value of brain imaging data, review the kinds of inferences one can draw from brain imaging studies, and consider the extent to which more recent methodological developments in analyzing neuroimaging data could accelerate progress in this rapidly evolving field.

MATHEMATICAL COGNITION AND DEVELOPMENT: BRAIN STRUCTURE AND FUNCTION

Although it is sometimes suggested that brain research should provide the scientific foundation for children's education in mathematics and in other academic areas, it is too early to directly apply findings from studies of brain processes during mathematical reasoning to classroom teaching and learning. Yet promising research emerging from the field of cognitive neuroscience is permitting investigators to begin forging links between neurobiological functions and mathematical cognition. We present a few highlights in the history of this research and then touch on an issue that has been overlooked for much of this history, that is, the difference between the developing and the mature brain (Ansari, 2010).

A Brief History

Early research on the relation between brain functioning and mathematical competencies was restricted to the sequelae of brain injury due to disease (e.g., stroke) or trauma (e.g., gunshot wounds) and largely focused on number and arithmetic. As an example and based on a series of case studies of the results of brain injury, Henschen and Schaller concluded, "In a general sense, akalkulia (sic) is the inability to form combinations of numbers" (Henschen & Schaller, 1925, p. 232) and is often associated with "cipher blindness" (poor recognition of numerals) and damage to the angular gyrus. Henschen and Schaller also observed that following these injuries "an educated person will revert to simple means of expression similar to those of ... stone folk" (Henschen & Schaller, 1925, p. 232); that is, they would thereafter rely on basic number words and the limited calculation abilities found in traditional populations. Their insight presaged Geary's (1995) distinction between evolved quantitative abilities and those dependent on formal education and Dehaene's number sense (2011; Dehaene, 1992; Dehaene & Cohen, 1995).

As is well known in this literature, Gerstmann (1940) defined a cluster of deficits associated with damage to the angular gyrus and adjacent regions of the occipital lobe that included calculation difficulties, along with finger agnosia (i.e., inability to recognize and differentiate different fingers), poor discrimination of the left and right side of the body, and often difficulties in understanding time (e.g., comprehending analog watches). The calculation difficulties included simple problems, as studied by Henschen and Schaller (1925), as well as a poor understanding of the meaning of numerals in different columns for more complex problems (e.g., 154+263, and not understanding that the "5" and "6" represent sets of 10). In his magnum opus, *Higher Cortical Functions in Man*, Luria (1980) described the same pattern of deficits associated with injury to the angular gyrus, and noted the importance of the prefrontal cortex for planning and enacting a problem-solving sequence. The latter would involve actually solving the problem 154+263, not simply understanding the magnitudes represented by the position of the columnar numerals; these processes were subsequently shown to be heavily dependent on working memory (Hitch, 1978). Luria also noted that temporal lobe lesions that disrupted language also disrupted arithmetical or other mathematical processes that were dependent on language, such as counting.

Dehaene and Cohen (1995; Dehaene, 1992) organized these early neuropsychological findings with their triple-code model of number and arithmetic processing. The model includes three forms of numerical and arithmetical representation: visual processing of numerals; a verbal word frame that supports the syntactic organization of the structure of complex number words and storage of arithmetic facts; and, an analog magnitude representation that supports the understanding of the magnitudes associated with the visual and verbal processing of numbers. The visual number system is largely supported by the ventral stream of the left hemisphere, that is, the occipito-temporal system that supports object recognition. The verbal word frame also shows a left-hemispheric bias, following the lateralization of language in most people, whereas the analog magnitude system is bilaterally represented and includes portions of the parietal cortex. The model has been extremely successful in organizing subsequent neuropsychological and brain imaging research not only on number and arithmetic processing but also other areas of mathematics (Schmithorst & Brown, 2004).

In the decade or so following the introduction and elaboration of the triple-code model, the field saw a spate of studies that used fMRI and related techniques focused on people's basic mental representations of number and quantity (Chochon, Cohen, van de Moortele, & Dehaene, 1999; Cohen Kadosh et al., 2005; Dehaene, Spelke, Pinel, Stanescu, & Tsivkin, 1999; Göbel, Calabria, Farné, & Rossetti, 2006; Halgren, Boujon, Clarke, Wang, & Chauvel, 2002; Pinel, Piazza, Le Bihan, & Dehaene, 2004; Temple & Posner, 1998; Vuilleumier, Ortigue, &

Brugger, 2004; Zorzi, Priftis, & Umiltá, 2002), as well as arithmetic (Gruber, Indefrey, Steinmetz, & Kleinschmidt, 2001; Rickard et al., 2000; Rivera, Reiss, Eckert, & Menon, 2005) and simple algebra (Anderson, Qin, Sohn, Stenger, & Carter, 2003; Qin et al., 2003, 2004). In most of these studies, researchers contrasted the brain regions activated when children (or adolescents) and adults solved the same arithmetic or algebra problems (Kawashima et al., 2004; Qin et al., 2003, 2004; Rivera et al., 2005); mapped changes in neural activity associated with practice at arithmetic (Delazer et al., 2003; Pauli et al., 1994); and differentiated the brain regions involved in arithmetic fact retrieval from those recruited for executing complex calculation procedures, such as trading across columns (Kong et al., 2005), as studied decades earlier by Luria (1980). In other studies, researchers have compared brain activity when the same quantities are presented in different notations (e.g., '8' versus 'eight'; Cohen Kadosh, Cohen Kadosh, Kaas, Henik, & Goebel, 2007; Piazza, Pinel, Bihan, & Dehaene, 2007).

There was of course some variation across studies in the brain areas engaged when solving different types of mathematical problems—due to differences in experimental procedures and specific problems presented across studies—but there are also intriguing consistencies. It has been repeatedly found that comparisons of number magnitudes (Pinel et al., 2004; Temple & Posner, 1998), quantitative estimation (Dehaene et al., 1999), use of a mental number line (Vuilleumier et al., 2004; Zorzi et al., 2002), and problem solving in arithmetic and algebra (Chochon et al., 1999; Qin et al., 2003; Rivera et al., 2005) activate several areas of the parietal cortex, including the bilateral intraparietal sulcus and angular gyrus. As extensively reviewed in the first volume of this series (Geary et al., 2015), the intraparietal sulcus is also active when nonhuman animals engage in numerical activities (Sawamura, Shima, & Tanji, 2002; Thompson, Mayers, Robertson, & Patterson, 1970) and it has been proposed that a segment of this sulcus supports the magnitude representation component of the triple-code model (Dehaene, Piazza, Pinel, & Cohen, 2003).

Progress in the field continues and has even accelerated with the widespread availability of fMRI technologies, advances in how the associated data are analyzed, and with the infusion of new researchers focused on mathematical cognition. The chapters of the current volume highlight some of the most exciting and important advances over the past decade. These new findings confirm many aspects of the early neuropsychological research on number and arithmetic processing, but also highlight the complexity of the brain systems underlying mathematical knowledge along with problem solving and issues to be addressed in the coming decade.

The Developing Brain

Gerstmann (1940) and Luria (1980) made a distinction between developing number and arithmetic competencies and fully developed competencies. Gerstmann, for instance, in describing the association of finger agnosia and arithmetic deficits noted fingers "supply the first aid in learning to calculate, as is commonly known. The first arithmetical operations of children are performed on the fingers of each hand" (Gerstmann, 1940, p. 401), and thus finger agnosia could result in later arithmetic deficits. Most of the case studies on which this early literature was built, nevertheless, were based on brain injury in adults. Indeed, the triple-code model (Dehaene, 1992; Dehaene & Cohen, 1995) and Geary's (1993) influential model of mathematical learning disabilities were based on these same literatures, largely because there was little else available, with the exception of a few studies of developmental dyscalculia (e.g., Temple, 1991).

As Ansari (2010) has aptly noted, trauma-induced disruptions of previously learned information (e.g., arithmetic facts) in the mature brain are not necessarily analogous to similar behavioral deficits in children who are in the process of learning this same information or for children who are having difficulties learning the information despite adequate instruction. In other words, even if the same behavior and cognition are observed in children and adults, such as retrieving addition facts from long-term memory, it should not be assumed that the same brain regions support this behavior and cognition.

Returning to our brief tour of early neuropsychological research, fMRI studies confirm the importance of the angular gyrus, especially in the left hemisphere, for the retrieval of arithmetic facts from long-term memory and engagement of a wider network of prefrontal and parietal regions when problems are solved procedurally (e.g., through counting; Grabner et al., 2009). Neuroimaging studies of children's arithmetical problem solving have also revealed engagement of this same network of brain regions, including the angular gyrus in the left and right hemisphere (Cho, Ryali, Geary, & Menon, 2011; Cho et al., 2012; De Smedt, Holloway, & Ansari, 2011; Qin et al., 2014). These studies have also identified the hippocampus as being critical for children's learning and retrieval of arithmetic facts, a brain region that was not identified in the studies of adults.

In retrospect, the identification of the hippocampus during children's fact learning is not surprising, given its importance to the formation of long-term associative memories (Squire & Zola-Morgan, 1991; Tse et al., 2007). Indeed, the learning of arithmetic facts, especially addition and multiplication, has a clear associative basis (Siegler & Shrager, 1984), and thus should be particularly dependent on the hippocampal memory system for initial learning; addition and multiplication facts are easier to memorize than subtraction and division facts because addition and multiplication are associative (e.g., $5+3=3+5$) whereas subtraction and division are not (e.g., $5-3\neq 3-5$). The hippocampus and surrounding areas support the initial formation of associative memories, and their engagement decreases as the memories are consolidated in the neocortex, eventually the left angular gyrus for arithmetic facts, as stated. Despite its clear theoretical importance in learning, the hippocampus was largely overlooked throughout the long history of neuropsychological and later brain imaging studies of number and arithmetic in adults, presumably as a result of an implicit assumption of no difference between the developed and developing brain (Ansari, 2010).

Longitudinal fMRI studies of developing children are critical to fully understanding differences between emerging competencies in the developing brain and developed competencies in the mature brain. In one such study, Qin et al. (2014) assessed 7- to 9-year-olds twice—about 1 year apart—while they solved simple addition problems. This age range was chosen because it is associated with the transition from use of procedural-based counting strategies for problem solving to the direct retrieval of answers. Overall, the transition to retrieval was related to decreased engagement of the frontal-parietal network that is associated with use of procedural strategies, and increased engagement of the hippocampus. Surprisingly, however, at an individual level, the children who showed the greatest gains in fact retrieval did not show a straightforward increase in hippocampal engagement relative to other children. Rather, the key seemed to be greater gains in the connectivity of the hippocampus with the prefrontal and parietal regions. A cross-sectional comparison of these children with adolescents and adults revealed decreased hippocampal engagement with age (and education), consistent with consolidation of long-term memories of basic facts. Critically, the consolidation of the long-term memories was associated with increased stability in the functional circuits recruited in the hippocampus and neocortex when solving each individual problem. In other words, memory consolidation and fluent fact retrieval appeared to be dependent on engagement of the same populations of hippocampal neurons when solving the same problem (see Menon, this volume).

The Qin et al. (2014) and related studies (e.g., Rivera et al., 2005; Vogel, Goffin, & Ansari, 2015) highlight the importance of taking Ansari's (2010) admonition seriously but still leave unaddressed the confound of the developing brain and level of expertise. The brain systems that support numerical and mathematic competencies may change with increasing levels of expertise, even in adults (e.g., Holloway, Battista, Vogel, & Ansari, 2013), and thus the longitudinal and cross-sectional changes in the brain systems underlying fact retrieval found by Qin et al. could be due to differences in the developing versus mature brain or to differences in level of expertise comparing children and adults, or some combination. For instance, is the engagement of the hippocampus and its functional connectivity to the prefrontal and parietal cortices the same comparing children and naïve adults learning the same information, or is the pattern of engagement different? The next step in fully understanding whether and how the developing brain learns differently than the mature brain will require consideration of levels of expertise, as well as age.

BRAIN IMAGING METHODS USED IN STUDYING MATHEMATICAL COGNITIVE DEVELOPMENT

As the reader will discover, every author in this volume discusses at least some findings resulting from the use of one or more methods of brain imaging. And although fMRI has been by far the most commonly used approach in the study of mathematical cognition to date (Cohen Kadosh, in press), other neuroimaging techniques that have been employed and are discussed in this volume include magnetic resonance imaging (MRI), positron emission tomography (PET), electroencephalography (EEG), magnetoencephalography (MEG), and near-infrared spectroscopy (NIRS). The interested reader is referred to Table 2 in Kucian (this volume) for descriptions of each of these procedures, how they measure brain activity, as well as their advantages and disadvantages. Kucian also provides detailed descriptions of advanced MRI techniques, including diffusion tensor imaging (DTI)—furnishes information about axonal fiber connections between brain regions by measuring white matter microstructure; and magnetic resonance spectroscopy (MRS)—a measure of brain chemistry that can be used as a noninvasive technique for studying metabolic changes in the brain. Furthermore, she describes the method of voxel-based morphometry (VBM)—an analytic technique that permits investigation of focal differences in brain structure. Finally, Sarkar and Cohen Kadosh (this volume) present detailed descriptions and reviews of various types of brain stimulation methods, including transcranial magnetic stimulation (TMS), transcranial direct-current stimulation (tDCS) and it variants, and transcranial electrical stimulation (tES).

Despite the increasing use of brain imaging techniques for studying developmental changes in the neural correlates of mathematical cognition, it is generally acknowledged that to date, such investigations have been comparatively rare (Kaufmann, Kucian, & von Aster, in press). Cohen Kadosh (in press) notes that this may be attributable to a variety of factors, such as (a) greater difficulty in recruiting children for brain imaging studies than college students, which he suggests may be due to the concerns of parents based on their lack of familiarity with imaging methods and their safety; or (b) more frequent unwanted movements by children than adults inside scanners for MRI studies that make the images difficult to interpret. Additional factors include fear in children—which can be aroused by the noise of the scanner, restlessness, or claustrophobia—as well as other methodological problems (see Kaufmann et al., in press).

For the most part, reports of the factors just described have been limited to sections of empirical papers indicating the number of participants excluded for any combination of these reasons. However, a recent study using fMRI with healthy children employed both a cross-sectional and longitudinal design to assess success rates on four tasks (verb-generation, syntactic prosody, picture-matching, and story-processing), where success was operationalized as the proportion of these tasks completed while in the scanner (Rajagopal, Byars, Schapiro, Lee, & Holland, 2014). In the cross-sectional study, with participants 5-18 years of age, 74% of the children completed all tasks. Success rates on individual tasks for 5- to 7-year-olds ranged from 34% to 67%, whereas for those ages 8-18 years, success rates ranged from 76% to 100%. For a subset of 5- to 7-year-old children studied longitudinally over a 10-year period, 89% completed all tasks in all 10 years. Among other benefits for fMRI researchers, the findings of this study have important implications for the design of imaging studies with children, suggesting that the sample sizes indicated by power analyses should be increased by 50% for children age 5-7 years and by 33% for youngster age 8-18 years (Rajagopal et al., 2014).

Despite the difficulties in recruiting children for brain imaging studies and the problems that can arise in the scanner, these kinds of investigations have more recently begun to increase at a rate that merits a collective review, analysis, and appraisal of developmental differences (cross-sectional) and changes (longitudinal) in the neural substrates of various domains of mathematical thinking and learning. The studies reviewed in this volume achieve these objectives admirably, and point the way toward future directions in this kind of research.

CRITICISMS OF fMRI

The reader of this volume will encounter reviews of studies that have made use of some version of the fMRI technique. As such, it is important to appreciate that a variety of criticisms have been leveled at the uses and interpretations of this approach, ranging from the assertion that the arbitrary decisions concerning the color-coding of different brain regions can be misleading to the potential for multiple statistical analyses involved in the generation of brain images to misrepresent the evidence. Although space limitations preclude a detailed analysis of these issues in this introductory chapter, fortunately, Farah (2014) has recently done just that by thoroughly and adroitly examining the evidence base for such claims. To begin with, she has classified such criticisms into four categories: *Inferential Distance and the Objects of Imaging, Metatheoretical Assumptions and Goals, Slippery Statistics*, and *The Undue Influence of Brain Images*. The following are, respectively, examples of each: (1) rather than gauging neural activity directly, brain images are generated only indirectly through measuring the flow of oxygen in the blood (Hardcastle & Stewart, 2002; the interested reader is referred to Aguirre, 2014 for an accessible tutorial on how neurovascular changes are measured); (2) the goal of fMRI is localizing particular psychological functions in the brain rather than explaining them—so-called neophrenology (Uttal, 2001; but see a description of the fMRI *adaptation paradigm* by Ansari, this volume, for a way around this criticism); (3) statistical analyses frequently consist of substituting estimated values for raw data to yield images that are consequently misleading (Shermer, 2008); and (4) the undue persuasive power of brain images, especially in applied contexts (Illes et al., 2010; McCabe & Castel, 2008). However, Farah's detailed review and analysis of these criticisms, summarized in Table 1, reveals that even though all of them possess a kernel of truth, they exaggerate these elements to the point of erroneously casting doubt on the usefulness or validity of the entire functional neuroimaging research enterprise. Her conclusion: "None of the criticisms reviewed here constitute reasons to reject or even drastically curtail the use of neuroimaging" (p. S28).

TABLE 1 Criticisms of Functional Brain Imaging

Criticism	Intended Scope	Evaluation
Inferential Distance and the Objects of Imaging		
Blood not brain	All fMRI	*Kernel of truth:* Neural activity per se is not imaged; the relation of BOLD to neural activity is not well understood.
Fabrications	All fMRI	*Kernel of truth:* Data representation requires arbitrary design decisions about, for example, scales and colors.
Metatheoretical Assumptions and Goals		
Localization vs. explanation	All fMRI	*Kernel of truth:* Some studies are aimed at localization per se, often for validation of method.
Cannot test psychological Hypotheses	All fMRI	*Kernel of truth:* There is no single decisive imaging finding.
Bias for modular hypotheses	All fMRI	*Kernel of truth:* Early approaches to image analysis were best suited to modular theories.
Wanton reverse inference	Particular studies	*Unqualified truth:* Without appropriate (often extensive) research, one cannot infer a psychological process from brain activation. Many studies have drawn invalid conclusions.
Slippery Statistics		
Statistics not reality	All fMRI	*Kernel of truth:* "Reality" is represented only through the filter of multiple statistical operations.
Multiple comparisons	Particular studies	*Unqualified truth:* Images lend themselves to large numbers of independent statistical tests, inviting spurious results if the tests are not appropriately constrained or corrected.
Circularity	Particular studies	*Unqualified truth:* Researchers must select regions or subsets of data by means independent of the subsequent test.
The Undue Influence of Brain Images		
Overly convincing	All fMRI	*Kernel of truth:* This phenomenon has been observed but does not appear robust; the aura of science more generally may be persuasive.
Overly appealing	All fMRI	*Kernel of truth:* This phenomenon is intuitively plausible, but there is little evidence to support it.

Note: BOLD stands for blood-oxygen-level-dependent.
Adapted from "Brain images, babies, and bathwater: Critiquing critiques of functional neuroimaging" by M. J. Farah, 2014, *Interpreting neuroimages: An introduction to the technology and its limits*, Special Report, *Hastings Center Report* 45, no. 2 (2014): s19-s30. http://dx.doi.org/10.1002/hast.295. Copyright 2014 by John Wiley & Sons. Adapted with permission.

Reverse Inference

One of the strongest criticisms of fMRI listed in Table 1 is worthy of a little more elaboration, namely, *Wanton reverse inference*. Reverse inference, as originally defined by Poldrack (2006), refers to the practice of reasoning backward from an observed activation of a specific brain region to a particular cognitive process that gave rise to it. Poldrack offered the following example to illustrate the kind of deductive logic comprising this inferential process for a given study:

1. In the current study, when task comparison X was administered, brain area A was active.
2. In other studies, when cognitive process C was purportedly engaged, brain area A was active.
3. Consequently, the activity of area A in the present study demonstrates engagement of cognitive process C by task comparison X.

Poldrack argued that such inferences are not deductively valid. Consequently, he recommended that cognitive neuroscientists be cautious about arriving at conclusions by means of this form of backward reasoning, especially when selectivity of activation in the region of interest either has not been established or has already been recognized as being weak. Nonetheless, he suggested that reverse inferences can provide some information of value, and that they can evoke novel hypotheses for testing in later studies. In a similar vein, Farah (2014) suggested that reverse inference is not intrinsically invalid as long as one has conscientiously determined the scope of cognitive processes that can activate a brain region under a specified set of conditions. Likewise, Mather, Cacioppo, and Kanwisher (2013) noted that "for the few cases of brain regions with very specific functions, and for which that specificity is widely replicated, these regions can indeed support solid reverse inferences" (p. 112).

In an exceptionally comprehensive analysis, Del Pinal and Nathan (2013) contended that successful reverse inferences do not require any assumption of a strong form of functional locationism. They also suggested that these kinds of inferences can be fruitful at early stages of psychological theorizing, as the neural bases of cognitive mechanisms can be differentiated with comparative ease during this period. At later, more advanced stages of theorizing, however, the cognitive mechanisms postulated are typically executed in overlapping brain regions, and thus are more difficult for imaging methods to discriminate. Furthermore, these authors pointed out that reverse inferences based on *patterns* of neural activation differ from those based on *locations* of activation.

Finally, as Farah (2014) has noted, a method known as *multi-voxel pattern analysis* (MVPA; Norman, Polyn, Detre, & Haxby, 2006) is essentially a combination of forward and reverse inference approaches (see descriptions of this technique by both Ansari, this volume, and De Smedt, this volume). Forward inference consists of dissociating/distinguishing between cognitive processes based on a qualitative difference in brain activity (Henson, 2006). According to Farah, "In MVPA studies, forward inference is first performed on all of the psychological states that reverse inference will later be used to select among. When the time comes for reverse inference, it is in effect a very large multiple-choice among the psychological states that were the subject of forward inference" (p. S24).

HOW BRAIN IMAGING CAN ADVANCE COGNITIVE THEORIZING

Theories serve as both a tool and a goal. With respect to the former, they provide heuristic value by framing relational propositions between abstract constructs from which specific empirical hypotheses can be derived, that can then be tested experimentally. As a goal, theories are constructed to provide explanations of the empirical phenomena that emerge from laboratory studies. Although these processes are reasonably straightforward when constructing and testing cognitive theories concerning behavioral data per se, it is not always immediately apparent how the measurement of neural activity can have a bearing on cognitive theorizing.

Mather et al. (2013) have argued that fMRI can inform theories of cognition by helping to answer four kinds of questions that go beyond simply trying to localize specific brain regions where particular cognitive processes are carried out: (1) *Which*, if any, mental processes are carried out in specialized brain regions—answering such a question is informative not because of the location per se (e.g., the intraparietal sulcus), but because it has implications for theories of the modularity of the brain, and therefore the mind; (2) if an imaging marker (i.e., neural activity in a highly specialized brain region) is discovered for a specific cognitive process, we can ask whether that process is engaged when participants perform a given task—this approach allows researchers to decode what kind of information is being processed; (3) how distinct are the representations from different tasks or stimulus classes—both the fMRI adaptation procedure referred to earlier (Grill-Spector & Malach, 2001) and the fMRI multi-voxel pattern analysis just mentioned permit the characterization of neural representations in specific brain regions by assessing which stimuli are treated the same and which are treated as different; and (4) do some pairs of tasks engage distinct or common processing mechanisms—experiments that reveal brain activation in the same or overlapping brain regions for two tasks can supply evidence in support of common mechanisms (within individual participants, and if suitable control conditions are employed).

A recent study by Klein et al. (2014) using fMRI data along with DTI probabilistic fiber tracking (for measuring white matter connectivity) illustrates how answering several of the questions posed by Mather et al. (2013) can inform the further development and refinement of the triple code, neurocognitive theory of mathematical cognition (Dehaene & Cohen, 1995) described earlier. The reader may recall that this model postulates three types of numerical codes that are supported by different brain regions, albeit with bidirectional connections (transcoding paths) between all three representations. To begin with, the use of fiber tracking by Klein et al. for measuring connectivity between the specific brain regions subserving the three different representational codes required the determination of so-called *seed points*—brain regions based on reliable anatomic-functional landmarks for starting the reconstruction of specific neuronal pathways. Choice of these areas was driven in part by prior meta-analytic studies of fMRI data (Arsalidou & Taylor, 2011) relevant to the neuroarchitecture specified by the triple-code model. Then Klein et al. (2014) reanalyzed some of their previous fMRI data based on a number bisection task and an exact/approximate addition task, and also employed the DTI procedure with another group of participants who were administered the same two tasks. The resulting evidence (a) suggested the involvement of the hippocampus as an additional brain region (i.e., not specified explicitly in the original triple-code model) essential for retrieving arithmetic facts, (b) specified the recruitment of four prefrontal cortical areas involved in number magnitude processing, and (c) identified the connections between these anatomic-functional manifestations of the representational codes supporting number magnitude processing and arithmetic fact retrieval, all of which led Klein et al. to propose a schematic update of the triple-code model.

BEHAVIORAL AND NEURO-GENETICS OF MATHEMATICAL COGNITION

As with neuroimaging, the 21st century has brought new and powerful technologies that provide insights into the role of genetics in creating mathematical competencies. Indeed, the enthusiasm and confidence in new genetic capabilities has led to rekindled public debates on the historical question of "nature versus nurture," as well as reinforced beliefs in genetic determinism that suggest huge swaths of the public are predestined to fail in one cognitive domain or another, including mathematical abilities.

Even before the awareness of genes, a few great thinkers held the belief that a biologically driven "master plan" mechanically unfolds a species' character and abilities. Aristotle (384-322 B.C.E.) introduced the seemingly modern concept of epigenesis in reference to the inborn developmental processes that direct the growth and characteristics of a species: "This much is clear in the first place, that, as in all other things, birth implies an antecedent beginning" (Aristotle, B.C.E.); "how will the foetus become greater by the addition of something else if that which is added remain unchanged? ...why not say that the semen from the very first is of such a kind that blood and flesh can be made out of it, instead of saying that it itself is blood and flesh?" (Aristotle, Platt, 1912).

The recognition of a paternal inheritance took hold early, but knowledge of the biological mechanisms of inheritance was slow to germinate. Although forgotten for several decades, Mendel's work published in 1865 set the stage for later biologists to differentiate the concepts of genotype and phenotype and for the field of genetics' reliance on quantitative analyses to demonstrate the impact of alleles (and genes) on phenotypes. His work also reinforced the belief that germinal matter determined the unfolding of adult characteristics in a systematic and unvarying process (Mendel, 1865). August Weismann (1834-1914), unaware of Aristotle's writings and Mendel's work with peas, proposed the "germ plasm theory" (working with multicellular organisms, but without the mass of empirical data Mendel had provided), suggesting that "biological inheritance is inconceivable except by way of germ plasm [now known as DNA] from parents to offspring" (Weismann, 1893). Extending his theory further, Weismann borrowed Darwin's (1859) theory of evolution and applied it to "germinal selection," thereby formally creating the concept of "genetic determinism."

Weismann (1893) was not alone; for many scholars of the 19th century, eugenics (the science of improving a population through planned or controlled breeding of specimens with desirable traits or characteristics)[1] and genetic determinism were taking hold. Francis Galton (1822-1911) coined the term "eugenics" and became a leading proponent of bettering the welfare of all members of society through selective breeding. As a young man, Galton excelled in mathematical thinking and turned his statistical concepts and interests (among other advances, Galton developed the concepts of "correlation" and "regression," and is known as the founder of psychometrics—the science of measurement of mental abilities) to the study of the relative roles of "nature" (biology) and "nurture" (environment) in human intelligence and inheritance, coining the phrase "nature-nurture." Before Galton, few had the ability to quantify specific cognitive abilities. He began by collecting data on the rates of "exceptionalism" in the families of eminent men and analyzed these in contrast to typical men to determine if qualities such as intelligence ran in families. In doing so, he demonstrated the heritable quality of mental abilities by degree of personal relatedness (Galton, 1869; see also Clark, 2014) and invented the methods of historiometry "the scientific discipline in which nomothetic hypotheses about human behavior are tested by applying quantitative analysis data concerning historical individuals" (Simonton, 1990, p. 3) and the use of surveys in behavioral research.

But Galton (1875) was dissatisfied with the statistical and scientific limitations of this work, and turned to the study of twin pairs to gain greater control of the variables of interest. "Twins have a special claim upon our attention; it is, that their history affords means of distinguishing between the effects of tendencies received at birth, and those that were imposed by the special circumstances of their after lives" (p. 156). In doing so, he created the most frequently used

1. In its original conceptualization, eugenics did not include forced sterilizations or exterminations of populations. By the early 20th century and the rise of political entities that distorted the principles with unethical and unacceptable practices, policies, and atrocities (e.g., the Nazis), eugenics fell out of favor.

method in behavioral genetics that is still used today[2] and provided some of the first human data addressing the "nature versus nurture" debate. Although he did not conduct studies with adopted twins, he proposed that such a method would significantly improve the science of inheritance, thus creating yet another method, one that was to become "the Rosetta Stone" for behavioral genetics research (Bouchard, 1999).

The young mathematical prodigy Ronald Fisher (1890-1962) entered college at Cambridge and quickly became intrigued by Mendel's quantitative principles of inheritance and, like Weismann before him, sought to harmonize these with Darwin's theory of evolution, but using biostatistical methods. In the process, he developed the concept of variance and the methodology of analysis of variance (Fisher, 1918). These and many other statistical advances Fisher published in later papers enabled a new understanding of how complex traits might be inherited or how genes might be linked, and revolutionized not only behavioral and population genetics, but most, if not all, behavioral sciences. These new statistical tools allowed the twin design even greater capacity to answer new questions and rapidly gained popularity among behavioral genetics researchers. With the reemergence of the nature versus nurture debate, twin studies flourished, focusing on the heritability of complex cognitive abilities such as intelligence (Bouchard, 1998; Devlin, Daniels, & Roeder, 1997), personality traits (Jang, Livesley, & Vernon, 1996), and physical characteristics (Silventoinen et al., 2003). A few studies have explored the heritability of mathematics and related cognitive abilities (Docherty et al., 2010; Husén, 1960; Thompson, Detterman, & Plomin, 1991; Tosto et al., 2014) and mathematics disability (Alarcón, DeFries, Light, & Pennington, 1997; Knopik & DeFries, 1999; Light & DeFries, 1995; Oliver et al., 2004; see Petrill and Kovas, this volume, for a detailed review and analysis of these topics).

2. There is debate whether Galton should be credited with this advance as, although he was explicitly aware of three types of perceptually different twin pairs, he was likely unaware of the different zygote types and status (monozygotic (MZ) or dizygotic (DZ)) of his research participants. Curtis Merriman (1875-1975) first elaborated the method of extracting heritability versus environmental influences by comparing identical (MZ) and fraternal (DZ) twins in a 1924 publication (Merriman, C. (1924). The intellectual resemblance of twins. *Psychological Monographs*, 33, 1-58). Hermann Siemens (1891-1961) independently but simultaneously published twin study findings in a book (Siemens, H. W. (1924). *Zwillingspathologie: Ihre Bedeutung; ihre Methodik, ihre bisherigen Ergebnisse [Twin pathology: Its meaning; its method; results so far]*. Berlin: Springer Verlag) that articulated this same method. Note that both these publications came some 50 years after Galton's. Recently, Liew and colleagues (Liew, S. H. M., Elsner, H., Spector, T. D., & Hammond, C. J. (2005). The first "classical" twin study? Analysis of refractive error using monozygotic and dizygotic twins published in 1922. *Twin Research and Human Genetics*,8(03), 198-200.) have suggested that a publication using the MZ-DZ comparison from Walter Jablonski (Jablonski, W. (1922). Ein Beitrag zur Vererbung der Refraktion menschlicher Augen [A contribution to the heredity of refraction in human eyes]. *Arch Augenheilk*, 91, 308-328) preceded even these two claims for the first publication.

It was not until 1953, when Francis Crick and James Watson discovered the structure of DNA (Watson & Crick, 1953) that the fields of behavioral and neuro-genetics were again profoundly transformed. For the first time, a specific phenotype could potentially be associated with a specific gene and a mechanism of action (protein and neural pathways). With the advent of molecular genetics came new research questions, new methodologies, new analytic approaches, and potentially, for the first time, new genetic intervention methods. Suddenly, with the discovery of the DNA structure and following decades of molecular biology advances, geneticists had the tools to link any measurable ability or behavior, including complex abilities, to a gene or genes. Many molecular genetic advances could now be combined with classic behavioral genetic methods and analyses to answer heritability questions on a new level.

The Human Genome Sequencing Study (International Human Genome Sequencing Consortium, Lander et al., 2001) began to crush the long-held notion of genetic determinism regarding cognitive abilities with findings that suggested that 99.99% of the genes of one person are identical to the genes of every other human, although different people can have different variations (alleles) of many of these genes. Epigenetic effects can now be tested for, are seen to have differing effects at various ages, and may even differ between monozygotic (identical) twins (Fraga et al., 2005). Combine these advances with the new neuroimaging technologies, and neurobehavioral geneticists have a scientific banquet at which to feast. Yet all of these findings lead to two unavoidable conclusions: (1) that phenotype is a function of genetics, epigenetics, and environment, in various complex mixtures, and (2) today's behavioral and neuro-genetics studies require close collaborations across multiple disciplines in the biobehavioral, medical, and molecular genetic sciences.

INTERPRETIVE CHALLENGES

While the classic behavioral genetics methodologies including twin, adoption, sibling, and family studies have profoundly changed our ability to tease out the relative contributions of genes and environment, they all share some common challenges in the interpretation and generalization of findings to which the readers should be attentive. For example, few investigators still use traditional family or sibling studies, as the ability to reliably separate environmental effects from genetic effects in these studies is poor; however, these methods can be very useful in the context of genomic testing when searching for a candidate gene.

In contrast, classic twin studies are still used to separate gene-environment effects. Most twin studies assume that twin pairs equally share an environment (this is known as the Equal Environment Assumption or EEA). If this is true, then the only difference between identical (MZ) and fraternal (DZ) twin pairs is their genes. Unfortunately, several studies have demonstrated that MZ pairs share more of a common environment than do DZ pairs (Jackson, 1960; Scarr, 1968) and that beginning from the time of gestation in the womb even MZ twin pairs begin to experience their environment differently (Bowles, Gintis, & Groves, 2005; Loehlin & Nichols, 1976; Plomin, Asbury, & Dunn, 2001). The result of these assumption errors is an overestimation of heritability. In response, modern twin studies focus on the assumption of trait-relevant EEA, but again, if the assumptions regarding which environmental factors are trait relevant or if they are not in fact equal, there may be an overestimation of heritability.

Most investigators searching for a separation of nature versus nurture assume that there is only one type of genetic effect, usually additive, for multiple variables of significance. But genetic effects can be additive, dominant-recessive, or epistatic (i.e., the effect of one nonallelic gene is influenced by the presence of other genes), and there can be interactions between gene-environment conditions. Likewise, the sometimes erroneous assumption of random parental mating will underestimate the shared genes of fraternal (DZ) twins and dull any small effect that might have been demonstrated. Additionally, most twin studies currently cannot and do not account for possible mosaicism (an incomplete mutation that affects some but not all cells in the body).

Adoption (or reared-apart) twin studies have similar risks if assumptions are ignored, but by completely eliminating the shared environmental effects they could increase the possibility of isolating genetic effects. Probably the best known adoption twin studies are the Minnesota Study of Twins Reared Apart (Bouchard, Lykken, McGue, Segal, & Tellegen, 1990), and the Swedish Adoption/Twin Study of Aging (Medlund, Cederlöf, Flodérus-Myrhed, Friberg, & Sörensen, 1976), and both have been challenged on various assumption concerns. Many of the twins in reared-apart twin studies actually were raised near to each other, often by other family members in the same town and with frequent contact between the twins, and may in fact have experienced very similar environments during childhood. The age of separation of twins may be an important factor for the assumption of different environments; no reared-apart studies have exclusively studied twins separated at birth. Often twin pairs develop strong relationships when they discover each other after some separation, and this close association may negate many reared-apart benefits for the study. Also, it is quite possible that even when twins are reared apart, their phenotypic similarities may create similar environmental responses to them, which is part of the gene-environment dynamic (Scarr & McCartney, 1983). Adoption twin studies are substantially more difficult for other reasons, including the need to locate appropriate participants for the study (locating the second twin may be quite difficult if the twins have not rediscovered each other) and in large enough numbers for meaningful statistical analyses, as well as the greater reliance on self-reported data; and the higher rate of missing historical data typically makes these studies substantially more expensive to conduct and subject to many scientific concerns.

In the past, some twin studies suffered from the inconclusive methods of confirming zygocity. Clearly, the accurate separation of twin pairs into MZ and DZ groups is critical to the assumptions of these studies. Although separations made by a survey of family members and self-reports have been shown to have a relatively high accuracy rate (91-98% accuracy), Christiansen and colleagues warn that accuracy of these determinations can vary by zygocity (lower accuracy for MZ), age/birth cohort (lower accuracy for twins in older birth cohorts (93%; $p < 0.04$)), and sex (the distribution of misclassified MZ and DZ pairs in men and women was statistically significant, $p = 0.009$, with males more likely to be incorrectly classified as DZ; Christiansen et al., 2003). When genetic material is unavailable for zygocity determination, use of a latent class analysis has been shown to be highly accurate (Heath et al., 2003). Today, investigators often use genetic testing to confirm zygocity of all twin pairs, but those investigators who do not leave their results vulnerable to questions of validity.

BEHAVIORAL GENETICS IN THE AGE OF MOLECULAR GENETICS AND NEUROSCIENCE

Much to the surprise of most geneticists (both behavioral and molecular), the completion of the human genome sequencing study did not result in rapid mapping of complex human abilities (phenotype) to specific genes. The smorgasbord of new molecular tools had not yielded the promised mapping of gene-to-behavior-to-brain location, nor did they elucidate the origins and cause-and-effect relationships we crave.

> *Ironically, it is precisely the aspect of the FW [Fisher-Wright] model that makes it so successful on a statistical level—the fractioning of genetic variance components into a collection of indistinguishable binomial effects—that makes it problematic as a model for the actual genetic etiology of traits. It is difficult to think of a biomedical disorder or behavioral trait with an etiology that is meaningfully described as an additive accumulation of a large number of very small independent causes, however useful such a model may be for modeling the occurrence of disorders and traits in pedigrees (Turkheimer, 2011, p. 229). What is missing from contemporary genomics is not heritability, but a meaningful link between statistical and etiological models of the transmission of complex traits (p. 232).*

How then can cognitive and behavioral neurogenetic science move forward with new discoveries? Both cognitive neuroscience and behavioral genetics have independently reached the conclusion that complex systems require complex solutions and that a key element of complexity is the unique, individual environmental effects on the system. Combined with the corollary or second shared realization that a single location will not yield the desired solution, both cognitive behavioral genetics and neuroscience have a shared trajectory. It is this simultaneous recognition of the powerful significance of individual factors that will lead to new and transformative discoveries.

How do sciences grounded in quantitative reasoning and population-level analyses advance a science that demands attention to individual factors and complex interactions? Both domains have already begun down the path of at least one solution—big data and informatics. Once combined, these data sets and complex algorithms hold great promise for these sciences to move beyond pure description to potential prescription. New analytic methods will also be required (and it seems fitting that new mathematical reasoning should contribute heavily to this endeavor).

Improved measurement is also in order. Detailed quantifying of cognitive development in mathematics, especially in the first years of life, and its interaction with other cognitive domains is needed. This quantification of early cognition needs to be matched with the quantifying of genetic and neurodevelopment, and a recognition of the contributions of unique, personal environmental factors. Only when we have developmental crosslinks will we be able to build and recognize a unified complex map of developmental cognition in any domain, such as mathematics. Only then will we be able to begin to harness the power of development as it rolls out in the individual.

SUMMARY AND CONCLUSIONS

In this chapter, we have examined some of the basic issues concerning the use of brain imaging methods and interpretations of the data they generate, as well as the ways in which these considerations may influence the study of mathematical cognitive development. We began by providing a brief history of brain imaging research in this field, followed by a discussion of the importance of adopting a developmental perspective. Then we reviewed various criticisms of fMRI, the dominant technique employed by the authors in this volume, and summarized the outcomes of a more prudent analysis that revealed these claims to have been mistakenly applied to the enterprise as a whole. Next, we described some specific ways in which neuroimaging can inform cognitive theories. Finally, we chronicled major developments in the field of genetics, described some of the interpretive challenges associated with behavioral genetic designs, and recommended that systematic quantification of mathematical cognitive development and its interaction with other cognitive domains be matched with the quantifying of genetic and neural development, taking into account the contributions of unique environmental factors experienced by individuals.

REFERENCES

Aguirre, G. K. (2014). Functional neuroimaging: Technological, logical, and social perspectives. In *Interpreting neuroimages: An introduction to the technology and its limits. Special report, Hastings Center Report* 45, No. 2: S8-S18. http://dx.doi.org/10.1002/hast.294.

Alarcón, M., DeFries, J. C., Light, J. G., & Pennington, B. F. (1997). A twin study of mathematics disability. *Journal of Learning Disabilities*, *30*, 617–623.

Anderson, J. R., Qin, Y., Sohn, M. H., Stenger, V. A., & Carter, C. S. (2003). An information-processing model of the BOLD response in symbol manipulation tasks. *Psychonomic Bulletin & Review*, *10*, 241–261.

Ansari, D. (2010). Neurocognitive approaches to developmental disorders of numerical and mathematical cognition: The perils of neglecting the role of development. *Learning and Individual Differences*, *20*, 123–129.

Aristotle. (350 B.C.E.) *Politics* – Book 7, Part XV, para. 2 (B. Jowett, Trans.). Retrieved from http://classics.mit.edu/Aristotle/politics.html.

Aristotle. (350 B.C.E.) *On the generation of animals* – Book 1, Part 18, para. 7 (A. Platt, Trans.) Retrieved from https://ebooks.adelaide.edu.au/a/aristotle/generation/book1.html.

Arsalidou, M., & Taylor, M. J. (2011). Is 2+2=4? Meta-analyses of brain areas needed for numbers and calculations. *NeuroImage*, *54*, 2382–2393.

Ashcraft, M. H., & Guillaume, M. M. (2009). Mathematical cognition and the problem size effect. In B. H. Ross (Ed.), *The psychology of learning and motivation: Vol. 51* (pp. 121–154). Burlington: Academic Press/Elsevier.

Bouchard, T. J. (1998). Genetic and environmental influences on adult intelligence and special mental abilities. *Human Biology*, *70*, 257–279.

Bouchard, T. (1999). Genes, environment, and personality. In S. J. Ceci & W. M. Williams (Eds.), *The nature-nurture debate: The essential readings* (pp. 97–103). Malden, MA: Blackwell.

Bouchard, T. J., Lykken, D. T., McGue, M., Segal, N. L., & Tellegen, A. (1990). Human psychological differences: The Minnesota study of twins reared apart. *Science*, *250*, 223–228.

Bowles, S., Gintis, H., & Groves, M. O. (Eds.), (2005). *Unequal chances: Family background and economic success*. Princeton, NJ: Princeton University Press.

Cho, S., Metcalfe, A. W. S., Young, C. B., Ryali, S., Geary, D. C., & Menon, V. (2012). Hippocampal-prefrontal engagement and dynamic causal interactions in the maturation of children's fact retrieval. *Journal of Cognitive Neuroscience*, *24*, 1849–1866.

Cho, S., Ryali, S., Geary, D. C., & Menon, V. (2011). How do you solve 7+8?: Decoding children's brain activity patterns during counting versus retrieval. *Developmental Science*, *14*, 989–1001.

Chochon, F., Cohen, L., van de Moortele, P. F., & Dehaene, S. (1999). Differential contributions of the left and right inferior parietal lobules to number processing. *Journal of Cognitive Neuroscience*, *11*, 617–630.

Christiansen, L., Frederiksen, H., Schousboe, K., Skytthe, A., von Wurmb-Schwark, N., Christensen, K., et al. (2003). Age- and sex-differences in the validity of questionnaire-based zygosity in twins. *Twin Research*, *6*, 275–278.

Clark, G. (2014). *The son also rises: Surnames and the history of social mobility*. Princeton, NJ: Princeton University Press.

Cohen Kadosh, R. (in press). Neuroscience. In R. Cohen Kadosh & A. Dowker (Eds.), *The Oxford handbook of numerical cognition*. Oxford: Oxford University Press.

Cohen Kadosh, R., Cohen Kadosh, K., Kaas, A., Henik, A., & Goebel, R. (2007). Notation-dependent and -independent representations of numbers in the parietal lobes. *Neuron*, *53*, 307–314.

Cohen Kadosh, R., Henik, A., Rubinsten, O., Mohr, H., Dori, H., van de Ven, V., et al. (2005). Are numbers special? The comparison systems of the human brain investigated by fMRI. *Neuropsychologia*, *43*, 1238–1248.

Darwin, C. (1859). *On the origin of species by means of natural selection*. London: John Murray.

De Smedt, B., Holloway, I. D., & Ansari, D. (2011). Effects of problem size and arithmetic operation on brain activation during calculation in children with varying levels of arithmetical fluency. *NeuroImage*, *57*, 771–781.

Dehaene, S. (1992). Varieties of numerical abilities. *Cognition, 44*, 1–42.

Dehaene, S. (2011). *The number sense: How the mind creates mathematics*. New York: Oxford University Press.

Dehaene, S., & Cohen, L. (1995). Towards an anatomical and functional model of number processing. *Mathematical Cognition, 1*, 83–120.

Dehaene, S., Piazza, M., Pinel, P., & Cohen, L. (2003). Three parietal circuits for number processing. *Cognitive Neuropsychology, 20*, 487–506.

Dehaene, S., Spelke, E., Pinel, P., Stanescu, R., & Tsivkin, S. (1999). Sources of mathematical thinking: Behavioral and brain-imaging evidence. *Science, 284*, 970–974.

Delazer, M., Domahs, F., Bartha, L., Brenneis, C., Lochy, A., Trieb, T., et al. (2003). Learning complex arithmetic—An fMRI study. *Cognitive Brain Research, 18*, 76–88.

Del Pinal, G., & Nathan, M. J. (2013). There and up again: On the uses and misuses of neuroimaging in psychology. *Cognitive Neuropsychology, 30*, 233–252.

Devlin, B., Daniels, M., & Roeder, K. (1997). The heritability of IQ. *Nature, 388*, 468–471.

Docherty, S. J., Davis, O. S. P., Kovas, Y., Meaburn, E. L., Dale, P. S., Petrill, S. A., et al. (2010). A genome-wide association study identifies multiple loci associated with mathematics ability and disability. *Genes, Brain and Behavior, 9*, 234–247.

Farah, M. J. (2014). Brain images, babies, and bathwater: Critiquing critiques of functional Neuroimaging. In *Interpreting neuroimages: An introduction to the technology and its limits. Special report, Hastings Center Report* 45, No. 2: S19-S30. http://dx.doi.org/10.1002/hast.295.

Fischer, M. H., & Shaki, S. (2014). Spatial associations in numerical cognition—From single digits to arithmetic. *The Quarterly Journal of Experimental Psychology, 67*, 1461–1483.

Fisher, R. A. (1918). The correlation between relatives on the supposition of mendelian inheritance. *Transactions of the Royal Society of Edinburgh, 52*, 399–433.

Fraga, M. F., Ballestar, E., Paz, M. F., Ropero, S., Setien, F., Ballestar, M., et al. (2005). Epigenetic differences arise during the lifetime of monozygotic twins. *Proceedings of the National Academy of Sciences of the United States of America, 102*, 10604–10609.

Galton, F. (1869). *Hereditary genius*. London: Macmillan.

Galton, F. (1875). The history of twins, as a criterion of the relative powers of nature and nurture. *Fraser's Magazine, 12*, 566–576.

Geary, D. C. (1993). Mathematical disabilities: Cognitive, neuropsychological, and genetic components. *Psychological Bulletin, 114*, 345–362.

Geary, D. C. (1995). Reflections of evolution and culture in children's cognition: Implications for mathematical development and instruction. *American Psychologist, 50*, 24–37.

Geary, D. C., Berch, D. B., & Mann Koepke, K. (Eds.), (2015). *Evolutionary origins and early development of number processing (Vol. 1, Mathematical Cognition and Learning)*. San Diego, CA: Elsevier Academic Press.

Gerstmann, J. (1940). Syndrome of finger agnosia, disorientation for right and left, agraphia and acalculia: Local diagnostic value. *Archives of Neurology and Psychiatry, 44*, 398–408.

Göbel, S. M., Calabria, M., Farné, A., & Rossetti, Y. (2006). Parietal rTMS distorts the mental number line: Simulating 'spatial' neglect in healthy subjects. *Neuropsychologia, 44*, 860–868.

Grabner, R. H., Ansari, D., Koschutnig, K., Reishofer, G., Ebner, F., & Neuper, C. (2009). To retrieve or to calculate? Left angular gyrus mediates the retrieval of arithmetic facts during problem solving. *Neuropsychologia, 47*, 604–608.

Grill-Spector, K., & Malach, R. (2001). fMR-adaptation: A tool for studying the functional properties of human cortical neurons. *Acta Psychologica, 107*, 293–321.

Gruber, O., Indefrey, P., Steinmetz, H., & Kleinschmidt, A. (2001). Dissociating neural correlates of cognitive components in mental calculation. *Cerebral Cortex, 11*, 350–359.

Halgren, E., Boujon, C., Clarke, J., Wang, C., & Chauvel, P. (2002). Rapid distributed fronto-parieto-occipital processing stages during working memory in humans. *Cerebral Cortex, 12*, 710–728.

Hardcastle, V. G., & Stewart, C. M. (2002). What do brain data really show? *Philosophy of Science, 69*, 572–585.

Heath, A. C., Nyholt, D. R., Neuman, R., Madden, P. A., Bucholz, K. K., Todd, R. D., et al. (2003). Zygosity diagnosis in the absence of genotypic data: An approach using latent class analysis. *Twin Research, 1*, 22–26.

Henschen, S. E., & Schaller, W. F. (1925). Clinical and anatomical contributions on brain pathology. *Archives of Neurology and Psychiatry, 13*, 226–249.

Henson, R. (2006). Forward inference using functional neuroimaging: Dissociations versus associations. *Trends in Cognitive Sciences, 10*, 64–69.

Hitch, G. J. (1978). The role of short-term working memory in mental arithmetic. *Cognitive Psychology, 10*, 302–323.

Holloway, I. D., Battista, C., Vogel, S. E., & Ansari, D. (2013). Semantic and perceptual processing of number symbols: Evidence from a cross-linguistic fMRI adaptation study. *Journal of Cognitive Neuroscience, 25*, 388–400.

Husén, T. (1960). Abilities of twins. *Scandinavian Journal of Psychology, 1*, 125–135.

Illes, J., Moser, M. A., McCormick, J. B., Racine, E., Blakeslee, S., Caplan, A., et al. (2010). Neurotalk: Improving the communication of neuroscience research. *Nature Reviews Neuroscience, 11*, 61–69.

Jackson, D. D. (1960). A critique of the literature on the genetics of schizophrenia. In D. D. Jackson (Ed.), *The etiology of schizophrenia* (pp. 37–87). Oxford, England: Basic Books.

Jang, K. L., Livesley, W. J., & Vernon, P. A. (1996). Heritability of the Big Five personality dimensions and their facets: A twin study. *Journal of Personality, 64*, 575–591.

Kaufmann, L., Kucian, K., & von Aster, M. (in press). Development of the numerical brain. In R. Cohen Kadosh & A. Dowker (Eds.), *The Oxford handbook of numerical cognition*. Oxford: Oxford University Press.

Kawashima, R., Taira, M., Okita, K., Inoue, K., Tajima, N., Yoshida, H., et al. (2004). A functional MRI study of simple arithmetic—A comparison between children and adults. *Cognitive Brain Research, 18*, 227–233.

Klein, E., Suchan, J., Moeller, K., Karnath, H. O., Knops, A., Wood, G., et al. (2014). Considering structural connectivity in the triple code model of numerical cognition: Differential connectivity for magnitude processing and arithmetic facts. *Brain Structure and Function, 219*, 1863–2653.

Knopik, V. S., & DeFries, J. C. (1999). Etiology of covariation between reading and mathematics performance: A twin study. *Twin Research, 2*, 226–234.

Kong, J., Wang, C., Kwong, K., Vangel, M., Chua, E., & Gollub, R. (2005). The neural substrate of arithmetic operations and procedure complexity. *Cognitive Brain Research, 22*, 397–405.

Lander, E. S., Linton, L. M., Birren, B., Nusbaum, C., Zody, M. C., Baldwin, J., et al. (2001). Initial sequencing and analysis of the human genome. *Nature, 409*, 860–921. http://dx.doi.org/10.1038/35057062.

Light, J. G., & DeFries, J. C. (1995). Comorbidity of reading and mathematics disabilities: Genetic and environmental etiologies. *Journal of Learning Disabilities, 28*, 96–106.

Loehlin, J. C., & Nichols, R. C. (1976). *Heredity, environment, and personality: A study of 850 sets of twins*. Austin, TX: University of Texas Press.

Luria, A. R. (1980). *Higher cortical functions in man* (2nd ed.). New York: Basic Books.

Mather, M., Cacioppo, J. T., & Kanwisher, N. (2013). How fMRI can inform cognitive theories. *Perspectives on Psychological Science, 8*, 108–113.

McCabe, D. P., & Castel, A. D. (2008). Seeing is believing: The effect of brain images on judgments of scientific reasoning. *Cognition, 107*, 343–352.

Medlund, P., Cederlöf, R., Floderus-Myrhed, B., Friberg, L., & Sörensen, S. (1976). A new Swedish twin registry containing environmental and medical base line data from about 14,000 same-sexed pairs born 1926-58. *Acta Medica Scandinavica Supplementum, 600*, 1–111.

Mendel, G. (1865). *Versuche über Pflanzen-Hybriden* ["Experiments in Plant Hybridization"], Mendel, Gregor. 1866. *Versuche über Plflanzenhybriden. Verhandlungen des naturforschenden Vereines in Brünn,* Bd. IV für das Jahr 1865, Abhandlungen, 3-47 (W. Bateson in 1901, corrections: R. Blumberg, Trans.). Retrieved from the MendelWeb project: http://www.esp.org/foundations/genetics/classical/gm-65.pdf.

Moyer, R. S., & Landauer, T. K. (1967). Time required for judgements of numerical inequality. *Nature, 215*, 1519–1520.

Norman, K. A., Polyn, S. M., Detre, G. J., & Haxby, J. V. (2006). Beyond mind-reading: Multi-voxel pattern analysis of fMRI data. *Trends in Cognitive Sciences, 10*, 424–430.

Oliver, B., Harlaar, N., Hayiou-Thomas, M. E., Kovas, Y., Walker, S. O., Petrill, S. A., et al. (2004). A twin study of teacher-reported mathematics performance and low performance in 7-year-olds. *Journal of Educational Psychology, 96*, 504–517.

Pauli, P., Lutzenberger, W., Rau, H., Birbaumer, N., Rickard, T. C., Yaroush, R. A., et al. (1994). Brain potentials during mental arithmetic: Effects of extensive practice and problem difficulty. *Cognitive Brain Research, 2*, 21–29.

Piazza, M., Pinel, P., Bihan, D. L., & Dehaene, S. (2007). A magnitude code common to numerosities and number symbols in human intraparietal cortex. *Neuron, 53*, 293–305.

Pinel, P., Piazza, D., Le Bihan, D., & Dehaene, S. (2004). Distributed and overlapping cerebral representations of number, size, and luminance during comparative judgments. *Neuron, 41*, 1–20.

Plomin, R., Asbury, K., & Dunn, J. (2001). Why are the children in the same family do different? Nonshared environment a decade later. *Canadian Journal of Psychiatry, 46*, 225–243.

Poldrack, R. A. (2006). Can cognitive processes be inferred from neuroimaging data? *Trends in Cognitive Sciences, 10*, 59–63.

Qin, Y., Carter, C. S., Silk, E. M., Stenger, V. A., Fissell, K., Goode, A., et al. (2004). The change of the brain activation patterns as children learn algebra equation solving. *Proceedings of the National Academy of Sciences of the United States of America, 101*, 5686–5691.

Qin, S., Cho, S., Chen, T., Rosenberg-Lee, M., Geary, D. C., & Menon, V. (2014). Hippocampal-neocortical functional reorganization underlies children's cognitive development. *Nature Neuroscience, 17*, 1263–1269.

Qin, Y., Sohn, M. H., Anderson, J. R., Stenger, V. A., Fissell, K., Goode, A., et al. (2003). Predicting the practice effects on the blood oxygenation level-dependent (BOLD) function of fMRI in a symbolic manipulation task. *Proceedings of the National Academy of Sciences of the United States of America, 100*, 4951–4956.

Rajagopal, A., Byars, A., Schapiro, M., Lee, G. R., & Holland, S. K. (2014). Success rates for functional MR imaging in children. *American Journal of Neuroradiology, 35*, 2319–2325.

Rickard, T. C., Romero, S. G., Basso, G., Wharton, C., Flitman, S., & Grafman, J. (2000). The calculating brain: An fMRI study. *Neuropsychologia, 38*, 325–335.

Rivera, S. M., Reiss, A. L., Eckert, M. A., & Menon, V. (2005). Developmental changes in mental arithmetic: Evidence for increased specialization in the left inferior parietal cortex. *Cerebral Cortex, 15*, 1779–1790.

Sawamura, H., Shima, K., & Tanji, J. (2002). Numerical representation for action in the parietal cortex of the monkey. *Nature, 415*, 918–922.

Scarr, S. (1968). Environmental bias in twin studies. *Eugenics Quarterly, 15*, 34–40.

Scarr, S., & McCartney, K. (1983). How people make their own environments: A theory of genotype greater than environment effects. *Child Development*, *54*, 424–435.

Schmithorst, V. J., & Brown, R. D. (2004). Empirical validation of the triple-code model of numerical processing for complex math operations using functional MRI and group Independent Component Analysis of the mental addition and subtraction of fractions. *NeuroImage*, *22*, 1414–1420.

Shermer, M. (2008). Why you should be skeptical of brain scans. *Scientific American Mind*, *19*, 66–71.

Siegler, R. S., & Shrager, J. (1984). Strategy choice in addition and subtraction: How do children know what to do? In C. Sophian (Ed.), *Origins of cognitive skills* (pp. 229–293). Hillsdale, NJ: Erlbaum.

Silventoinen, K., Sammalisto, S., Perola, M., Boomsma, D. I., Cornes, B. K., Davis, C., et al. (2003). Heritability of adult body height: A comparative study of twin cohorts in eight countries. *Twin Research*, *6*, 399–408.

Simonton, D. K. (1990). *Psychology, science, and history: An introduction to historiometry*. New Haven: Yale University Press.

Squire, L. R., & Zola-Morgan, S. (1991). The medial temporal lobe memory system. *Science*, *253*, 1380–1386.

Temple, C. M. (1991). Procedural dyscalculia and number fact dyscalculia: Double dissociation in developmental dyscalculia. *Cognitive Neuropsychology*, *8*, 155–176.

Temple, E., & Posner, M. I. (1998). Brain mechanisms of quantity are similar in 5-year old children and adults. *Proceedings of the National Academy of Sciences of the United States of America*, *95*, 7836–7841.

Thompson, L. A., Detterman, D. K., & Plomin, R. (1991). Associations between cognitive abilities and scholastic achievement: Genetic overlap but environmental differences. *Psychological Science*, *2*, 158–165.

Thompson, R. F., Mayers, K. S., Robertson, R. T., & Patterson, C. J. (1970). Number coding in association cortex of the cat. *Science*, *168*, 271–273.

Tosto, M. G., Hanscombe, K. B., Haworth, C. M. A., Davis, O. S. P., Petrill, S. A., Dale, P. S., et al. (2014). Why do spatial abilities predict mathematical performance? *Developmental Science*, *17*, 462–470.

Tse, D., Langston, R. F., Kakeyama, M., Bethus, I., Spooner, P. A., Wood, E. R., et al. (2007). Schemas and memory consolidation. *Science*, *316*, 76–82.

Turkheimer, E. (2011). Still missing. *Research in Human Development*, *8*, 227–241.

Uttal, W. R. (2001). *The new phrenology: The limits of localizing cognitive processes in the brain*. Cambridge, MA: MIT Press.

Vogel, S. E., Goffin, C., & Ansari, D. (2015). Developmental specialization of the left parietal cortex for the semantic representation of Arabic numerals: An fMR-adaptation study. *Developmental Cognitive Neuroscience*, *12*, 61–73.

Vuilleumier, P., Ortigue, S., & Brugger, P. (2004). The number space and neglect. *Cortex*, *40*, 399–410.

Watson, J. D., & Crick, F. H. C. (1953). A structure for deoxyribose nucleic acid. *Nature*, *171*, 737–738.

Weismann, A. (1893). *The Germ-Plasm*, Charles Scribner's Sons (prepared by Robert Robbins for Electronic Scholarly Publishing). Retrieved from http://www.esp.org/books/weismann/germ-plasm/facsimile/.

Zbrodoff, N. J., & Logan, G. D. (2005). What everyone finds: The problem-size effect. In J. I. Campbell (Ed.), *Handbook of mathematical cognition* (pp. 331–345). New York: Psychology Press.

Zorzi, M., Priftis, K., & Umiltá, C. (2002). Neglect disrupts the mental number line. *Nature*, *417*, 138.

Part I

Neural substrates

Chapter 2

Number Symbols in the Brain

Daniel Ansari

Department of Psychology and Brain and Mind Institute, The University of Western Ontario, London, Ontario, Canada

INTRODUCTION

Over the past three decades, our understanding of how humans process numerical quantity (the total number of items in a set) has increased dramatically. Through experimental work, it has been demonstrated that very young infants can discriminate between arrays containing different numbers of dots (for a review, see Libertus & Brannon, 2009). Moreover, work with a variety of animal species has revealed that, like human infants, animals can also discriminate between numerical quantities (Geary, Berch, & Mann Koepke, 2015). This research has led to the powerful suggestion that humans share with animals a sense of numerical quantity that is possibly innate (e.g., Brannon, 2006; Dehaene, 1997; Feigenson, Dehaene, & Spelke, 2004).

The finding that numerical discrimination performance can be predicted using the same parameters in nonhuman animals as well as human infants, children, and adults provides further support for the notion that a common representation of numerical quantity exists across species and developmental time. Specifically, across species and over developmental time, the accuracy of numerical discrimination performance can be predicted by the numerical ratio of the numerical quantities being compared. In other words, numerical discrimination performance decreases as the ratio of the quantities being compared increases (e.g., Buckley & Gillman, 1974; Halberda & Feigenson, 2008; Moyer & Landauer, 1967). The effect of ratio on numerical discrimination performance has been taken to suggest that the representation of numerical quantity is analog and therefore approximate. The representation of numerical quantity that humans share with other species and that can be detected from infancy onward is not exact. Instead, numbers that are close to one another share overlapping representations, and this overlap increases as a function of their size. That is, larger numbers overlap more than relatively smaller ones, leading to the effect of ratio on discrimination performance. In view of the fact that ratio effects have been found in nonhuman animals, human infants, children, and adults, it has

Development of Mathematical Cognition. http://dx.doi.org/10.1016/B978-0-12-801871-2.00002-2

been argued that an analog, approximate representation of numerical quantity exhibits both phylogenetic and ontogenetic continuity.

In addition to behavioral studies, neuroimaging studies of infants, adults, and children as well as single-cell recording studies in awake behaving animals have provided evidence consistent with the notion that there exists an approximate system for the representation of numerical quantity (e.g., Brannon, 2006; Nieder & Dehaene, 2009; Nieder & Miller, 2004; Nieder, Freedman, & Miller, 2002). Specifically, noninvasive neuroimaging studies with infants, children, and adults have consistently demonstrated that the right intraparietal sulcus (IPS, see Figure 1) is activated during numerical discrimination tasks (Cantlon, Brannon, Carter, & Pelphrey, 2006; Hyde, Boas, Blair, & Carey, 2010; Izard, Dehaene-Lambertz, & Dehaene, 2008). The common activation of this brain region at different points in development lends support to the notion that the core system for numerical quantity processing exhibits developmental continuity.

Furthermore, putative "number neurons" have been identified in the prefrontal and parietal cortex of nonhuman primates using invasive single-cell recording. These neurons respond to specific quantities while the monkey (*Macaca mulatta*) is performing a dot discrimination task. Moreover, the neurons exhibit the signature of the analog, approximate representation of the numerical quantity described. Consider, for example, a neuron that fires preferentially when the animal sees four objects on the screen. This neuron will also fire during the presentation of three and five dots, but less so than to four dots and even less to two and six dots. Put differently, these neurons exhibit analog, ratio-dependent tuning to a specific number of objects (Nieder & Miller, 2003, 2004; Nieder et al., 2002).

The aforementioned evidence is but a brief overview of a rich set of empirical data garnered at multiple levels of analysis, using a diverse collection of methodologies, to demonstrate the existence of numerical quantity representations that are shared across species and can be detected in humans from

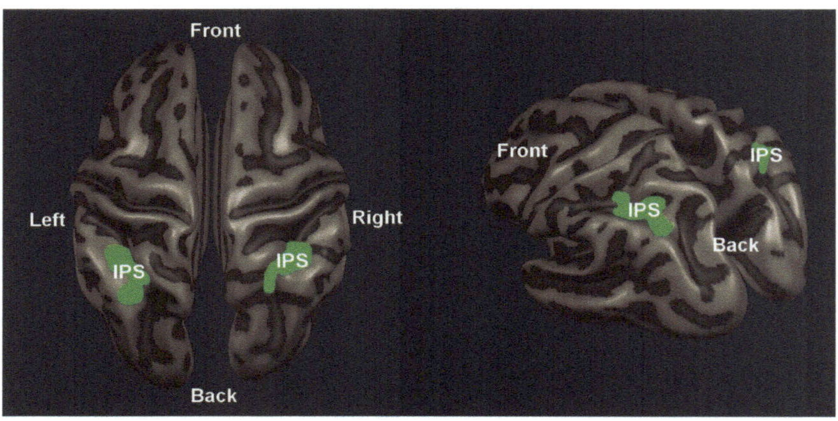

FIGURE 1 Reconstruction of the human brain, showing the intraparietal sulcus (IPS) displayed in green.

infancy onward. Note, however, that this research has been concerned with what is commonly referred to as *nonsymbolic representations of numerical quantity*, such as arrays of dots, numbers of beeps, collection of objects, and the like. Unlike nonhuman animals, humans use symbols (and by symbols I refer here to the culturally acquired representations of numerical quantity), such as number words and Arabic numerals to represent numerical quantity. These symbols have been invented by humans over the course of human history (Menninger, 1992). Like letters, these symbols for numerical quantity are non-iconic, that is the relationship between the sound of a number word or the shape of an Arabic numeral and the quantity they represent is completely arbitrary and needs to be learned, typically through schooling.

Therefore, from a cognitive neuroscience perspective, the acquisition of symbolic representations of numerical magnitude must be accompanied by changes in the brain and the gradual construction of brain regions (or networks of brain regions) that are engaged during the processing of numerical symbols. The acquisition of reading, which involves the mapping of letters onto speech sounds, is often referred to as a process of "cultural brain plasticity," whereby the association between letters and speech sounds becomes represented in the regions of the brain typically associated with auditory processing (e.g., Van Atteveldt, Formisano, Goebel, & Blomert, 2004). A similar process of brain plasticity must ensue in order to allow humans to go from processing symbolic representation of numerical magnitude as an arbitrary sequence of sounds (number words) or visual shapes (numerals) to effortlessly use them for comparing quantities, computing sums, or as ordinal labels.

Against this background, the aim of the present chapter is to review recent findings on how numerical symbols are processed in the human brain. More specifically, this chapter reviews evidence from cognitive neuroscience that has addressed the following questions: (1) Which brain regions are engaged during the processing of numerical symbols? (2) Do brain regions become specialized for processing numerical symbols over the course of learning and development? (3) Are there brain regions specialized for the perceptual processing of number symbols? Put differently, is there a visual number form area? (4) Do symbolic and nonsymbolic representations of quantity activate similar brain networks?

It should be noted at the outset that the focus of this chapter (as should be evident from the preceding guiding questions) is on what we have learned from neuroimaging studies about the brain circuitry underlying symbolic number processing rather than from behavioral studies alone.

WHICH BRAIN REGIONS ARE ENGAGED DURING THE PROCESSING OF NUMERICAL SYMBOLS?

Evidence from Comparison Tasks

Functional neuroimaging methods, such as functional magnetic resonance imaging (fMRI), have been used to identify the brain regions engaged when human adults process numerical symbols. Most such studies have used a numerical discrimination/comparison paradigm. In this task, participants are presented with either two numerical symbols and have to choose the one that represents the larger numerical quantity or with one numerical symbol and are asked to judge whether this symbol is numerically larger or smaller than a reference symbol (e.g., "5"). These studies have consistently revealed that the comparison of numerical symbols is associated with activation in and around the IPS in both the left and right hemispheres (e.g., Ansari, Fugelsang, Dhital, & Venkatraman, 2006; Kaufmann et al., 2005; Pinel, Dehaene, Riviere, & Le Bihan, 2001; Pinel et al., 1999). Furthermore, these studies have revealed that the amount of activation of these brain regions during the discrimination of numerical symbols is predicted by the numerical distance/ratio between the numbers. In other words, parietal brain regions are more strongly engaged when the numerical distance between the two numerical symbols that are being compared is relatively small (i.e., the numerical ratio is relatively large). The finding of distance/ratio effects on brain activation during numerical symbol comparison tasks has been taken as evidence to suggest that the analog, approximate representation of numerical quantity (described in the introduction) underlies the patterns of brain activation observed during symbolic number comparison.

Response-Selection Confounds

While comparison tasks have been widely used to index the representation and processing of numerical symbols, interpreting their neuronal correlates is not straightforward. Specifically, these tasks involve response selection (indicating which number is larger by choosing one of two buttons to press), thus confounding response selection with symbolic number processing. This is particularly problematic, given that a large body of neuroimaging studies has demonstrated that the IPS is engaged during response selection even when the tasks used to measure brain activation do not involve numerical stimuli (Göbel, Johansen-Berg, Behrens, & Rushworth, 2004; Gobel & Rushworth, 2004). In an effort to address this potential confound, Göbel and colleagues compared the neural correlates of symbolic and nonnumerical comparison. Specifically, adult participants were asked to perform three types of comparison: (1) numerical comparison of both single and double-digit numerals to a reference number; (2) judging whether or not a numerical symbol contained a vertical line (e.g., as in 5 but not 3); and (3) deciding whether an arbitrary, nonnumerical symbol contained a vertical line (Göbel et al., 2004). The comparison of the brain activation underlying these different conditions did not reveal any regions that were more activated during the numerical (i.e., is the numerical symbol presented smaller or larger than the reference number?) compared to the nonnumerical tasks (i.e., does the symbol contain a vertical line or not?). The absence of any activation specific to number comparison raises doubts about the extent to which comparison tasks can reveal the neural correlates of symbolic number processing independently of the neural mechanisms engaged during general response-selection and task-general stimulus processing.

Evidence from fMRI Adaptation Studies

In an effort to measure the neural correlates of symbolic number processing independently of response selection, researchers have used a methodology in which the brain response to stimuli can be measured in the absence of an active task. This methodology is commonly referred to as fMRI adaptation, fMRI habituation, or repetition suppression fMRI (for a review see Grill-Spector, Henson, & Martin, 2006). The logic behind fMRI adaptation is that neuronal populations that are sensitive to a particular category of stimuli will reduce their response to those stimuli (i.e., become adapted/habituated) when they are repeated and will recover from this repetition-induced suppression of the signal when the stimulus is changed.

Using this methodology, Notebaert, Nelis, and Reynvoet (2010) investigated the neural correlates underlying the processing of both single and double-digit Arabic numerals. In their experiment, single ("6") or double-digit ("32") numerals were repeatedly presented in different font types and spatial locations on a computer screen. This train of repeated numerals was interspersed with "numerical deviants" that differed from the numerals that were used for adaptation (see Figure 2). In particular, for both the small and large number adaptation conditions, 6 numerical deviants were presented among the train of repeated stimuli (6 or 32). These 6 numerical deviants fell into three ratio conditions (ratio between the adaptation number and the deviant). In this way the effect of numerical ratio (smaller/larger) on the magnitude of the response to deviants could be estimated. The participant's task was to press a button whenever the color of the numerals changed from white to red.

Using this design, Notebaert and colleagues (2010) investigated which brain regions responded to the numerical deviants and whether the magnitude of this brain response depended on the ratio between the adaptation number and the deviant. Their analysis revealed that a region of the left IPS responded to the numerical deviants in both the single and double-digit conditions. Moreover, this response scaled with the numerical ratio in such a way that the magnitude of the response increased as the ratio between the numbers decreased (see Figure 3). Put differently, when participants were adapted to the numeral 6, the response to deviants 3 and 12 (ratio of 0.5) was larger than the responses to deviants 4 and 9 (ratio of 0.66) and deviants 5 and 8 (ratios of 0.83 and 0.75).

These findings suggest that the parietal cortex (particularly the left parietal cortex) responds to numerical symbols in a ratio-dependent manner even in the absence of response selection. Thus, while comparison tasks revealing parietal activation may be confounded by response selection, the association between symbolic number processing and the IPS remains even when response selection is removed, as is the case in fMRI adaptation studies. Strikingly, the left IPS was the only region found to be associated with a ratio-dependent response to the numerical deviants. While comparison tasks frequently reveal a network of fronto-parietal regions associated with number comparison, this adaptation

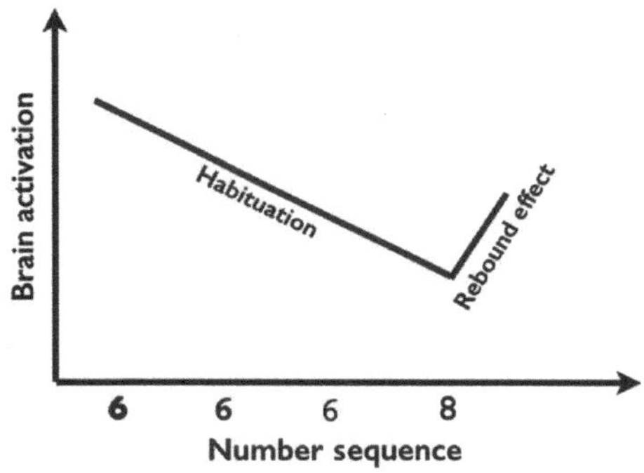

FIGURE 2 Predicted repetition-related decreases in brain activation followed by signal recovery after a change in the number presented. The font size of both habituation and deviant numbers were varied over the course of the experiment so that habituation could not occur to a specific font type or size.

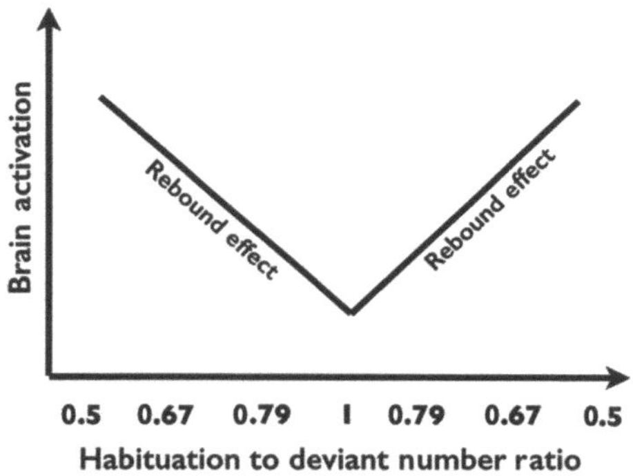

FIGURE 3 Predicted ratio-dependent rebound effect. The smaller the ratio between the adaptation and deviant number the larger the predicted amount of rebound (signal recovery).

study suggests that once response-selection confounds are removed, only parietal regions exhibit sensitivity to numerical quantity, at least in arithmetically competent adults. Furthermore, Notebaert and colleagues found a highly similar, ratio-dependent response to the numerical deviants both in the single and double-digit conditions, suggesting a similar coding for these symbolic representations in the left IPS.

Semantic or Perceptual Processing of Number Symbols in the IPS?

While the previously discussed fMRI adaptation results by Notebaert and colleagues (2010) demonstrate that responses to symbolic numbers can be detected in the IPS independently of response selection, it is not clear from these results whether such a response is strictly semantic in nature (i.e., based on the cardinal value of the numeral) or whether it is related to perceptual features of the stimuli. This question is particularly important in the context of a series of recent behavioral studies showing that physical similarity influences responses to number symbols over and above numerical ratio, especially in paradigms in which participants do not have to explicitly process the meaning of the numerical symbols. In other words, it has been shown that numbers close to one another are more physically similar than numbers further apart in the number sequence (Cohen, 2009; Cohen, Warren, & Blanc-Goldhammer, 2013).

This raises the question of how to more directly investigate whether the IPS is involved in semantic rather than perceptual processing in the absence of response selection. The most direct way of answering this question is to compare brains that have a semantic representation of numerical symbols to those that do not. This, of course, is difficult given that adults (in literate societies) have a semantic representation of the numerical symbols used (most commonly Arabic numerals). One option may be to take a developmental approach and compare children who know the meaning of Arabic numerals with those who do not yet understand the quantities represented by these symbols. Such an approach is complicated for a number of reasons. First of all, children who have and have not yet acquired a semantic representation of Arabic numerals are likely to differ in age, thereby confounding age with the comparison of interest (having or not having a semantic representation of Arabic numerals). Moreover, it is likely that the semantic knowledge young children have of Arabic numerals is on a continuum, thereby making it nearly impossible to neatly categorize children into those who do and do not have a semantic representation of Arabic numerals.

An alternative approach is to find a number symbol system that is used by some individuals but not others. Consider, for example, Chinese ideographs. These are symbolic representations of numerical quantity that are learned by readers of Chinese (see Table 1). In this way they are symbolic, noniconic (with the exception of 1-3), semantic representations of number for Chinese readers but not for individuals who do not read Chinese. Thus, if responses to the symbolic numbers in the IPS reflect the processing of the meaning (semantics) of numerical symbols, then only Chinese readers should exhibit such a brain response during the processing of Chinese ideographs.

This hypothesis was tested by Holloway, Battista, Vogel, and Ansari (2013) using an fMRI adaptation design. Specifically, these authors tested

TABLE 1 Numerals and Ideographs

Numerals	3	4	5	6	8	9	12
Ideographs	三	四	五	六	八	九	十二

both Chinese-English and English-non-Chinese bilinguals. Both groups were adapted to both Chinese ideographs and Arabic numerals using the fMRI adaptation method employed by Notebaert and colleagues (2010), which was discussed in detail earlier. The results revealed that both groups exhibited ratio-dependent responses to the numerical deviants in the Arabic numeral condition in the left IPS, thereby directly replicating the results by Notebaert et al. Critically, for the Chinese ideograph condition, only the Chinese-English bilinguals exhibited ratio-dependent responses in the IPS. Moreover, in contrast to the left-lateralized responses to the Arabic numeral deviants, activation correlated with the Chinese ideograph deviants most strongly in the right IPS.

These data lead to a number of conclusions. First of all, they show that numerical symbols are associated with activation of the IPS only when participants have a semantic representation of their numerical meanings. Second, the data point to interesting hemispheric differences in the responses to different formats of symbolic number. Even though both groups exhibited ratio-dependent responses in the left IPS in the Arabic numeral condition, the brain activation correlated with the Chinese ideographs was found to be strongest in the right IPS. There are several possible explanations for these hemispheric differences. On the one hand, it is possible that there are format-specific representations that vary as a function of the hemisphere (Cohen Kadosh, Cohen Kadosh, Kaas, Henik, & Goebel, 2007). Alternatively, as argued by Holloway et al., there are hemispheric differences as a function of the extent to which numerical symbols are used routinely. Although all the Chinese-bilinguals could name the numerical ideographs, none of them indicated that they used these routinely to represent number or calculate. Thus, it is possible that symbolic numbers that are used in everyday life and form the basis for higher-level numerical and mathematical cognition are more strongly lateralized to the left hemisphere and that the early learning of such symbols—before fluent processing of them—is more strongly lateralized to the right hemisphere.

This account sets up two developmental predictions: (1) the degree of engagement of the left IPS during Arabic numeral processing should increase over developmental time, and (2) individual differences in the amount of left IPS recruitment during symbolic number processing should correlate with individual differences in the higher-level usages of those symbols, such as symbolic mental arithmetic. These predictions will be evaluated against the background of developmental neuroimaging data, which are the focus of the next section of this chapter.

NUMERICAL SYMBOLS IN THE BRAIN—EVIDENCE FROM DEVELOPMENTAL STUDIES

Given that we are not born with an understanding of symbolic representations of numerical quantity (at least those that have been invented over the course of cultural history), the brain response to numerical symbols should be expected to change over the course of learning and development as children go from initially perceiving numerical symbols as meaningless shapes or sounds to having a rich representation of their meanings. There exists a small, but gradually growing body of studies that have investigated age-related changes in brain activation during symbolic number processing. Most of these studies have used comparison tasks and have compared groups of children to groups of adults. These studies have revealed that (a) the engagement of the IPS during symbolic number comparison increases over developmental time, and (b) the prefrontal regions tend to be more strongly activated in children compared to adults (Ansari, Garcia, Lucas, Hamon, & Dhital, 2005; Kaufmann et al., 2006). These findings have been interpreted as evidence for an experience-related specialization of the semantic representation of symbolic number in the parietal cortex. The decreasing engagement of the prefrontal cortex, found in some but not all studies, is thought to reflect a decreasing demand on executive functions, such as working memory and attention, subserved by areas of the frontal lobes, as children gain greater fluency and automaticity in processing numerical symbols. In other words, as the semantic representation in the parietal cortex becomes increasingly refined and more precise, there is less demand on domain-general functions of the prefrontal cortex during symbolic number comparison.

As we have discussed, comparison studies potentially confound symbolic number processing with response selection. In developmental neuroimaging studies this confound is exacerbated by the fact that children and adults will often differ significantly in terms of their reaction times and accuracy rates. Children tend to be slower and less accurate than adults. This complicates the interpretation of any age-related differences in brain activation as these may be driven by differences in the speed and accuracy of task performance rather than reflecting true developmental changes. Put differently, differences in age-related experiences and behavioral performance are confounded with one another. There are several solutions to addressing this confound. One of them is to match the groups of children and adults on the behavioral performance. This approach does eliminate group differences in behavioral performance and thus allows for a clearer interpretation of any age-related changes. However, this methodology can lead to selection biases, such as selecting fast and accurate children and matching them with relatively slow and inaccurate adults. Another approach is to use a passive task, such as the fMRI adaptation tasks previously discussed.

In a recent study, Vogel, Goffin, and Ansari (2014) adopted this latter approach and used fMRI adaptation to investigate the neural correlates of symbolic number processing in a group of 6- to 14-year-old children. The methods were nearly identical to those used in studies with adults by Notebaert et al. (2011) and Holloway et al. (2013), as discussed (see Figures 2 and 3). Using this paradigm, Vogel et al. investigated the ratio-dependent response to symbolic number deviants in two ways. First of all, the authors probed whether there were brain regions that exhibited a ratio-dependent response to the numerical deviants in an age-invariant manner. In other words, they investigated whether any brain regions respond to the symbolic number deviants across the entire group of children. This analysis revealed a region in the right IPS that was found to exhibit a ratio-dependent response to the numerical deviants but did not correlate with the children's chronological age. Second, to investigate whether any brain regions exhibited age-related differences in their ratio-dependent response, the authors correlated chronological age with brain activation in response to the numerical deviants. This analysis revealed that an area in the left IPS was significantly, and positively, correlated with chronological age. Put differently, older children activated the left IPS relatively more than their younger peers.

These results have a number of implications for our understanding of how the processing of symbolic numbers in the brain changes with age and experience. Specifically, the data reveal hemispheric differences in the extent to which the IPS undergoes age-related changes for symbolic number processing. The finding that the right IPS, at least between the ages of 6-14 years, exhibits age-invariant responses to symbolic numbers may suggest that this region has a less precise, more rudimentary, representation of symbolic numerical magnitude. This interpretation is consistent with findings showing that the right IPS responds to nonsymbolic numerical magnitudes from infancy onward and that brain activation during nonsymbolic fMRI adaptation is similar in 4-year-old children and adults in the right but not the left IPS (Cantlon et al., 2006; Hyde et al., 2010; Izard et al., 2008). Against this background, one might speculate that symbolic and nonsymbolic quantities are linked in the right IPS, whereas the left IPS activation reflects processes that are specific to the representations of numerical symbols, such as numerical order. Furthermore, the interpretation that the right IPS houses a less-elaborated representation of numerical symbols is also consistent with the cross-linguistic findings by Holloway et al., as discussed. Recall that they found that the less-used Chinese ideographs were associated with ratio-dependent responses in the right IPS, whereas the fMRI adaptation to the more commonly used Arabic numerals was correlated with activation in the left IPS.

Developmental differences in the response of the right and left IPS during number processing, similar to those reported by Vogel et al. (2014), were recently reported by Emerson and Cantlon (2014) who employed a longitudinal neuroimaging design. Specifically, children between the ages of 4 and 9 years were scanned at two time points that were separated by 1-2 years. In an fMRI experiment, children were presented with stimuli that displayed an array of dots and an Arabic numeral and were asked to indicate whether the dots and numerals were matching in quantity or not. When Emerson and Cantlon examined longitudinal changes in the brain response correlated with this symbolic-nonsymbolic matching task, they found that activation in the right, but not left IPS were correlated across the time points, suggesting continuity in the response of the right IPS. In contrast, changes in the left IPS were associated with behavioral changes in the precision with which children were performing the task. These data go further than those presented by Vogel et al. by linking developmental changes in the left IPS to age-related changes in behavioral performance. In other words, the results reported by Emerson and Cantlon (2014) reveal that the left but not right IPS is linked to changes in children's number processing abilities. Furthermore, as this is a longitudinal neuroimaging study, it offers greater insight into actual developmental changes in the relative activation of the left and the right IPS during number processing. In the context of the present focus on the neural correlates of symbolic number processing, it should be noted that the study by Emerson and Cantlon employed a matching task using symbolic and nonsymbolic stimuli. Therefore, for the purposes of the present chapter, it remains ambiguous whether the findings by Emerson and Cantlon are specifically related to symbolic number development or, as is more likely, to the neural correlates of the connection between symbolic and nonsymbolic representations of numerical quantity.

A similar result linking activation in the left IPS during symbolic number processing and children's numerical competence was also reported by Bugden, Price, McLean, and Ansari (2012). In this study, we examined the relationship between brain activation during symbolic number processing and individual differences in children's arithmetic achievement. More specifically, children between the ages of 8 and 10 years were asked to complete a symbolic number comparison task (which of two Arabic numerals is numerically larger) while their brain activation was measured using fMRI. In addition, outside the MRI scanner, children completed standardized tests of reading and arithmetic fluency (using the Woodcock-Johnson III Math and Reading Fluency subtests; Woodcock, McGrew, & Mather, 2001). A whole brain analysis revealed that children's left IPS was significantly modulated by the numerical ratio between the number symbols. That is, greater activation in the left IPS was revealed when children were comparing two numbers separated by a relatively large compared to a small ratio. More importantly, a whole brain correlation analysis to investigate the relationship between individual differences in the effect of numerical ratio on brain activation and performance on the mathematical fluency test (which asks children to solve as many arithmetic problems as they can in 3 min) revealed a significant, positive correlation in the left IPS. Put differently, those children who scored more highly on the mathematics fluency test were also those who exhibited a larger effect of numerical ratio in the left IPS. Furthermore, this correlation was found to be specific to children's performance on the arithmetic fluency test, as no such correlation was revealed between reading fluency and individual differences in brain activation. These findings then, like those by Emerson and Cantlon (2014), establish a link between individual differences in behavioral markers of numerical and mathematical proficiency and variability in children's recruitment of the left, but not the right IPS during symbolic number processing.

Taken together, developmental neuroimaging studies investigating the neural correlates of symbolic number processing have revealed that the left IPS becomes increasingly engaged as a function of children's chronological age and their numerical and mathematical competencies. In view of this one might speculate that the more children have specialized their left IPS for symbolic number processing, the better prepared they are to use symbolic numbers in higher-level tasks, such as solving arithmetic problems. It remains for future longitudinal studies to investigate this hypothesis more directly.

Most of this evidence points to the left IPS as a critical region involved in the development of symbolic number processing and in the association between individual differences in symbolic number processing and higher-level numerical abilities, such as arithmetic. However, it is important not to focus solely on a single region when considering the neuronal correlates of symbolic number processing. In view of this, a recent study by Emerson and Cantlon (2012) is particularly instructive. Rather than using the common approach of relating activation of single brain regions to individual differences in numerical and mathematical achievement, these authors investigated the extent to which connectivity (indexed by a correlation) between brain regions and between-subject variability therein was associated with variability in achievement. Their findings indicated that individual differences in the extent to which activation in parietal and prefrontal regions were correlated with one another was positively associated with standardized measures of early math achievement in a sample of 4- to 11-year-old children. Importantly, individual differences in fronto-parietal connectivity were associated with standardized measures of early mathematics achievement but not with variability in children's intelligence.

The findings of Emerson and Cantlon (2012) are important as they highlight the critical role that networks of brain regions play. Moreover, their data suggest that it is not the relative amount of activation of the parietal cortex alone that is linked to individual differences in children's numerical and mathematical achievement, but rather the connectivity between parietal and other brain regions (such as the prefrontal cortex) that plays a critical role in symbolic number development. In this context, it should be noted that a related study by Cantlon and Li (2013) also investigated connectivity between brain regions and their relationship to standardized measures of children's mathematical proficiency. However, since that study did not specifically investigate the neural networks involved in symbolic number processing, it will not be discussed further here.

PERCEPTUAL REPRESENTATION OF NUMBER SYMBOLS IN THE BRAIN

Thus far I have focused on the brain mechanisms underlying the processing of the meaning of numerical symbols. However, the brain does not only process symbols semantically. Before a symbol's meaning can be accessed, it must first be perceptually identified. Numerical symbols are ubiquitous in literate societies. It is therefore reasonable to hypothesize that there might be a brain region in the visual cortex that becomes specialized for the visual recognition of numerical symbols, much like brain regions that are specialized for the perceptual processing of other highly frequent perceptual stimuli, such as faces and words (Kanwisher & Yovel, 2006; McCandliss, Cohen, & Dehaene, 2003). Indeed, the Triple-Code Model by Dehaene and Cohen (1995), which is arguably the most influential neuropsychological model of numerical processing, predicts the existence of a visual number form area in the ventral visual cortex. In contrast to the large number of empirical studies on the semantic processing of numerical symbols and the role of the IPS, comparatively few studies have tested the Triple-Code Model's hypothesis that there exists a brain region specialized for the processing of numerical symbols in the ventral stream.

To test whether a brain region exists that exhibits category-specific activation for the processing of numerical symbols (rather than activation related to the processing of symbols generally), it is critical to contrast numerical symbol processing with activation related to control conditions consisting of the presentation of nonnumerical symbols, such as letters. The majority of functional imaging studies that have contrasted Arabic numeral and letter processing have been unable to find support for the existence of a dedicated visual number form area in the ventral visual cortex (Cantlon, Pinel, Dehaene, & Pelphrey, 2011; Chochon, Cohen, van de Moortele, & Dehaene, 1999; Price & Ansari, 2011; Stanescu-Cosson et al., 2000). Although a study by Park, Hebrank, Polk, and Park (2012) did reveal greater activation for Arabic numerals compared to letters in the right ventral stream (lateral occipital cortex), the same region did not exhibit a difference in activation between letters and meaningless stimuli (e.g., pseudowords).

While there is not strong evidence in support of the existence of a dedicated visual number form area in the ventral visual stream, a recent study has suggested that the left angular gyrus may play a category-specific role in the processing of Arabic numerals. More specifically, Price and Ansari (2011), used fMRI to contrast brain activation in response to the perception of single-digit Arabic numerals with activation related to both letter and scrambled[1] stimuli. Although this study did not reveal category-specific activation in the ventral stream, a region of the ventral left angular gyrus was found to be significantly more activated during the processing of Arabic numerals compared to both letters and scrambled stimuli. In studies of mathematical processing, the left angular gyrus has frequently been associated with the retrieval of arithmetic facts (e.g., Grabner et al., 2007, 2008; Zamarian, Ischebeck, & Delazer, 2009). In view of its association with the retrieval of arithmetic facts as well as the visual processing of Arabic numerals, it has been hypothesized that the left angular gyrus is involved in the mapping of numerical symbols (be they arithmetic expressions or single-digit numerals) to their semantic referents. Interestingly, the left ventral angular gyrus has been associated with the conceptual identification of visual inputs (Seghier, 2013; Seghier, Fagan, & Price, 2010). In view of this, Price and Ansari (2011) argue that category-specific activation of the left angular gyrus cannot be purely perceptual and that it is

1. The scrambled stimuli were created by segmenting and rearranging the intact letters and digits into a unified but completely novel shape. In other words, each letter and each digit had an unrecognizable scrambled counterpart with equivalent number of pixels and approximately similar numbers of angles and curves.

not a visual number form area. Instead they contend that the engagement of the left angular gyrus may reflect the brain's initial identification of Arabic numerals as stimuli that have meaning.

In view of the aforementioned, functional neuroimaging studies have not been successful at identifying a brain region that exhibits category-specific activation in response to numerical symbols. Of course, as with any null result, the absence of category-specific activation for numerical symbols in the ventral stream does not preclude the possibility that a visual number form area exists. Functional neuroimaging methods have many limitations that might explain such a null result. For example, the resolution of functional neuroimaging methods, such as fMRI, is relatively coarse. In addition, there could be image acquisition artifacts that prevent the detection of signals within brain areas that may house a visual number form area.

Recent evidence from a different, invasive, method to measure brain activation in humans suggests that fMRI may not be capable of detecting the visual number form area. More specifically, Shum et al. (2013) used intracranial electrophysiology to record from regions of the inferior temporal gyrus while participants viewed Arabic numerals as well as control stimuli (scrambled numerals, letters, scrambled letters, foreign numerals, number words as well as nonnumber words). Recordings from seven patients undergoing surgery for epileptic seizures revealed significantly higher responses in the high-frequency broadband in the inferior temporal gyrus for Arabic numerals compared to all the control stimuli. These findings provide support for the Triple-Code Model's prediction for a visual number form area in the ventral visual cortex.

In addition to revealing a visual number form area, Shum et al. (2013) report fMRI evidence that explains why previous studies may not have been able to reveal a visual number form area. Specifically, in a separate fMRI experiment with six healthy participants, they found that the region that was found to exhibit category-specific responses to Arabic numerals in the intracranial electrophysiology experiment lies within a region that exhibits signal dropout during the acquisition of fMRI data. In other words, the visual number form area lies within a region of the brain from which it is hard to detect signals using fMRI (a so-called fMRI eclipse zone). This finding could account for the null results for a visual number form area in the fMRI studies that have sought to test for such a region. In view of this, it will be critical for future studies to replicate the electrophysiological evidence and to seek ways to increase the amount of signal measurable from this region using fMRI.

ARE SYMBOLIC AND NONSYMBOLIC QUANTITY REPRESENTATIONS LINKED IN THE BRAIN?

When children first encounter numerical symbols (e.g., 4, 7, "ten"), these do not carry any meaning for them. So how do numerical symbols become representations of numerical quantity? It has frequently been hypothesized (e.g., Dehaene, 2007; Piazza, 2010) that symbols acquire numerical meaning by being connected to the approximate, nonsymbolic representations of numerical quantity that humans share with other species and can be measured in very young infants, as described earlier.

The hypothesis that there is a strong link between symbolic and nonsymbolic representations of numerical quantity provides a set of testable hypotheses for cognitive neuroscience. Specifically, if symbolic and nonsymbolic representations are linked with one another, they should (a) be represented in similar brain areas/networks of brain region, (b) demonstrate cross-format coding—that the processing of symbolic representation should lead to cross-activation of nonsymbolic representations and vice versa, and (c) display similar neural representational features of symbolic and nonsymbolic numerical quantity.

To test whether symbolic and nonsymbolic representations of numerical quantity are correlated with activation in similar brain networks, several researchers have used a method referred to as conjunction analysis. This statistical tool enables researchers to examine which brain regions are activated by both symbolic and nonsymbolic representations of numerical quantity. The results from these studies have revealed that areas of the IPS are activated by both symbolic and nonsymbolic numerical quantity (Cantlon et al., 2008; Holloway & Ansari, 2010; Holloway, Price, & Ansari, 2010). The laterality of these effects is somewhat unclear, with some authors reporting common activation in the right IPS and others revealing joint activation in the left IPS. As previously discussed, there is evidence to suggest that before children have acquired a fluent symbolic representation of numerical quantity, nonsymbolic numerical quantity processing is correlated with activation in the right IPS (Hyde et al., 2010; Izard et al., 2008), while increasing expertise with symbolic number is correlated with activation in the left IPS. Thus, if there is a neural connection between symbolic and nonsymbolic representations of numerical quantity, it is likely to originate in the right IPS.

Another method by which to test whether there is an association between nonsymbolic and symbolic representations of numerical quantity in the brain is to examine whether the presentation of one representational format also leads to the activation of the other (i.e., cross-format activation). Using fMRI adaptation, Piazza, Pinel, Le Bihan, and Dehaene (2007) tested whether adaptation to symbolic number quantities would also activate nonsymbolic representations and vice versa. More specifically, participants were first adapted to either a symbolic (Arabic numerals) or nonsymbolic (dot arrays) numerical quantity. Following adaptation they were presented with numerical deviants in the other format. Responses to these deviants were found in a bilateral network of frontal parietal regions. Importantly, the degree of deviant-related response in the IPS was related to the numerical distance between the number to which the brain was adapted and the deviant. In other words, the amount of deviant-related activation increased as the numerical distance between the adaptation and deviant numbers increased. Importantly, this distance effect occurred both when the change from notation was from symbolic to nonsymbolic and vice versa.

This cross-format numerical distance effect suggests that symbolic and nonsymbolic representations of numerical quantity tap into a common approximate, neural code of quantity representation. As the authors put it, the findings:

> *"support the idea that symbols acquire meaning by linking neural populations coding symbol shapes to those holding nonsymbolic representations of quantities."*
>
> (Piazza et al., 2007, p. 293)

Whereas the findings reported by Piazza et al. (2007) support the notion of a format-independent representation of numerical quantity in the parietal cortex, more recent findings have questioned the conclusion drawn. Also using fMRI adaptation to study cross-format processing in the brain, Cohen Kadosh et al. (2011) found that a change of format in the absence of a change in numerical quantity (e.g., a change from 12 dots to the symbolic representation "12") led to activation in the same IPS regions that responded to a change in quantity. Furthermore, format changes led to greater activation in the IPS than did quantity changes, which is at odds with the existence of a common, amodal quantity code.

Importantly, the use of both conjunction analysis and fMRI adaptation to examine commonalities in the neuronal correlates of symbolic and nonsymbolic numerical quantity processing suffer from a number of limitations regarding the inferences that can be drawn about the presence or lack of a strong link between symbolic and nonsymbolic quantity processing in the brain. Because fMRI has a relatively coarse spatial resolution and involves averaging the activation across points of observation (called voxels; three-dimensional pixels), it is entirely possible that the common activation of a brain region for symbolic and nonsymbolic numerical quantity processing does not imply equivalence of processes engaged by each condition. In other words, it is possible that the activation of a brain region by two experimental conditions is driven by two different underlying mechanisms.

To address this issue, cognitive neuroscientists have begun to use novel, multivariate analytic techniques. Rather than averaging activation across voxels, these new tools analyze the pattern of activation of multiple voxels and compare and correlate the pattern of activation between conditions. This method is often referred to as Multi-Voxel Pattern Analysis (MVPA). Bulthé, De Smedt, and Op de Beeck (2014) used this approach to understand whether the patterns of different symbolic and nonsymbolic numerical quantities can be distinguished from one another both with and across formats in both the IPS and other brain regions. Their results reveal that while it is possible to distinguish the patterns of activation elicited by different quantities within formats (e.g., distinguishing 4 dots from 6 dots or the numeral 5 from the numeral 9), there was no evidence for cross-format classification. Specifically, the authors found that the patterns that distinguish symbolic quantities from one another cannot be applied to distinguishing nonsymbolic quantities and vice-a-versa. This evidence points to a lack of overlap in the distributed (across voxels) representations of symbolic and nonsymbolic numerical quantities.

Similar evidence was obtained in a study by Damarla and Just (2013), who also showed good within-format classification of numerical quantities, but poor cross-format classification (i.e., the multi-voxel patterns of different numbers could be distinguished within but not across formats of numerical quantity representation). Another study by Eger et al. (2009) using MVPA to classify the patterns of symbolic and nonsymbolic numerical quantities revealed modest cross-format classification (just over 50% correct) but only in one direction: the multi-voxel patterns for dot arrays could be used to predict the activation of corresponding Arabic numerals, but the patterns of activation elicited during the processing of the numerals could not be used to predict the patterns correlated with processing dot arrays.

Similarly, through examining the correlation between multi-voxel pattern activity (an approach referred to as Representational Similarity Analysis), Lyons, Ansari, and Beilock (2014) found no correlation between the patterns of individual symbolic and nonsymbolic numbers.

Taken together, recent neuroimaging methods using cutting-edge multivariate neuroimaging analysis techniques have begun to question the commonly held view that symbolic representations of numerical quantity acquire their meaning by becoming associated with the preexisting, approximate, nonsymbolic representations of quantity (Dehaene, 2007; Piazza, 2010). An alternative explanation is that number symbols are initially linked to small quantities (~1-4) that can be enumerated exactly and without counting (a process referred to as subitizing; Trick & Pylyshyn, 1994). In other words, early in development, the number words 1-4 become associated with nonsymbolic numbers in the subitizing range, but larger numbers are not linked to nonsymbolic representations of numerical quantity (Carey, 2001).

Future, developmental neuroimaging studies should test this hypothesis by studying young children who are in the initial stages of learning the quantities represented by numerical symbols (number words) and examining the extent to which brain activation patterns of nonsymbolic representations of numbers in the subitizing range are linked to the first number words that children learn and, if so, to what extent the strength of this association changes over developmental time. In view of the developmental neuroimaging studies discussed earlier, one might also speculate that the initial association with small quantities is subserved by the right IPS, while the continued development of symbolic number representations (in the absence of a strong link to nonsymbolic representations of quantity) is associated with the left IPS.

DIFFERENCES IN CARDINAL AND ORDINAL PROCESSING OF NUMBER SYMBOLS IN THE BRAIN

Beyond being representations of quantities, numerical symbols are part of an ordered sequence. Put differently, symbolic numbers do not only convey "how many" (cardinality) but also "what position" (ordinality). Furthermore, positional information provides relational information, such as which number comes before and which comes after. To date, neuroimaging studies (see earlier) of symbolic number processing have focused almost entirely on the question of symbols as representations of numerical quantity (the "how much" question).

Recent evidence by Lyons and Beilock (2013) demonstrates that a thorough investigation of the neural correlates of ordinality may help to significantly advance both our understanding of symbolic number representations in the brain and the differences in the way in which the brain codes for symbolic and nonsymbolic representations of numerical quantity. Specifically, Lyons and Beilock (2013) asked adult participants to (a) judge which of two stimuli is numerically greater (the "how many" question) and (b) indicate whether a series of three numerical quantities were in increasing, decreasing, or mixed numerical order (e.g., 1-2-3 being in order and 2-4-1 being in a mixed order). These two conditions were presented using both Arabic numerals and dot arrays. The judgment of "how much" (cardinality) was found to be associated with activation in areas of the IPS for both symbolic and nonsymbolic number processing. For ordinal processing, however, symbolic and nonsymbolic formats diverged significantly with respect to their neural correlates. While for nonsymbolic ordering, activation was observed in the same brain region as for cardinal processing, symbolic ordering was associated with activation in the left premotor cortex, but not the IPS. These findings suggest that processing the ordinality and cardinality of symbolic representations of numerical quantity is associated with different brain regions, suggesting that they may represent two very distinct aspects of symbolic number processing. Furthermore, the fact that no such differences between ordinal and cardinal judgments were found for the nonsymbolic condition may suggest that ordinality is a key aspect of symbolic number processing that differentiates it from nonsymbolic numerical quantity processing.

Against this background, future studies should investigate further the neural correlates of the ordinal and cardinal processing of number symbols as well as developmental and experience-related changes and individual differences in their neuronal correlates.

CONCLUSIONS AND FUTURE DIRECTIONS

Numerical symbols, such as number words and numerals, have been invented over the course of human cultural history. Therefore, brain representations of symbolic numerical quantity need to be constructed over the course of learning—a process sometimes referred to as "cultural brain plasticity" (Ansari, 2012). The present chapter provides an overview of what is known about (a) the brain circuitry associated with symbolic number processing; (b) experience-related and developmental changes in the neural correlates of symbolic number processing and their relationship to individual differences in mathematical achievement; (c) semantic and perceptual processing of numerical symbols; and (d) how symbols are connected to nonsymbolic representations of numerical quantity in the brain.

It is evident from this review that the last two decades have witnessed tremendous progress in our understanding of how the brain processes numerical symbols. A large body of empirical research studies using noninvasive neuroimaging methods have consistently shown that regions of the IPS are associated with the processing of numerical symbols, especially the processing of the meaning of numerical symbols (i.e., that they are representations of numerical quantity). In addition, a growing body of studies is pointing to the left IPS as a critical region for the processing of numerical symbols. Specifically, it has been demonstrated that the left IPS is associated with symbolic number processing even after confounds, such as response selection, are removed. Moreover, it has been shown that the left IPS is a region whose activation during symbolic number processing increases over developmental time and is associated with individual differences in calculation fluency. The increasing engagement of the left IPS over developmental time might point to the emergence of a brain representation for symbolic representations of numerical quantity that is different from the processing of nonsymbolic numerical quantity in the right IPS which can already be detected in young infants and children who have not yet acquired fluent, symbolic number processing skills.

Even though much progress has been made in understanding the brain regions (and their networks) that are correlated with processing the meaning of numerical symbols, the literature reviewed here remains unclear regarding a number of critical questions. First of all, functional neuroimaging studies have yet to reveal a region in the visual cortex that is specialized for the visual, asemantic processing of numerical symbols (a visual number form area). The reasons for such a null result should be explored more fully in future studies. A complete account of the neural correlates of symbolic number processing should be able to map out the brain regions involved in both semantic and perceptual processing of symbolic number. In turn, this will enable an examination of how different brain regions subserving these functions change differentially over the course of learning numerical symbols and vary between individuals with different levels of numerical and mathematical proficiency.

Another area in need of further exploration is the brain's representation of numerical order. As discussed, recent evidence by Lyons and Beilock (2013) points to different brain mechanisms underlying cardinal and ordinal judgments of symbolic numbers. Future studies can both supplement and elucidate these findings by employing longitudinal designs to explore whether ordinality and cardinality are subserved by different developmental trajectories in the brain and whether these are associated with individual differences in higher-level numerical and mathematical competence.

Perhaps the most critical issue for future research into the neurocognitive systems underlying symbolic number processing pertains to the way in which numerical symbols acquire meaning. The leading and most straightforward prediction is that symbols are grounded (mapped to) the approximate, nonsymbolic representations of numerical quantity that we share with other species. While this prediction is certainly plausible, the neuroimaging evidence reviewed here does not support it well. Instead, most findings point to weak connections between symbolic and nonsymbolic numerical quantity representations in the brain. Therefore the "symbol-grounding" problem (Harnad, 1990) with respect to numerical symbols has not been solved. Cognitive neuroscience studies that employ developmental and/or training designs may shed further light on this central issue.

ACKNOWLEDGMENTS

Daniel Ansari is supported by operating grants from the Natural Sciences and Engineering Council of Canada (NSERC), the Canadian Institutes of Health Research (CIHR), the Canada Research Chairs Program as well as an E.W.R. Steacie Memorial Fellowship from the Natural Sciences and Engineering Council of Canada (NSERC).

REFERENCES

Ansari, D. (2012). Culture and education: New frontiers in brain plasticity. *Trends in Cognitive Sciences*, *16*, 93–95. http://dx.doi.org/10.1016/j.tics.2011.11.016.

Ansari, D., Fugelsang, J. A., Dhital, B., & Venkatraman, V. (2006). Dissociating response conflict from numerical magnitude processing in the brain: An event-related fMRI study. *NeuroImage*, *32*(2), 799–805. Retrieved from http://www.ncbi.nlm.nih.gov/entrez/query.fcgi?cmd=Retrieve&db=PubMed&dopt=Citation&list_uids=16731007.

Ansari, D., Garcia, N., Lucas, E., Hamon, K., & Dhital, B. (2005). Neural correlates of symbolic number processing in children and adults. *Neuroreport*, *16*(16), 1769–1773. Retrieved from http://www.ncbi.nlm.nih.gov/entrez/query.fcgi?cmd=Retrieve&db=PubMed&dopt=Citation&list_uids=16237324.

Brannon, E. M. (2006). The representation of numerical magnitude. *Current Opinion in Neurobiology*, *16*, 222–229.

Buckley, P. B., & Gillman, C. B. (1974). Comparisons of digits and dot patterns. *Journal of Experimental Psychology*, *103*(6), 1131–1136. Retrieved from http://www.ncbi.nlm.nih.gov/entrez/query.fcgi?cmd=Retrieve&db=PubMed&dopt=Citation&list_uids=4457588.

Bugden, S., Price, G. R., McLean, D. A., & Ansari, D. (2012). The role of the left intraparietal sulcus in the relationship between symbolic number processing and children's arithmetic competence. *Developmental Cognitive Neuroscience*, *2*, 1–10. http://dx.doi.org/10.1016/j.dcn.2012.04.001.

Bulthé, J., De Smedt, B., & Op de Beeck, H. P. (2014). Format-dependent representations of symbolic and non-symbolic numbers in the human cortex as revealed by multi-voxel pattern analyses. *NeuroImage*, *87*, 311–322. http://dx.doi.org/10.1016/j.neuroimage.2013.10.049.

Cantlon, J. F., Brannon, E. M., Carter, E. J., & Pelphrey, K. A. (2006). Functional imaging of numerical processing in adults and 4-y-old children. *PLoS Biology*, *4*(5), e125. Retrieved from http://www.ncbi.nlm.nih.gov/entrez/query.fcgi?cmd=Retrieve&db=PubMed&dopt=Citation&list_uids=16594732.

Cantlon, J. F., & Li, R. (2013). Neural activity during natural viewing of Sesame Street statistically predicts test scores in early childhood. *PLoS Biology*, *11*(1), e1001462. http://dx.doi.org/10.1371/journal.pbio.1001462.

Cantlon, J. F., Libertus, M. E., Pinel, P., Dehaene, S., Brannon, E. M., & Pelphrey, K. A. (2008). The neural development of an abstract concept of number. *Journal of Cognitive Neuroscience*. http://dx.doi.org/10.1162/jocn.2008.21159.

Cantlon, J. F., Pinel, P., Dehaene, S., & Pelphrey, K. A. (2011). Cortical representations of symbols, objects, and faces are pruned back during early childhood. *Cerebral Cortex (New York, NY: 1991)*, *21*(1), 191–199. http://dx.doi.org/10.1093/cercor/bhq078.

Carey, S. (2001). Cognitive foundations of arithmetic: Evolution and ontogenesis. *Mind and Language*, *16*(1), 37–55. http://dx.doi.org/10.1111/1468-0017.00155.

Chochon, F., Cohen, L., van de Moortele, P. F., & Dehaene, S. (1999). Differential contributions of the left and right inferior parietal lobules to number processing. *Journal of Cognitive Neuroscience*, *11*(6), 617–630. Retrieved from http://www.ncbi.nlm.nih.gov/entrez/query.fcgi?cmd=Retrieve&db=PubMed&dopt=Citation&list_uids=10601743.

Cohen, D. J. (2009). Integers do not automatically activate their quantity representation. *Psychonomic Bulletin & Review*, *16*(2), 332–336. http://dx.doi.org/10.3758/PBR.16.2.332.

Cohen Kadosh, R., Bahrami, B., Walsh, V., Butterworth, B., Popescu, T., & Price, C. J. (2011). Specialization in the human brain: The case of numbers. *Frontiers in Human Neuroscience*, *5*, 62. http://dx.doi.org/10.3389/fnhum.2011.00062.

Cohen Kadosh, R., Cohen Kadosh, K., Kaas, A., Henik, A., & Goebel, R. (2007). Notation-dependent and -independent representations of numbers in the parietal lobes. *Neuron*, *53*(2), 307–314.

Cohen, D. J., Warren, E., & Blanc-Goldhammer, D. (2013). Cross-format physical similarity effects and their implications for the numerical cognition architecture. *Cognitive Psychology*, *66*(4), 355–379. http://dx.doi.org/10.1016/j.cogpsych.2013.03.001.

Damarla, S. R., & Just, M. A. (2013). Decoding the representation of numerical values from brain activation patterns. *Human Brain Mapping*, *34*(10), 2624–2634. http://dx.doi.org/10.1002/hbm.22087.

Dehaene, S. (1997). *The number sense*. New York, NY: Oxford University Press.

Dehaene, S. (2007). Symbols and quantities in parietal cortex: Elements of a mathematical theory of number representation and manipulation. In P. Haggard, Y. Rossetti, & M. Kawato (Eds.), *Sensorimotor foundations of higher cognition*: *Vol. XXII* (pp. 527–574). Boston: Harvard University Press.

Dehaene, S., & Cohen, L. (1995). Towards an anatomical and functional model of number processing. *Mathematical Cognition*, *1*, 83–120.

Eger, E., Michel, V., Thirion, B., Amadon, A., Dehaene, S., & Kleinschmidt, A. (2009). Deciphering cortical number coding from human brain activity patterns. *Current Biology*, *19*(19), 1608–1615. http://dx.doi.org/10.1016/j.cub.2009.08.047.

Emerson, R. W., & Cantlon, J. F. (2012). Early math achievement and functional connectivity in the fronto-parietal network. *Developmental Cognitive Neuroscience*, *2*(Suppl. 1), S139–S151. http://dx.doi.org/10.1016/j.dcn.2011.11.003.

Emerson, R. W., & Cantlon, J. F. (2014). Continuity and change in children's longitudinal neural responses to numbers. *Developmental Science*. http://dx.doi.org/10.1111/desc.12215.

Feigenson, L., Dehaene, S., & Spelke, E. (2004). Core systems of number. *Trends in Cognitive Sciences*, *8*(7), 307–314. Retrieved from http://www.ncbi.nlm.nih.gov/entrez/query.fcgi?cmd=Retrieve&db=PubMed&dopt=Citation&list_uids=15242690.

Geary, D. C., Berch, D. B., & Mann Koepke, K. (Eds.), (2015). *Evolutionary origins and early development of number processing*. San Diego, CA: Academic Press.

Göbel, S. M., Johansen-Berg, H., Behrens, T., & Rushworth, M. F. (2004). Response-selection-related parietal activation during number comparison. *Journal of Cognitive Neuroscience*, *16*(9), 1536–1551. Retrieved from http://www.ncbi.nlm.nih.gov/entrez/query.fcgi?cmd=Retrieve&db=PubMed&dopt=Citation&list_uids=15601517.

Gobel, S. M., & Rushworth, M. F. (2004). Cognitive neuroscience: Acting on numbers. *Current Biology*, *14*(13), R517–R519. Retrieved from http://www.ncbi.nlm.nih.gov/entrez/query.fcgi?cmd=Retrieve&db=PubMed&dopt=Citation&list_uids=15242633.

Grabner, R. H., Ansari, D., Koschutnig, K., Reishofer, G., Ebner, F., & Neuper, C. (2008). To retrieve or to calculate? Left angular gyrus mediates the retrieval of arithmetic facts during problem solving. *Neuropsychologia*. http://dx.doi.org/10.1016/j.neuropsychologia.2008.10.013.

Grabner, R. H., Ansari, D., Reishofer, G., Stern, E., Ebner, F., & Neuper, C. (2007). Individual differences in mathematical competence predict parietal brain activation during mental calculation. *NeuroImage*, *38*(2), 346–356. Retrieved from http://www.ncbi.nlm.nih.gov/entrez/query.fcgi?cmd=Retrieve&db=PubMed&dopt=Citation&list_uids=17851092.

Grill-Spector, K., Henson, R., & Martin, A. (2006). Repetition and the brain: Neural models of stimulus-specific effects. *Trends in Cognitive Sciences*, *10*(1), 14–23.

Halberda, J., & Feigenson, L. (2008). Developmental change in the acuity of the "Number Sense": The Approximate Number System in 3-, 4-, 5-, and 6-year-olds and adults. *Developmental Psychology*, *44*(5), 1457–1465. http://dx.doi.org/10.1037/a0012682.

Harnad, S. (1990). The symbol grounding problem. *Physica D*, *42*, 335–346.

Holloway, I. D., & Ansari, D. (2010). Developmental specialization in the right intraparietal sulcus for the abstract representation of numerical magnitude. *Journal of Cognitive Neuroscience*, *22*, 2627–2637.

Holloway, I. D., Battista, C., Vogel, S. E., & Ansari, D. (2013). Semantic and perceptual processing of number symbols: Evidence from a cross-linguistic fMRI adaptation study. *Journal of Cognitive Neuroscience*, *25*, 388–400. http://dx.doi.org/10.1162/jocn_a_00323.

Holloway, I. D., Price, G. R., & Ansari, D. (2010). Common and segregated neural pathways for the processing of symbolic and nonsymbolic numerical magnitude: An fMRI study. *NeuroImage*, *49*, 1006–1017.

Hyde, D. C., Boas, D. A., Blair, C., & Carey, S. (2010). Near-infrared spectroscopy shows right parietal specialization for number in pre-verbal infants. *NeuroImage*, *53*(2), 647–652. http://dx.doi.org/10.1016/j.neuroimage.2010.06.030.

Izard, V., Dehaene-Lambertz, G., & Dehaene, S. (2008). Distinct cerebral pathways for object identity and number in human infants. *PLoS Biology*, *6*(2). http://dx.doi.org/10.1371/journal.pbio.0060011, e11.

Kanwisher, N., & Yovel, G. (2006). The fusiform face area: A cortical region specialized for the perception of faces. *Philosophical Transactions of the Royal Society of London, Series B: Biological Sciences*, *361*(1476), 2109–2128. http://dx.doi.org/10.1098/rstb.2006.1934.

Kaufmann, L., Koppelstaetter, F., Delazer, M., Siedentopf, C., Rhomberg, P., Golaszewski, S., et al. (2005). Neural correlates of distance and congruity effects in a numerical Stroop task: An event-related fMRI study. *NeuroImage*, *25*(3), 888–898. Retrieved from http://www.ncbi.nlm.nih.gov/entrez/query.fcgi?cmd=Retrieve&db=PubMed&dopt=Citation&list_uids=15808989.

Kaufmann, L., Koppelstaetter, F., Siedentopf, C., Haala, I., Haberlandt, E., Zimmerhackl, L. B., et al. (2006). Neural correlates of the number-size interference task in children. *Neuroreport*, *17*(6), 587–591. Retrieved from http://www.ncbi.nlm.nih.gov/entrez/query.fcgi?cmd=Retrieve&db=PubMed&dopt=Citation&list_uids=16603917.

Libertus, M. E., & Brannon, E. M. (2009). Behavioral and neural basis of number sense in infancy. *Current Directions in Psychological Science*, *18*(6), 346–351. http://dx.doi.org/10.1111/j.1467-8721.2009.01665.x.

Lyons, I. M., Ansari, D., & Beilock, S. L. (2014). Qualitatively different coding of symbolic and nonsymbolic numbers in the human brain. *Human Brain Mapping*, http://dx.doi.org/10.1002/hbm.22641.

Lyons, I. M., & Beilock, S. L. (2013). Ordinality and the nature of symbolic numbers. *The Journal of Neuroscience: The Official Journal of the Society for Neuroscience*, *33*(43), 17052–17061. http://dx.doi.org/10.1523/JNEUROSCI.1775-13.2013.

McCandliss, B. D., Cohen, L., & Dehaene, S. (2003). The visual word form area: Expertise for reading in the fusiform gyrus. *Trends in Cognitive Sciences*, *7*(7), 293–299. Retrieved from http://www.ncbi.nlm.nih.gov/entrez/query.fcgi?cmd=Retrieve&db=PubMed&dopt=Citation&list_uids=12860187.

Menninger, K. (1992). Number words and number symbols: A cultural history of numbers. *American Anthropologist, 72*, 480.

Moyer, R. S., & Landauer, T. K. (1967). Time required for judgements of numerical inequality. *Nature, 215*(109), 1519–1520. Retrieved from http://www.ncbi.nlm.nih.gov/entrez/query.fcgi?cmd=Retrieve&db=PubMed&dopt=Citation&list_uids=6052760.

Nieder, A., & Dehaene, S. (2009). Representation of number in the brain. *Annual Review of Neuroscience, 32*, 185–208. http://dx.doi.org/10.1146/annurev.neuro.051508.135550.

Nieder, A., Freedman, D. J., & Miller, E. K. (2002). Representation of the quantity of visual items in the primate prefrontal cortex. *Science, 297*, 1708–1711.

Nieder, A., & Miller, E. K. (2003). Coding of cognitive magnitude: Compressed scaling of numerical information in the primate prefrontal cortex. *Neuron, 37*(1), 149–157. Retrieved from http://www.ncbi.nlm.nih.gov/entrez/query.fcgi?cmd=Retrieve&db=PubMed&dopt=Citation&list_uids=12526780.

Nieder, A., & Miller, E. K. (2004). A parieto-frontal network for visual numerical information in the monkey. *Proceedings of the National Academy of Sciences of the United States of America, 101*(19), 7457–7462.

Notebaert, K., Nelis, S., & Reynvoet, B. (2010). The magnitude representation of small and large symbolic numbers in the left and right hemisphere: An event-related fMRI study. *Journal of Cognitive Neuroscience, 23*(3), 622–630. http://dx.doi.org/10.1162/jocn.2010.21445.

Notebaert, K., Nelis, S., & Reynvoet, B. (2011). The magnitude representation of small and large symbolic numbers in the left and right hemisphere: An event-related fMRI study. *Journal of Cognitive Neuroscience, 23*(3), 622–630.

Park, J., Hebrank, A., Polk, T. A., & Park, D. C. (2012). Neural dissociation of number from letter recognition and its relationship to parietal numerical processing. *Journal of Cognitive Neuroscience, 24*(1), 39–50. http://dx.doi.org/10.1162/jocn_a_00085.

Piazza, M. (2010). Neurocognitive start-up tools for symbolic number representations. *Trends in Cognitive Sciences, 14*(12), 542–551. http://dx.doi.org/10.1016/j.tics.2010.09.008.

Piazza, M., Pinel, P., Le Bihan, D., & Dehaene, S. (2007). A magnitude code common to numerosities and number symbols in human intraparietal cortex. *Neuron, 53*, 293–305.

Pinel, P., Dehaene, S., Riviere, D., & Le Bihan, D. (2001). Modulation of parietal activation by semantic distance in a number comparison task. *NeuroImage, 14*(5), 1013–1026. Retrieved from http://www.ncbi.nlm.nih.gov/entrez/query.fcgi?cmd=Retrieve&db=PubMed&dopt=Citation&list_uids=11697933.

Pinel, P., Le Clec, H. G., van de Moortele, P. F., Naccache, L., Le Bihan, D., & Dehaene, S. (1999). Event-related fMRI analysis of the cerebral circuit for number comparison. *Neuroreport, 10*(7), 1473–1479. Retrieved from http://www.ncbi.nlm.nih.gov/entrez/query.fcgi?cmd=Retrieve&db=PubMed&dopt=Citation&list_uids=10380965.

Price, G. R., & Ansari, D. (2011). Symbol processing in the left angular gyrus: Evidence from passive perception of digits. *NeuroImage, 57*(3), 1205–1211. http://dx.doi.org/10.1016/j.neuroimage.2011.05.035.

Seghier, M. L. (2013). The angular gyrus: Multiple functions and multiple subdivisions. *The Neuroscientist: A Review Journal Bringing Neurobiology, Neurology and Psychiatry, 19*(1), 43–61. http://dx.doi.org/10.1177/1073858412440596.

Seghier, M. L., Fagan, E., & Price, C. J. (2010). Functional subdivisions in the left angular gyrus where the semantic system meets and diverges from the default network. *The Journal of Neuroscience: The Official Journal of the Society for Neuroscience, 30*(50), 16809–16817. http://dx.doi.org/10.1523/JNEUROSCI.3377-10.2010.

Shum, J., Hermes, D., Foster, B. L., Dastjerdi, M., Rangarajan, V., Winawer, J., et al. (2013). A brain area for visual numerals. *The Journal of Neuroscience: The Official Journal of the Society for Neuroscience*, *33*(16), 6709–6715. http://dx.doi.org/10.1523/JNEUROSCI.4558-12.2013.

Stanescu-Cosson, R., Pinel, P., van De Moortele, P. F., Le Bihan, D., Cohen, L., & Dehaene, S. (2000). Understanding dissociations in dyscalculia: A brain imaging study of the impact of number size on the cerebral networks for exact and approximate calculation. *Brain*, *123*(Pt 1), 2240–2255. Retrieved from http://www.ncbi.nlm.nih.gov/entrez/query.fcgi?cmd=Retrieve&db=PubMed&dopt=Citation&list_uids=11050024.

Trick, L. M., & Pylyshyn, Z. W. (1994). Why are small and large numbers enumerated differently? A limited-capacity preattentive stage in vision. *Psychological Review*, *101*(1), 80–102. Retrieved from http://www.ncbi.nlm.nih.gov/pubmed/8121961.

Van Atteveldt, N., Formisano, E., Goebel, R., & Blomert, L. (2004). Integration of letters and speech sounds in the human brain. *Neuron*, *43*(2), 271–282. Retrieved from http://www.ncbi.nlm.nih.gov/entrez/query.fcgi?cmd=Retrieve&db=PubMed&dopt=Citation&list_uids=15260962.

Vogel, S. E., Goffin, C., & Ansari, D. (2014). Developmental specialization of the left parietal cortex for the semantic representation of Arabic numerals: An fMR-adaptation study. *Developmental Cognitive Neuroscience*, *12C*, 61–73. http://dx.doi.org/10.1016/j.dcn.2014.12.001.

Woodcock, R. W., McGrew, K. S., & Mather, N. (2001). *Woodcock-Johnson III tests of achievement*. Itasca, IL: Riverside Publishing.

Zamarian, L., Ischebeck, A., & Delazer, M. (2009). Neuroscience of learning arithmetic-evidence from brain imaging studies. *Neuroscience and Biobehavioral Reviews*, *33*(6), 909–925. http://dx.doi.org/10.1016/j.neubiorev.2009.03.005.

Chapter 3

Neural and Behavioral Signatures of Core Numerical Abilities and Early Symbolic Number Development

Daniel C. Hyde, Yi Mou
Department of Psychology, University of Illinois Urbana-Champaign, Champaign, Illinois, USA

INTRODUCTION

More than 100 years ago, Jevons (1871) experimentally tested a question that philosophers had been debating for centuries: How many items can be held in the mind at once? To do this, he tossed a variable number of beans into a pan and estimated, as quickly as possible, the number of beans that fell. He found that he was perfectly accurate up to 4 items (see Figure 1(A)). In contrast, as the number of beans increased beyond 4, so did error in his estimates (see Figure1(A)). Based on these data, Jevons argued that fundamentally different cognitive processes were operating to enumerate smaller and larger sets of items. Many studies have since replicated and expanded on Jevons's findings, calling the phenomenon of accurately and seemingly instantaneously enumerating small numbers of items (up to 4) *subitizing* to distinguish it from the approximate estimation or iterative counting processes which operate to enumerate larger sets of items (>4) (e.g., Kaufman, Lord, Reese, & Volkmann, 1949; Mandler & Shebo, 1982; Revkin, Piazza, Izard, Cohen, & Dehaene, 2008; Taves, 1941; Trick & Pylyshyn, 1994; Warren, 1897).

Substantial debate regarding the interpretation of these data, however, has continued until the present day (Cordes, Gelman, Gallistel, & Whalen, 2001; Hyde, 2011; Revkin et al., 2008; Trick & Pylyshyn, 1994). The working hypothesis, having been put forth by others, is that this dissociation suggests nonverbal numerical cognition is supported by at least two qualitatively distinct cognitive systems (see Feigenson, Dehaene, & Spelke, 2004; Hyde, 2011; Mou & vanMarle, 2014 for review). Others argue, however, that despite the temptation to interpret the dissociation as evidence of two systems, the data are more

Development of Mathematical Cognition. http://dx.doi.org/10.1016/B978-0-12-801871-2.00003-4

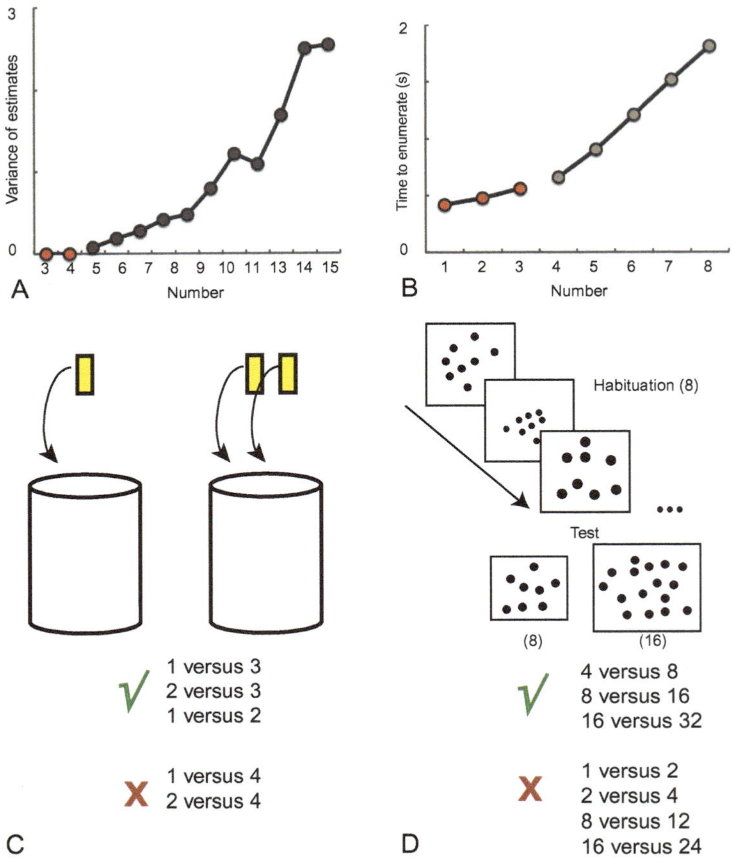

FIGURE 1 Examples of small/large number processing dissociations in behavioral work. (A) Results from Jevons (1871) enumeration task. (B) Results from Trick and Pylyshyn (1994) enumeration task. (C) Schematic depiction of infant manual choice task and results over several studies (Feigenson, Carey, & Hauser, 2002; vanMarle, 2013). (D) Schematic depiction of infant number habituation task and results from several studies (includes results from Xu, 2003, Xu, et al, 2005; Xu & Spelke, 2000). Data for (B) extracted from original figure using WebPlotDigitizer (http://arohatgi.info/WebPlotDigitizer).

consistent with a single cognitive system underlying nonverbal numerical abilities. Here we will argue that the dissociation between small and large number processing provides a window into the numerical brain and its development. To do so, we will briefly review the theory and behavioral evidence for and against the two systems view of numerical cognition. Next, we will show how recent advances in cognitive neuroscience have shed light on the cognitive and neural distinction between the two systems. We will then discuss the various hypotheses about how the two systems contribute to symbolic numerical and mathematical abilities. Finally, we will discuss recent results that begin to narrow the hypothesis space regarding the contribution of core systems to the development of symbolic numerical and mathematical systems across human cultures and within human ontogeny.

TWO SYSTEMS FOR NONVERBAL NUMERICAL COGNITION

A decade ago Feigenson et al. (2004) proposed the *two core systems* account of numerical cognition to explain dissociations between small and large number processing seen across human development and in other nonhuman animals (for further discussion see Posid & Cordes, 2015, Volume 1; Starr & Brannon, 2015, Volume 1; vanMarle, 2015, Volume 1). This account posits that humans possess at least two, evolutionarily ancient cognitive systems that allow for nonverbal numerical cognition: the parallel individuation (PI) system and the approximate number system (ANS). These systems are proposed to support nonverbal numerical cognition through distinct types of mental representation, each with its distinct utilities and limits (Feigenson et al., 2004). Furthermore, these systems are thought to underlie the development of symbolic systems of number and mathematics (e.g., Carey, 2009; Dehaene, 1997; Gallistel & Gelman, 2000).

Parallel Individuation System

The PI (also known as the object-tracking system or the object file system) allows us to select, pay attention to, track, and remember individual objects in space and over time (Kahneman & Treisman, 1984; Kahneman, Treisman, & Gibbs, 1992; Pylyshyn, 1994; Pylyshyn & Storm, 1988; Trick & Pylyshyn, 1994). This system is deeply intertwined with our abilities to mentally represent, conceptualize, and organize objects (Cheries, Mitroff, Wynn, & Scholl, 2009; Scholl, 2001). As such, the representations of this system are not inherently "numerical." Rather, reference to the PI system is an appeal to the aspects of object representation that allow us to represent the numerical identity, or distinctness, of individual items. The PI represents a set of objects by assigning a single mental symbol to each item in the set (rather than a single symbol to summarize the numerosity of the set: Gallistel, 2007; Hyde, 2011). PI can also be used to carry out some basic numerical computations such as comparison or simple arithmetic through the employment of one-to-one correspondence procedures (Carey, 2009). For example, the PI system can detect which of two sets has more items by matching back-and-forth symbols for individual items from the two sets (Carey, 2009; Feigenson et al., 2004). Such computations draw heavily on the general perceptual and cognitive resources of attention and working memory. As such, the PI system is constrained by the limits of these resources such that only a small number of items, up to 3 sets of 3 to 4 individual items, can be represented and/or compared by the system simultaneously (see Cowan, 2001; Leslie, Xu, Tremoulet, & Scholl, 1998; Scholl, 2002 for reviews).

Thus, engagement of the PI can be identified by its signature sensitivity to a capacity limit of about 3 to 4 items, as is the case in subitizing.

Approximate Number System

The ANS allows us to nonverbally represent the numerical magnitude of a set of objects (Dehaene, 1997; Gallistel, 1990; Gallistel & Gelman, 2000, 2005; Gibbon, 1977; Meck & Church, 1983). The output of this system, more broadly, is one of many types of estimates the brain makes to rapidly summarize the complex and uncertain visual environment (Alvarez & Cavanagh, 2004; Burr & Ross, 2008). Psychophysical studies of nonverbal numerical abilities show that, in the absence of exact verbal counting, the uncertainty or imprecision in ANS representations increases logarithmically as numerosity increases (Dehaene, 2003; Dehaene & Changeux, 1993; Dehaene, Izard, Spelke, & Pica, 2008). This systematicity gives rise to strict behavioral limitations on comparison and estimation: the ability to discriminate two numerosities is determined by their numerical ratio rather than their absolute difference (Dehaene, 1997; Moyer & Landauer, 1967; Parkman, 1971; Feigenson et al., 2004; Revkin et al., 2008). For example, it is easier to discriminate 5 from 10 (1:2 ratio) than 20 from 25 (4:5 ratio) even though the absolute difference is the same in both comparisons because mental representations of larger numerosities are more imprecise, resulting in more overlap in the mental representations of 20 and 25 than of 5 and 10. Thus, engagement of the ANS can be identified by its signature sensitivity to the numerical ratio between numerosities.

Representations generated by the ANS can also be entered into numerical computations such as addition, subtraction, multiplication, and division (Cantlon & Brannon, 2007; Cordes, Gallistel, Gelman, & Latham, 2007; see Gallistel & Gelman, 2005 for review). However, because ANS representations themselves are imprecise, the numerical computations over these representations are also approximate (see McCrink, 2015, Volume 1 for review). For example, human adults, children and nonhuman primates are able to estimate the sum of two sequentially presented dot arrays (e.g., 5 + 10) and choose which of two test arrays is closer to the correct answer (e.g., Barth et al., 2006; Cantlon & Brannon, 2007). As in the case of comparing approximate numerosities, these participants' approximate arithmetic ability is dependent on the numerical ratio between the actual sum and distractor arrays.

Behavioral Evidence for Distinct Systems of Numerical Cognition

Evidence for two systems comes largely from work showing that the different signature limits operate over small and large number processing (Feigenson et al., 2004; Hyde, 2011 for reviews). The best known dissociation comes from the enumeration literature, where human adults are very accurate and fast at enumerating a small number of items (subitizing), but are increasingly slower and more error prone as the number increases above 4 (Jevons, 1871; Kaufman et al., 1949; Revkin et al., 2008; Taves, 1941; Trick & Pylyshyn, 1994) (see Figure 1(A) and (B)). However, from the enumeration literature alone, it is unclear whether dissociations between small and large numbers reflect differential experience between the numerical ranges, differences in difficulty to enumerate small and large numbers, or even whether they reflect two systems at all. For example, some work shows that temporarily impairing verbal processing abilities available for a numerical task, through verbal shadowing routines (whereby participants are asked perform some verbal task jointly with the numerical task), eliminates the stark processing differences between small and large numbers, suggesting that dissociations may be mediated by verbal/symbolic number systems (e.g., Balakrishnan & Ashby, 1992; Cordes et al., 2001, 2007; Whalen, Gallistel, & Gelman, 1999). Similarly, some studies of enumeration in nonhuman primates (without a system of symbolic number or spoken language) show no break in performance between small and large numbers as seen in human, suggesting continuity between processing small and large numbers (Beran, 2007; Brannon & Terrace, 1998). Furthermore, other models of enumeration explain differences in accuracy and/or reaction time between small and large numbers as consistent with the ANS operating over all numbers with increased precision at the lower end of the number range (see Gallistel & Gelman, 2005 for review).

Other work from the infant behavioral literature shows even more striking dissociations between small and large number processing while avoiding some of the limitations of the adult enumeration work (Feigenson & Carey, 2003; Posid & Cordes, 2015, Volume 1; Starr & Brannon, 2015, Volume 1; vanMarle, 2015, Volume 1). For example, after watching an experimenter place items sequentially into an opaque box, 12- to 14-month-old children manually search reliably longer in the box if all but one of the items are seen being removed than if all of the items are seen being removed. However, they only do so if the number of items initially hidden is less than 4. Once more than 3 items are hidden, infants do not search longer when one is left in the box compared to when they are all removed (Feigenson & Carey, 2003). Similarly, after watching an experimenter place a certain number of food items in one opaque bucket and a different number of food items in another opaque bucket, 10- to 12-month-old infants reliably chose to retrieve the food items from the bucket containing more food items in cases of 1 versus 2, 1 versus 3, and 2 versus 3 food items. However, once more than 3 food items are hidden in one of the buckets (2 vs. 4 or 1 vs. 4), infants retrieve equally often from both buckets (Feigenson & Carey, 2005; Feigenson et al., 2002; vanMarle, 2013) (see Figure 1(C)). Results from manual search paradigms suggest that infants' capacity limit for representing and comparing individual items on the basis of number is 3, consistent with the use of the PI system. Once the capacity is surpassed, this ability to make numerical comparisons in these tasks dramatically fails.

However, infants, like adults, can also represent larger numerosities with the ANS (vanMarle & Wynn, 2011; Xu & Spelke, 2000; Xu et al., 2005; see Mou & vanMarle, 2014; Starr & Brannon, 2015, Volume 1; vanMarle, 2015, Volume 1 for review). For example, after being repeatedly shown novel images of 8 dots until boredom sets in, 6-month-old infants increase visual attention (i.e., they look longer) to a novel test image of 16 but not to a novel test image of 8 dots that varies equally from previous images in other nonnumerical visual parameters such as cumulative area (Xu & Spelke, 2000; Xu et al., 2005) (see Figure 1(D)). However, at this age, the ability to discriminate numerosities is limited to those that differ by a 1:2 ratio (8 vs. 16; 16 vs. 32) and failures to discriminate are seen when closer ratios are presented (8 vs. 12; 6 vs. 9) (Lipton & Spelke, 2003; Wood & Spelke, 2005; Xu & Spelke, 2000; Xu et al., 2005) (see Figure 1(D)). The ratio-dependent discrimination is also observed in manual choice tasks, such as the aforementioned, where 10- to 12-month-olds reliably choose the larger set of hidden food items in 4 versus 8 and 5 versus 10 (a 1:2 ratio), and in 4 versus 6 and 6 versus 9 (a 2:3 ratio), but not 6 versus 8 (a 3:4 ratio) (vanMarle, 2012, 2013; vanMarle & Wynn, 2011). The critical ratio limit at which two numerosities can be discriminated decreases with age from about 1:3 at birth to somewhere between 7:8 and 10:11 in adulthood (Barth, Kanwisher, & Spelke, 2003; Halberda & Feigenson, 2008; Izard, Sann, Spelke, & Streri, 2009; Libertus & Brannon, 2010; Lipton & Spelke, 2003, 2004; vanMarle & Wynn, 2009; Wood & Spelke, 2005; Xu & Spelke, 2000; Xu et al., 2005).

Together, these results show a dissociation in early development that is indicative of engagement of distinct representational systems over small and large numbers. In tasks where tracking the exact number of individual items is required, infants show a capacity limit of about 3–4 and fail to engage approximate number representations to make numerical comparisons across the small/large boundary. In contrast, within other tasks that may encourage approximate, rather than exact, numerical discrimination, infants are strictly limited by the numerical ratio relationship between the two numerosities to be compared rather than a capacity limit. What is striking about the infant work is that they appear to engage either the PI system or ANS, even when engagement of the other system could yield better performance (and where, presumably, adults would easily succeed), suggesting automatic engagement of distinct systems in different numerical contexts (Feigenson et al., 2004; Xu, 2003).

Unlike the adult data, dissociations between small and large numbers in infants cannot be explained by symbolic number knowledge or differential experience with enumerating small compared to large numbers because such experience is not yet present in infancy. Nevertheless, other infant studies observe signatures of the ANS for both small and large numerosities (see Mou & vanMarle, 2014 for review). For example, 5- to 7-month-olds can successfully discriminate 2 versus 4 tones (vanMarle & Wynn, 2009), 2 versus 4 moving groups of squares (Wynn, Bloom, & Chiang, 2002), and 2 versus 8 visual dots

(Cordes & Brannon, 2009). In addition, when 6-month-olds were presented with two side-by-side displays, with one containing 2 or 4 items presented alternately and the other containing constant number of items (e.g., 4), infants looked longer at the changing display, suggesting they detected the numerical changes between 2 and 4 (Starr, Libertus, & Brannon, 2013a). Under some conditions, nonhuman primates also compare and order numerosities across the small/large boundary with ratio-dependent discriminability, suggesting engagement of the ANS for both small and large numbers (see Brannon & Merritt, 2011; Starr & Brannon, 2015, Volume 1 for review). These studies suggest that, in some contexts, the ANS may operate over all numbers. From the behavioral evidence alone, it is unclear why all numbers appear to be represented by the ANS under some conditions and small numbers appear to be represented distinctly from larger numbers by the PI system under other conditions.

THE COGNITIVE NEUROSCIENCE OF TWO CORE SYSTEMS OF NUMBER

Recent advances in cognitive neuroscience have shed light on the dissociations between small and large number processing. Specifically, neuroscience methods have allowed small and large number processing to be directly compared using the same dependent measure and experimental paradigm across age groups, while eliminating several common confounds (see discussion later). As a result, the use of these methods has provided key insights into the nature and development of these systems in the mind and brain.

Establishing the Neural Signatures of Two Systems

A series of recent event-related potential (ERP) studies have provided important insights into the nature and timing of brain activity associated with small and large number processing, as well as the underlying cognitive and brain systems. Several of these studies have employed a number adaptation paradigm similar to that used in fMRI (Cantlon, Brannon, Carter, & Pelphrey, 2006; Grill-Spector & Malach, 2001; Piazza, Izard, Pinel, Le Bihan, & Dehaene, 2004), and infant habituation studies (e.g., Lipton & Spelke, 2003; vanMarle & Wynn, 2009; Xu & Spelke, 2000; Xu et al., 2005). In number adaptation experiments, participants are presented with a stream of novel images containing sets of items (e.g., dots). Critically, a majority of the images contain the same number of objects. Unbeknownst to the participant, occasional test images appear within the stream that either contained the same number of items as the context images (no change) or a different number of items. The position, size, and spacing of the items in the arrays vary from image to image.

The logic of this paradigm is that if a brain region(s) is sensitive to number, then the repeated presentation of the same number will lead to a diminished response, or adaptation, in that region despite constant changes on nonnumerical dimensions of the stimulus such as size, spacing, and spatial position (Cantlon et al., 2006; Piazza et al., 2004; Xu & Spelke, 2000). Furthermore, occasional changes in number should be associated with changes in brain activity proportional to the ratio of adaptation number to test number. If other dimensions of the stimuli are properly controlled, one can isolate response(s) specific to the dimension of interest from more general response to stimulus change. The number adaptation paradigm has the added benefit of being able to be recorded without a demanding cognitive task and with no number-related instruction. As such, the use of this methodology eliminates potential concerns of difficulty differences between large and small numbers or explicit engagement of a symbolic number system common in many classic tasks (Cordes et al., 2001; Kaufman et al., 1949; Revkin et al., 2008).

Hyde and Spelke (2009) employed such a paradigm to ask whether small and large numerosities were processed by the ANS or by distinct cognitive systems. They asked adult participants to simply fixate on the computer screen and pay attention to what was presented. During half of the blocks participants were presented with adaptation arrays of 1, 2, and 3 items (small numbers) with occasional test arrays of 1, 2, or 3 items. For the other half of blocks, context arrays of 8, 16, or 24 items (large numbers) were presented along with occasional test arrays of 8, 16, and 24 items. As such, over the entire experiment, participants were presented with four different ratio change conditions (no change or a 1:1 ratio; small change or a 2:3 ratio; medium change or a 1:2 ratio; and large change or a 1:3 ratio) in both the small and large number contexts. While participants passively watched these displays, the ongoing EEG was recorded using a 128-channel high-density system. After data collection, ERPs were derived in response to the test images. ERPs were analyzed by the numerosity of the image itself (cardinal value averaged across contexts) as well as the numerical ratio change (relative to context). The prediction was that if both large and small numbers were spontaneously represented by the ANS, then the brain response should be modulated by the ratio of change in both the small and large number ranges. However, if small numbers were spontaneously processed differently than large numbers, then a qualitatively different pattern may emerge: sensitivity to the total number of objects presented in an array for small numbers rather than extent of ratio change.

Results revealed a double dissociation in the functional brain response to small and large numbers as predicted by the two core systems hypothesis (Hyde & Spelke, 2009; also see Hyde & Spelke, 2012 and Hyde & Wood, 2011). Small number images modulated early visual attention by the total number of items (or cardinal value) presented in a given array, regardless of the numerical context (adaptation context) it was presented in. More specifically, an early negative component called the N1 (see Hillyard, Mangun, Luck, & Heinze, 1990), peaking around 150 ms over widespread posterior scalp sites, became more negative with the number of items presented in the small number images (1–3 items) (see Figure 2(A)). In contrast, mid-latency parietal processing of large number

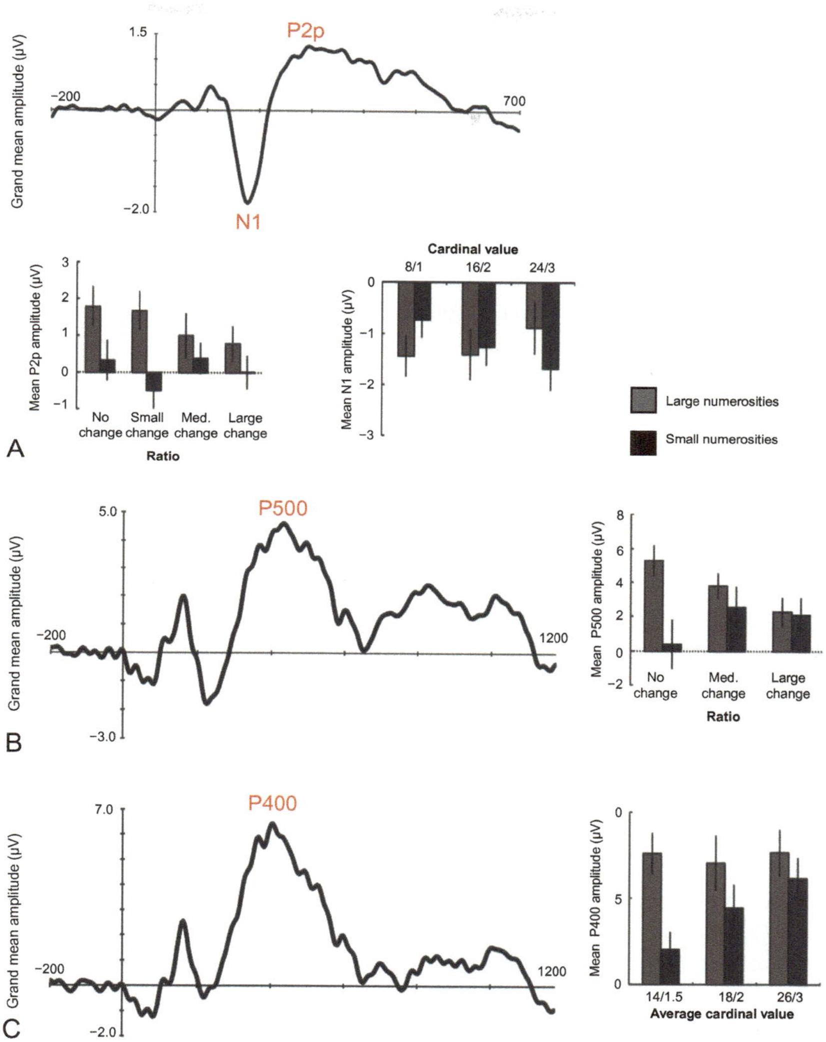

FIGURE 2 Examples of ERP results showing a double dissociation between small and large number processing. (A) Representative results from small and large number adaptation study with adults (Experiment 2, Hyde & Spelke, 2009). Waveform represents the average response to all conditions over parietal scalp sites. Bar graphs represent the extracted mean amplitudes over parietal scalp sites in response to ratio change on the P2p component and cardinal value on the N1. (B) Results from infant ERP study of number processing (Hyde & Spelke, 2011) showing the effects of ratio change on P500. Waveform represents the average response to all conditions. Bar graphs represent the extracted mean amplitudes in response to ratio change for both small and large numbers. (C) Results from infant ERP study of number processing (Hyde & Spelke, 2011) showing the effects of cardinal value on P400. Waveform represents the average response to all conditions. Bar graphs represent the extracted mean amplitudes in response to the average cardinal values presented within the different condition blocks for both small and large numbers.

images was dependent on the numerical ratio relationship between adaptation and test number. More specifically, a mid-latency positivity called the P2p (see Dehaene, 1996), peaking around 200–250 ms over bilateral parietal scalp sites, became more positive as the ratio of change between adaptation and test number decreased (or the distance between adaptation and test number decreased) (see Figure 2(A)). The dissociation between small and large numbers could not be attributed to differences in difficulty between small and large conditions since participants were simply asked to pay attention and no explicit response was required.

More broadly, P2p has been associated with numerical processing in a variety of explicit contexts such as both symbolic and nonsymbolic number comparison tasks (Dehaene, 1996; Temple & Posner, 1998). The results of Hyde and Spelke (2009) demonstrated further that this posterior parietal component was also spontaneously sensitive to nonsymbolic number change. P2p is thought to reflect brain activity related to the abstract numerical meaning or mental representation of number underlying the physical stimulus presented (see Hyde, 2011). Furthermore, sensitivity of the functional response to numerical distance or ratio change suggests it to be a marker of ANS engagement.

In contrast, N1 has classically been associated with the earlier distribution of visual-spatial attention to objects and/or locations (see Luck, 2005 for a review). For example, a greater (more negative) N1 response is observed when an object appears in an attended location compared to an unattended location (Hillyard & Annlo-Vento, 1998; Hillyard et al., 1990). Similar modulation of N1 has been observed in a variety of other enumeration and object-tracking ERP studies (Drew & Vogel, 2008; Ester, Drew, Klee, Vogel, & Awh, 2012; Libertus, Woldorff, & Brannon, 2007; Nan, Knosche, & Luo, 2006; Pagano & Mazza, 2012; Mazza & Caramazza, 2011; Temple & Posner, 1998). Recently, Mazza, Pagano, and Caramazza (2012) reported that while both the earlier N1 and a later posterior component called N2pc were modulated by the number of items (for small numbers of items 1–4) in an enumeration task, only the N2pc (but not N1) modulation reflected the number of items to be enumerated when additional distractor items were present. As such, they concluded that N1 modulation by number likely reflects the initial visual-spatial indexing of objects, with strict limits on the number of items that can be indexed at once, while further binding of item location with item features is reflected in downstream processing (of N2pc and the following sustained activity). N1 differences have also been observed between number in different formats (digit vs. word), further suggesting that it does not reflect number-specific activation, but distribution of attention to objects more broadly (Libertus et al., 2007; Temple & Posner, 1998). Based on this evidence, it appears that N1 is a good marker of initial engagement of the PI system.

Using the established ERP signatures of the PI system and the ANS, Hyde and Wood (2011) found that differences in the N1 modulation (previously observed between small and large numbers) actually reflect the extent to which early attention selects individual items or selects the set as a single item. In other words, dissociations reflect more general properties and limits of attention rather than systems specialized for particular number ranges (small and large). To show this, they systematically manipulated the extent to which spatial attention could individuate (or not) items in small, nonsymbolic object arrays (1–3 items) only. In one experiment, participants were adapted to small numbers of objects that were spaced sufficiently or insufficiently (crowded) for visual attention to individuate the items (e.g., as in Intriligator & Cavanagh, 2001). In another experiment of the same study, participants were adapted to small nonsymbolic object arrays under high or low spatial attention load (e.g., as in Fougnie & Marois, 2006). They found that when items were sufficiently spaced or when spatial attention was not tied up in another task, N1 was sensitive to the number of items in an image, as in previous studies (Hyde & Spelke, 2009). However, when spatial attention was taxed by an additional task or when spacing was beyond the limits of the attentional system to individuate items (because of crowding), N1 was not sensitive to the number of items in the image. Instead, the later P2p became sensitive to the ratio of change between adaptation and test images for small numerosities, as had been observed for large numbers in previous experiments (Hyde & Spelke, 2009). This study experimentally confirmed intuitions raised by Ansari, Lyons, van Eimeren, and Xu (2007) that differences between small and large number processing reflect differences in attentional allocation. Furthermore, it suggests that early attentional processes determine whether a set of objects will be treated as distinct items through PI or as a single set with an approximate numerical magnitude through the ANS. As such, it could be reasoned that it is only when multiple items are treated as a single attentional item that approximate number is calculated (Hyde, 2011).

Distinct Brain Mechanisms of Two Systems

In addition to further revealing the nature and relationship between mental representations of the PI system and the ANS, cognitive neuroscience research has also made progress in identifying the brain mechanisms underlying these systems. Neuroimaging of numerical cognition has overwhelmingly focused on the question of whether the mechanisms for the ANS overlap with those for symbolic numerical abilities (Piazza, Pinel, Le Bihan, & Dehaene, 2007; see Cantlon, 2015, Volume 1). The evidence points to a network of fronto-parietal regions, including the intraparietal sulcus (IPS), left angular gyrus (AG), bilateral superior parietal lobe (SPL), and prefrontal cortex (PFC) (see Dehaene, 1997; Dehaene, Piazza, Pinel, & Cohen, 2003; Nieder & Dehaene, 2009 for reviews; see also Ansari, this volume) as important for mentally representing both symbolic and nonverbal numerical information, as well as for performing mathematical calculations. Of these regions, the IPS of the parietal lobe is thought to underlie the abstract representation of number. Less work has investigated whether and to what extent the mechanisms of number processing of the

PI system and those of the ANS are different. Nevertheless, the evidence thus far suggests that distinct brain mechanisms operate to individuate items and approximate the numerosity of sets.

Early neuropsychological patient studies (a review of lesion studies is beyond the scope of this chapter; see Butterworth, 1999; Dehaene & Cohen, 1994), along with behavioral dissociations cited previously suggested the possibility the PI system and the ANS could be cortically as well as cognitively dissociated in the brain. Whereas early neuroimaging studies on the issue were largely inconclusive (Piazza, Giacomini, Le Bihan, & Dehaene, 2003; Piazza, Mechelli, Butterworth, & Price, 2002; Sathian et al., 1999), recent studies have found more definitive evidence for differences in the neural basis of small and large number processing. For example, Ansari et al. (2007) used fMRI to measure the brain response as adult participants compared either small or large numbers of items. They observed extensive overlap in bilateral IPS for large and small number comparison. In addition, they observed greater activation in the temporal-parietal junction (TPJ) for small compared to large nonsymbolic comparison. In fact, TPJ activity was suppressed in large nonsymbolic number comparison and greater suppression was associated with better performance. They concluded that these differences reflect differential reliance on attention for small and large number processing, similar to the conclusion later reached and expanded on by Hyde and Wood (2011). Hyde and Spelke (2012) have applied source-modeling techniques to estimate the cortical origin of event-related scalp potentials underlying the neural signatures of the two systems (Hyde & Spelke, 2009). Using such techniques, they observed that the ERP signatures of the PI system and the ANS had distinct cortical origins. In particular, the ratio-dependent signature of the ANS was estimated to arise from right intraparietal regions, whereas the cardinal value sensitivity of the PI system was estimated to arise from bilateral extrastriate visual regions, as well as left parietal and right TPJ regions. Most recently, Cutini, Scatturin, Basso Moro, and Zorzi (2014) used functional near-infrared spectroscopy (fNIRS) to show that even within an IPS region where overlapping activation is observed, distinct functional patterns of activity arise for large and small number processing. Together the results suggest that PI processes and approximate number likely rely on some distinct brain mechanisms.

Continuity in Neural Signatures over Development

One limitation in studying adult number processing is that participants have a wealth of experience that could lead them to identify small number arrays as sets with particular symbolic numerosities more readily than large number arrays. Thus, it is unclear from adult data alone whether dissociations between small and large numbers result from two core systems or whether such dissociations arise after differential experience. Even in cases of passive viewing where the influence of the symbolic system itself is likely decreased, it is possible that dissociations between small and large number processing indicative of two systems arise as a result of differential experience in processing or recognizing small compared to large numerosities. To determine whether the dissociation between the two systems arises as a result of experience, some have turned to testing preverbal infants.

Hyde and Spelke (2011) measured ERPs in response to arrays of objects with both large and small numerosities in 6-month-olds. To do this, they employed a number alternation paradigm (e.g., Libertus & Brannon, 2010; Starr et al., 2013a), where alternating images of different numerosities were presented over different blocks. In this particular study, three blocks of images were presented: a no change block where novel images of the same number of items (of different sizes and arrangements) were presented during the entire block, a small change block where images alternated in numerosity by a 1:2 ratio, and a large change block where images alternated in numerosity by a 1:3 or 1:4 ratio. Furthermore, half of the infants in the study saw large number images (e.g., 8, 16, and 32) and half the infants saw small number images (e.g., 1, 2, and 3). ERPs were derived from the response to all images presented and ERP data were analyzed both with respect to the ratio relationship presented in the block as well as the actual cardinal values used in the block.

Similar to the adult ERP data, Hyde and Spelke (2011) observed a double dissociation between small and large number processing in preverbal infants (see Figure 2(B) and (C)). Processing of large numerosities was sensitive to the ratio change within a given block but not to the actual cardinal values used in the images. Specifically, a widespread, mid-latency, posterior parietal component descriptively labeled the P500 was modulated by the numerical ratio presented in the block. That is, P500 amplitude increased as ratio decreased (see Figure 2(B)). Small number processing, in contrast, was sensitive to the actual cardinal values used within the block. More specifically, an earlier positive component peaking only over left and right lateral posterior scalp sites, descriptively termed the P400, was sensitive to the average cardinal value presented in each block. That is, P400 was greater to blocks where greater cardinal values were presented (see Figure 2(C)). The double dissociation in the functional brain response of infants mirrored the sensitivity patterns seen in adults, providing strong evidence for continuity in the two core systems of number and the brain systems that underlie them.

Change in Core Numerical Processing over Development

Some progress has also been made in determining the brain mechanisms underlying the development of the ANS. A majority of studies have investigated this issue in children using fMRI. For example, Cantlon et al. (2006) carried out a number adaptation study with adults and 4-year-old children using fMRI. Both groups engaged bilateral regions of the IPS for processing the numerosity of object arrays. Activity for number further dissociated from activity for shape changes, suggesting specialization of these regions in 4-year-olds. Some differences between adults and children were also observed. First, the degree of activity in and around the IPS was greater in adults than in children. Second, activity in children was predominantly right lateralized while activity in adults was more bilateral. Ansari and Dhital (2006) have observed similar differences between adults and children using an active nonsymbolic number comparison task. Together, these results suggest both continuity and change in the brain mechanisms associated with approximate number processing.

Only a few studies to date have attempted to measure the neural mechanisms of number processing in infancy before exposure to symbolic number systems or spoken language. Izard, Dehaene-Lambertz, and Dehaene (2008) used source-modeling techniques to estimate the neural origin and specificity of scalp electrophysiology associated with number processing in 3-month-old infants. Using such techniques, they observed a right-lateralized fronto-parietal network for processing numerical changes that dissociated from a more ventral response to changes in the shape of objects in the array. Hyde, Boas, Blair, and Carey (2010) used fNIRS, a method with higher spatial resolution than ERPs, to investigate functional brain organization for number in the preverbal infant brain. To do this, they employed an adaptation method similar to previous fMRI studies with adults and children (Cantlon et al., 2006; Piazza et al., 2004) and ERP studies with adults, children, and infants (Hyde & Spelke, 2011; Temple & Posner, 1998). They observed a right-lateralized response to number. Furthermore, as has been observed in children and adults, they found that the response to number dissociated with the response to an equally perceptually salient change in the shape of the objects, suggesting the right IPS is specialized for number before the acquisition of a symbolic number system.

Most recently, Edwards, Wagner, Simon, and Hyde (in press) have confirmed, using the number alternation paradigm, that number processing is initially right lateralized and specialized in the infant brain. Using an fNIRS optical probe that covered a much broader surface of the posterior scalp than that of previous studies (Hyde et al., 2010), they observed that only a right parietal region responded to number and that the response to number dissociated spatially from a more general bilateral occipital and middle parietal visual-attentional response. Together these studies of infant approximate number processing suggest that the brain is specialized for number before encountering symbolic numbers or receiving formal instruction with them. Furthermore, this work suggests that although some continuity in specialization for approximate number is seen in right lateralized brain regions, number-specific activity becomes bilateral over development.

Little if any work has explicitly investigated the neural basis of the PI system in developmental populations. Izard et al. (2008) included a small number processing condition in their ERP source-modeling study, but did not analyze these data separately from large numbers to estimate the cortical source of ERPs. Hyde and Spelke (2011) observed a topographical difference in ERP signatures of small and large number processing. However, given the coarse spatial resolution of scalp ERPs, this difference can be taken only as preliminary. More broadly, Wilcox, Haslup, and Boas (2010) have studied the neural basis of object processing in the infant brain using fNIRS. Over several studies they have found that regions of the posterior parietal cortex are sensitive to spatio-temporal information used for object individuation while more ventral temporal lobe areas are responsive to featural information when used for individuation (Wilcox et al., 2010; Wilcox, Stubbs, Hirshkowitz, & Boas, 2012). However, to date, no one has investigated potential dissociations between small and large number processing in the infant brain using a neuroimaging method with high spatial resolution. Future work with infants using emerging developmental neuroimaging techniques should further address this question.

THE RELATIONSHIP BETWEEN CORE SYSTEMS AND SYMBOLIC NUMBER ABILITIES

Core system(s) of nonverbal numerical cognition have long been hypothesized to form the basis for further symbolic numerical and mathematical knowledge (Carey, 2009; Chen & Li, 2014; Dehaene, 1997; Fazio, Bailey, Thompson, & Siegler, 2014; Feigenson et al., 2004; Gallistel & Gelman, 1992; Geary, 1995; Leslie, Gelman, & Gallistel, 2008; McCrink, 2015, Volume 1; Starr & Brannon, 2015, Volume 1; vanMarle, Chu, Li, & Geary, 2014; but see Rips, Asmuth, & Bloomfield, 2006; Rips, Bloomfield, & Asmuth, 2008). Although empirical evidence is now emerging in support of this proposal and the number of investigations into this matter has recently escalated, consensus is far from evident.

Approximate Number System and Symbolic Number and Mathematics Abilities

Most of the work to date has focused on the relationship of the ANS to symbolic numerical and mathematical abilities (see Starr & Brannon, 2015, Volume 1 for further discussion). Early work highlighted associations between symbolic and nonverbal systems of number (see Piazza, 2010 for an extensive review). For example, researchers observed that behavioral comparison of symbolic numbers showed signatures of the ANS. That is, comparing the numerical magnitudes of Arabic digits that were closer together (65 vs. 71) was harder than comparing digits that were numerically more distant (65 vs. 91) (Moyer & Landauer, 1967; Van Opstal, Gevers, De Moor, & Verguts, 2008), paralleling effects of nonsymbolic numerosity comparison. Similarly, comparisons of symbolic and nonsymbolic numbers show common patterns of brain response in overlapping brain areas (Libertus et al., 2007; Piazza et al., 2007; Pinel, Piazza, Le Bihan, & Dehaene, 2004; Temple & Posner, 1998).

More recent efforts have shown a relationship between approximate number abilities and mathematics achievement (e.g., Halberda, Mazzocco, & Feigenson, 2008) (see Figure 3(A)). A majority of these studies report significant correlations between the ability to make approximate, nonsymbolic number comparisons and previous, current, and/or future mathematics achievement scores (e.g., Bonny & Lourenco, 2013; Chen & Li, 2014; Chu, vanMarle, & Geary, 2013; Fazio et al., 2014; Gilmore, McCarthy, & Spelke, 2010; Halberda & Feigenson, 2008; Halberda, Ly, Wilmer, Naiman, & Germine, 2012; Libertus, Feigenson, & Halberda, 2011). This type of research, however, only establishes an association or correlation between the ANS and the symbolic number system, the direction of which is unclear. There is some reason to remain skeptical, given that other studies have either not found a relationship between the ANS and mathematics achievement or found that the relationship is mediated by more general cognitive abilities (e.g., Fuhs & McNeil, 2013; Gilmore et al., 2013; Lyons, Ansari, & Beilock, 2012; Sasanguie, De Smedt, Defever, & Reynvoet, 2012; see De Smedt, Noël, Gilmore, & Ansari, 2013 for a review). Others have suggested that the relationship between the ANS and mathematics achievement, at least in young children, is mediated by initial learning of the symbolic number system (vanMarle et al., 2014). Similarly, Lyons, Price, Vaessen, Blomert, and Ansari (2014) assessed the relationship between several number processing tasks and mathematics achievement and found that no evidence that the ANS uniquely predicts mathematics achievement, whereas symbolic number processing does. Finally, at least one published study suggests the opposite direction of causation: symbolic number acquisition changes ANS precision (Piazza, Pica, Izard, Spelke, & Dehaene, 2013).

Others have used experimental training procedures to inform the relationship between the ANS and symbolic numerical and mathematics abilities (e.g., Hyde et al., 2014; Park & Brannon, 2013; also see Starr & Brannon, 2015, Volume 1 for further discussion). Several studies with adults and at least one study with children have now shown that training the ANS leads to gains in exact, symbolic arithmetic (e.g., Hyde et al., 2014; Park & Brannon, 2013). For example, Hyde et al. (2014) showed that practicing approximate nonsymbolic number comparison or approximate nonsymbolic addition problems improved subsequent exact, symbolic addition performance in first-grade children, while active control conditions involving the comparison or addition of other nonnumerical magnitudes (brightness or line lengths) did not (see Figure 3(B)). Similarly, Park and Brannon (2013) have shown that individual gains in approximate numerical addition training were associated with gains in symbolic arithmetic ability. Although this work suggests a causal direction with the ANS changing symbolic mathematics ability, it falls short of demonstrating that it is this relationship that drives actual numerical development in childhood.

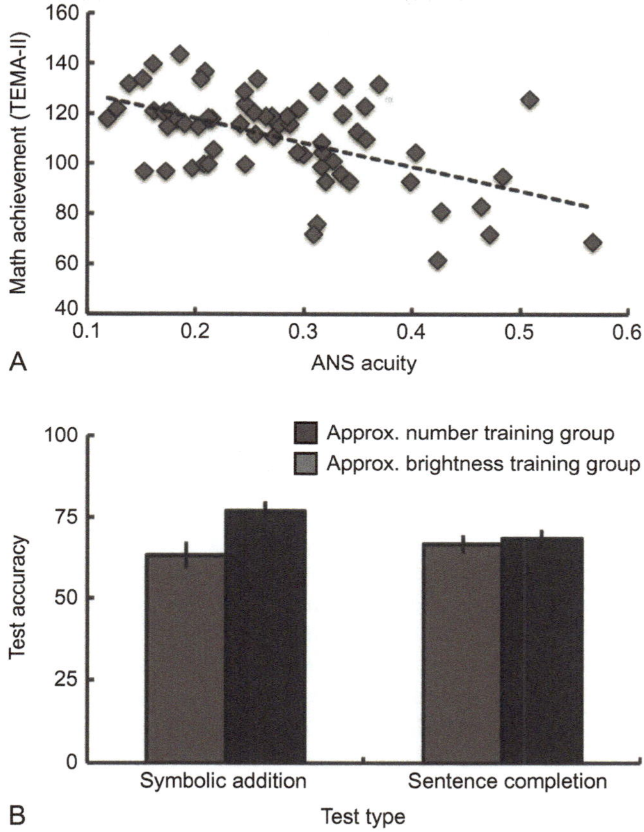

FIGURE 3 Examples of empirical evidence for a relationship between approximate number system and symbolic mathematics. (A) Results from Halberda and colleagues (2008) showing a correlation between approximate number system acuity and mathematics achievement. (B). Results from Hyde and colleagues (2014) showing an effect of approximate number training on mathematics test performance (Experiment 2) compared to a control training condition. Data for 3A extracted from original figure using WebPlotDigitizer (http://arohatgi.info/WebPlotDigitizer).

The Relationship of Core Systems to Early Number Concept Development

Almost all current theories of early numerical development posit a causal role of nonverbal core systems in initial, symbolic number concept development (for a detailed review of the neuroimaging evidence pertaining to matter, see Ansari, this volume). Theories differ substantially, however, in their characterization of which and to what extent core systems contribute to the process. One major aspect in which theories differ is the role to which they assign the core systems in this development. Some propose that it is through the mapping of ANS representations onto number words that children initially acquire a symbolic number system (Gallistel & Gelman, 1992, 2000). Others posit a special role for the PI system in early symbolic number acquisition (Carey, 2009; Le Corre & Carey, 2007). Still others hypothesize a contribution of both systems to the conceptual development process (Spelke, 2011; Spelke & Tsivkin, 2001; vanMarle, Chu, Mou, & Geary, under review). Finally, some posit no role at all for core systems (Rips et al., 2008). Despite the prominent roles core systems of numbers play in most current theories of early symbolic number development, only a few empirical studies to date have actually investigated the issue.

One such study used a longitudinal approach to measure visual attention to numerosity change in infancy and its relationship to later mathematics achievement scores in preschool (Starr, Libertus, & Brannon, 2013b). They observed that attention to numerosity change at 6 months of age correlated to ANS acuity and mathematics achievement at 3.5 years, even after controlling for general intelligence, suggesting that early individual differences in the ANS led to more advantaged mathematics development. Importantly, no relationship was observed between visual attention to nonnumerical changes and mathematics, suggesting the relationship was number-specific.

In contrast, Carey and colleagues have proposed that the ANS aids in later numerical and mathematical development but plays no role in initial symbolic number concept development (i.e., initial understanding of the exact cardinal values of number words) (e.g., Carey, 2009; Le Corre & Carey, 2007). In support of this position, Le Corre and Carey (2007) provided evidence that approximate number representations were only mapped onto the number words after those number words were understood. Thus, they concluded that the ANS does not play a role in the initial symbolic number concept acquisition. To do this, they showed that children in their sample understood the exact meanings of number words, on average, about 6 months before they showed signatures of approximate number representations over larger numbers (6–10 items). The authors concluded that the patterns of early symbolic number concept acquisition suggest a strong role for the PI system (combined with quantifiers in language—which they term enriched PI; see Carey, 2009 for a full account). That is, the fact that children learn the first few number words in a belabored, piecemeal, and largely sequential fashion and that this process transitions at around 3 or 4 items, provides qualitative evidence for involvement of PI in the learning process (Gelman, 1972, 1993; Wynn, 1990, 1992).

Experiments with children earlier in the learning process provide a different picture. Huang, Spelke, and Snedeker (2010) briefly trained subset knowers (children who only understood the meanings of the first few number words) on the next number on the count list beyond their knower-level (the highest number for which a child understands its exact, numerical meaning). They found that children who knew the meanings of 1 and 2 and were briefly trained on the meaning of 3, were able to identify new arrays of 3 objects (in a choice between 3 objects and another number they did not yet understand) only if those objects were of the same category type as those on which they were trained. In contrast, children who knew the meanings of the numbers 1–3 and were briefly trained on 4 were able to identify new arrays of 4 objects across novel categories as well. Furthermore, the responses of 3-knowers to 4 (but not 2-knowers to 3) were indicative of an approximate rather than an exact meaning of 4, suggesting that the initial conceptualization of 4 was based at least in part on the ANS. The results with 3-knowers also contrasted with those of 2-knowers, whose response to 3 was not indicative of an approximate meaning. They interpreted these results as evidence that the PI system and the ANS both contribute to early number concept development. However, they qualified that the PI system aids selectively in learning the first few numbers, while the ANS operates to make inferences about larger number word meanings, ultimately leading children to understand counting.

VanMarle, Chu, Mou, and Geary (2015) conducted the most comprehensive study to date on the role of core systems in early symbolic number concept development by measuring individual differences in ANS, object tracking, and counting ability in a large sample of children over the first year of preschool. The ANS was measured with a standard nonsymbolic numerical comparison task (Halberda et al., 2008), while the PI system was measured by a novel object-tracking task. In the object-tracking task, children watched as a small number of objects (0, 1, or 2) were sequentially hidden in a box and then a single object was either added or removed. Children were then shown the contents of the box and had to judge whether the number of objects in the box was correct or incorrect based on the transformation they observed (also see vanMarle, 2015, Volume 1 for a description). VanMarle and colleagues (2015) observed that individual differences in both the tracking task with small numbers of objects and approximate numerical comparison correlated with cardinal number knowledge at the beginning of the school year but only approximate number comparison performance correlated with individual differences in cardinal number knowledge at the end of the school year. Based on these results, they concluded that the PI system (object-tracking system in their terms) plays a role in the initial stages of conceptual development, but that its contribution is short-lived. In contrast, they claimed that that the contribution of the ANS is more enduring over the entire developmental process and, therefore, more fundamental to symbolic number concept development.

One major limitation of this study, however, is that the object-tracking task, designed to tap the nonverbal PI system, could be alternatively solved by using a verbal counting strategy (e.g., counting the number of objects going into the box rather than nonverbally tracking). If this were the case, then the object-tracking task with small numbers was really a counting task with small numbers. The consequence of this confound is that short-lived correlations between the PI system and early cardinal knowledge may reflect a correlation between the ability to count a small number of objects and early (but not later) numerical development rather than a nuanced relationship between the PI system and numerical development.

It appears, then, that the evidence to date suggests that the PI system is related to early phases of symbolic number development and that the ANS is, at the very least, related to later symbolic number abilities. It is still unclear whether children employ the ANS to learn the meanings of the numbers 1–3, and the evidence to date concerning whether the ANS is deployed in learning the meanings of larger number words (4 and on) is debatable. Ideas differ vastly in how these data should be interpreted in relation to conceptual development. Even though there appear to be clearly documented steps that most children take to learn the meanings of number words (Gallistel & Gelman, 1992; Gelman & Gallistel, 1978; Wynn, 1992), there is no consensus in the field as to what point in this process the child can be said to possess a symbolic number system. Opinions range from when they are able to employ counting successfully to being able to explicitly state mathematical axioms (Carey, 2009; Rips et al., 2008). This discord, in our opinion, reflects a deeper theoretical debate as to whether the acquisition of a symbolic number system constitutes true conceptual change or not (see Carey, 2009 for a discussion). Some posit that the acquisition of a symbolic number system requires new concepts to be formed (Carey, 2009; Spelke, 2011; Spelke & Tsivkin, 2001) while others see symbolic number development as an enrichment of previously held concepts (Dehaene, 1997; Gallistel & Gelman, 1992, 2000; vanMarle, 2015, Volume 1). Those who advocate that symbolic number system development involves conceptual change must clearly define and agree upon a point in the process at which children can be said to possess a symbolic number system. Those who advocate that symbolic number system acquisition does not involve conceptual change must argue that the requisite concepts are present before behavioral competency with the symbolic number system and, therefore, must define at what point in the conceptual development process symbolic system acquisition should be granted. Despite what appears to be clear evidence of the involvement of both systems in early symbolic number concept development, an understanding of the role that core systems of numbers contributes to numerical development will not be complete until common ground rules for what develops are established.

CONCLUSIONS

An enduring question in psychology has been how the mind creates the uniquely human endeavor of mathematics. Decades of work with human infants, children, adults, and nonhuman animals suggests primitive, nonverbal systems of number are present before language-based symbolic number systems. Over the last few decades, a succession of well-conceived and brilliantly designed experiments have shown how these systems can be used to mentally represent number, as well as to perform sophisticated numerical computations before the onset of spoken language or formal education. Recent work in cognitive neuroscience has confirmed intuitions long posited by behavioral scientists that these abilities are based on at least two distinct cognitive and brain systems: a PI system for representing individual items and an ANS for representing the numerical magnitude of sets. These brain systems appear to be functional from very early infancy and persist across the lifespan.

More recently, researchers have taken on the challenge to determine how these primitive, nonverbal systems are related to the uniquely human symbolic number concepts and mathematics. There is already strong evidence for a relationship between core systems and symbolic numerical and mathematical abilities. However, much debate still surrounds the exact role these nonverbal cognitive systems play in early symbolic number and mathematics concept development. The emerging empirical evidence to date suggests that the PI system contributes to the earliest number concept formation, while the ANS contributes to later and more sophisticated numerical abilities and mathematics achievement. However, the extent to which the contribution of each is fundamental for conceptual development in the realm of number remains unclear.

ACKNOWLEDGMENTS

We would like to thank Ilaria Berteletti for her helpful comments on earlier versions of this chapter. Writing of this chapter was supported by an NSF grant to D.C.H (NSF DRL 1252445).

REFERENCES

Alvarez, G. A., & Cavanagh, P. (2004). The capacity of visual short-term memory is set both by visual information load and by number of objects. *Psychological Science*, *15*, 106–111.

Ansari, D., & Dhital, B. (2006). Age-related changes in the activation of the intraparietal sulcus during nonsymbolic magnitude processing: An event-related functional magnetic resonance imaging study. *Journal of Cognitive Neuroscience*, *18*, 1820–1828.

Ansari, D., Lyons, I. M., van Eimeren, L., & Xu, F. (2007). Linking visual attention and number processing in the brain: The role of the temporo-parietal junction in small and large symbolic and nonsymbolic number comparison. *Journal of Cognitive Neuroscience*, *19*, 1845–1853.

Balakrishnan, J. D., & Ashby, F. G. (1992). Subitizing: Magical numbers or mere superstition? *Psychological Research*, *54*, 80–90.

Barth, H., Kanwisher, N., & Spelke, E. (2003). The construction of large number representations in adults. *Cognition*, *86*, 201–221.

Barth, H., La Mont, K., Lipton, J., Dehaene, S., Kanwisher, N., & Spelke, E. (2006). Nonsymbolic arithmetic in adults and young children. *Cognition*, *98*, 199–222.

Beran, M. J. (2007). Rhesus monkeys (*Macaca mulatta*) enumerate large and small sequentially presented sets of items using analog numerical representations. *Journal of Experimental Psychology: Animal Behavior Processes*, *33*, 42–54.

Bonny, J. W., & Lourenco, S. F. (2013). The approximate number system and its relation to early math achievement: Evidence from the preschool years. *Journal of Experimental Child Psychology*, *114*, 375–388.

Brannon, E. M., & Terrace, H. S. (1998). Ordering of numerosities 1-9 by monkeys. *Science*, *282*, 746–749.

Brannon, E. M., & Merritt, D. (2011). Evolutionary foundations of the approximate number system. In S. Dehaene & E. M. Brannon (Eds.), *Space, time, and number in the brain: Searching for the foundations of mathematical thought* (pp. 207–224). New York: Elsevier.

Burr, D. C., & Ross, J. (2008). A visual sense of number. *Current Biology*, *18*, 425–428.

Butterworth, B. (1999). *The mathematical brain*. London, UK: Macmillan.

Cantlon, J. F. (2015). Analog origins of numerical concepts. In D. C. Geary, D. B. Berch, & K. Mann Koepke (Eds.), *Evolutionary origins and early development of number processing* (pp. 225–251). San Diego, CA: Academic Press.

Cantlon, J. F., & Brannon, E. M. (2007). Basic math in monkeys and college students. *PLoS Biology*, *5*, 2912–2919.

Cantlon, J. F., Brannon, E. M., Carter, E. J., & Pelphrey, K. A. (2006). Functional imaging of numerical processing in adults and 4-year-old children. *PLoS Biology*, *4*, e125.

Carey, S. (2009). *The origin of concepts*. New York: Oxford University Press.

Chen, Q., & Li, J. (2014). Association between individual differences in nonsymbolic number acuity and math performance: A meta-analysis. *Acta Psychologica*, *148*, 163–172.

Cheries, E. W., Mitroff, S. R., Wynn, K., & Scholl, B. J. (2009). Do the same principles constrain persisting object representations in infant cognition and adult perception? The cases of continuity and cohesion. In B. M. Hood, & L. R. Santos (Eds.), *The origins of object knowledge* (pp. 107–134). New York: Oxford University Press.

Chu, F. W., vanMarle, K., & Geary, D. C. (2013). Quantitative deficits of preschool children at risk for mathematical learning disability. *Frontiers in Psychology*, *4*, 195. http://dx.doi.org/10.3389/fpsyg.2013.00195.

Cordes, S., & Brannon, E. M. (2009). Crossing the divide: Infants discriminate small from large numerosities. *Developmental Psychology*, *45*, 1583–1594.

Cordes, S., Gallistel, C. R., Gelman, R., & Latham, P. (2007). Nonverbal arithmetic in humans: Light from noise. *Perception and Psychophysics*, *69*, 1185–1203.

Cordes, S., Gelman, R., Gallistel, C. R., & Whalen, J. (2001). Variability signatures distinguish verbal from nonverbal counting for both large and small numbers. *Psychonomic Bulletin & Review*, *8*, 698–707.

Cowan, N. (2001). The magical number 4 in short-term memory: A reconsideration of mental storage capacity. *Behavioral and Brain Sciences*, *24*, 87–185.

Cutini, S., Scatturin, P., Basso Moro, S., & Zorzi, M. (2014). Are the neural correlates of subitizing and estimation dissociable? An fNIRS investigation. *NeuroImage*, *85*, 391–399.

De Smedt, B., Noël, M. P., Gilmore, C., & Ansari, D. (2013). The relationship between symbolic and nonsymbolic numerical magnitude processing and the typical and atypical development of mathematics: A review of evidence from brain and behavior. *Trends in Neuroscience and Education*, *2*, 48–55.

Dehaene, S. (1996). The organization of brain activations in number comparison: Event-related potentials and the additive-factors method. *Journal of Cognitive Neuroscience*, *8*, 47–68.

Dehaene, S. (1997). *The number sense: How the mind creates mathematics*. New York: Oxford University Press.

Dehaene, S. (2003). The neural basis of Weber-Fechner's law: Neuronal recordings reveal a logarithmic scale for number. *Trends in Cognitive Sciences*, *7*, 145–147.

Dehaene, S., & Changeux, J. (1993). Development of elementary numerical abilities: A neuronal model. *Journal of Cognitive Neuroscience*, *5*, 390–407.

Dehaene, S., & Cohen, L. (1994). Dissociable mechanisms of subitizing and counting: Neuropsychological evidence from simultanagnosic patients. *Journal of Experimental Psychology: Human Perception and Performance*, *20*, 958–975.

Dehaene, S., Izard, V., Spelke, E., & Pica, P. (2008). Log or linear? Distinct intuitions of the number scale in Western and Amazonian indigene cultures. *Science*, *320*, 1217–1220.

Dehaene, S., Piazza, M., Pinel, P., & Cohen, L. (2003). Three parietal circuits for number processing. *Cognitive Neuropsychology*, *20*, 487–506.

Drew, T., & Vogel, E. K. (2008). Neural measures of individual differences in selecting and tracking multiple moving objects. *Journal of Neuroscience*, *28*, 4183–4191.
Edwards, L. A., Wagner, J. B., Simon, C. E., & Hyde, D. C. (in press). Functional brain organization for number processing in preverbal infants. *Developmental Science*.
Ester, E. F., Drew, T. W., Klee, D., Vogel, E. K., & Awh, E. (2012). Neural measures reveal a fixed item limit in subitizing. *Journal of Neuroscience*, *32*, 7169–7177.
Fazio, L. K., Bailey, D. H., Thompson, C. A., & Siegler, R. S. (2014). Relations of difference types of numerical magnitude representations to each other and to mathematics achievement. *Journal of Experimental Child Psychology*, *123*, 53–72.
Feigenson, L., & Carey, S. (2003). Tracking individuals via object files: Evidence from infants' manual search. *Developmental Science*, *6*, 568–584.
Feigenson, L., & Carey, S. (2005). On the limits of infants' quantification of small object arrays. *Cognition*, *97*, 295–313.
Feigenson, L., Carey, S., & Hauser, M. D. (2002). The representations underlying infants' choice of more: Object files versus analog magnitudes. *Psychological Science*, *13*, 150–156.
Feigenson, L., Dehaene, S., & Spelke, E. S. (2004). Core systems of number. *Trends in Cognitive Sciences*, *8*, 307–314.
Fougnie, D., & Marois, R. (2006). Distinct capacity limits for attention and working memory. Evidence from attentive tracking and visual working memory paradigms. *Psychological Science*, *17*, 526–534.
Fuhs, M. W., & McNeil, N. M. (2013). ANS acuity and mathematics ability in preschoolers from low-income homes: Contributions of inhibitory control. *Developmental Science*, *16*, 136–148.
Gallistel, C. R. (1990). *The organization of learning*. Cambridge, MA: Bradford Books/MIT Press.
Gallistel, C. R. (2007). Commentary on Le Corre & Carey. *Cognition*, *105*, 439–445.
Gallistel, C. R., & Gelman, R. (1992). Preverbal and verbal counting and computation. *Cognition*, *44*, 43–74.
Gallistel, C. R., & Gelman, R. (2000). Nonverbal numerical cognition: From reals to integers. *Trends in Cognitive Sciences*, *4*, 59–65.
Gallistel, C. R., & Gelman, R. (2005). Mathematical cognition. In K. Holyoak, & R. Morrison (Eds.), *The Cambridge handbook of thinking and reasoning* (pp. 559–588). New York: Cambridge University Press.
Geary, D. C. (1995). Reflections of evolution and culture in children's cognition: Implications for mathematical development and instruction. *American Psychologist*, *50*, 24–37.
Gelman, R. (1972). The nature and development of early number concepts. In H. W. Reese (Ed.), *Advances in Child Development* (pp. 115–167). New York: Academic Press.
Gelman, R. (1993). A rational-constructivist account of early learning about numbers and objects. In D. Medin (Ed.), *Learning and motivation* (pp. 61–96). New York: Academic Press.
Gelman, R., & Gallistel, C. R. (1978). *The child's understanding of number*. Cambridge, MA: Harvard University Press.
Gibbon, J. (1977). Scalar expectancy theory and Weber's law in animal timing. *Psychological Review*, *84*, 279–325.
Gilmore, C., Attridge, N., Clayton, S., Cragg, L., Johnson, S., Marlow, N., et al. (2013). Individual differences in inhibitory control, not nonverbal number acuity, correlate with mathematics achievement. *PLoS One*, *8*, e67374. http://dx.doi.org/10.1371/journal.pone.0067374.
Gilmore, C. K., McCarthy, S. E., & Spelke, E. S. (2010). Nonsymbolic arithmetic abilities and mathematics achievement in the first year of formal schooling. *Cognition*, *115*, 394–406.
Grill-Spector, K., & Malach, R. (2001). fMR-adaptation: A tool for studying the functional properties of human cortical neurons. *Acta Psychologica*, *107*, 293–321.

Halberda, J., & Feigenson, L. (2008). Developmental change in the acuity of the "number sense": The approximate number system in 3-, 4-, 5-, 6-year-olds and adults. *Developmental Psychology, 44*, 1457–1465.

Halberda, J., Ly, R., Wilmer, J. B., Naiman, D. Q., & Germine, L. (2012). Number sense across the lifespan as revealed by a massive internet-based sample. *Proceedings of the National Academy of Science of the United States of America, 109*, 11116–11120.

Halberda, J., Mazzocco, M., & Feigenson, L. (2008). Individual differences in nonverbal number acuity predict maths achievement. *Nature, 455*, 665–668.

Hillyard, S. A., & Annlo-Vento, L. (1998). Event-related brain potentials in the study of visual selective attention. *Proceedings of the National Academy of Science of the United States of America, 95*, 781–787.

Hillyard, S. A., Mangun, G. R., Luck, S. J., & Heinze, H. J. (1990). Electrophysiology of visual attention. In E. R. John, T. Harmony, L. Prichep, M. Valdez, & P. Valdez (Eds.), *Machinery of the mind* (pp. 186–205). Boston: Birkhausen.

Huang, Y. T., Spelke, E. S., & Snedeker, J. (2010). When is *four* far more than *three*? Children's generalization of newly acquired number words. *Psychological Science, 21*, 600–606.

Hyde, D. C. (2011). Hypothesis and theory article: Two systems of nonsymbolic numerical cognition. *Frontiers in Human Neuroscience, 5*, 150. http://dx.doi.org/10.3389/fnhum.2011.00150.

Hyde, D. C., Boas, D. A., Blair, C., & Carey, S. (2010). Near-infrared spectroscopy shows right parietal specialization for number in pre-verbal infants. *NeuroImage, 53*, 647–652.

Hyde, D. C., Khanum, S., & Spelke, E. S. (2014). Brief nonsymbolic, approximate number practice enhances subsequent exact symbolic arithmetic in children. *Cognition, 131*, 92–107.

Hyde, D. C., & Spelke, E. S. (2009). All numbers are not equal: An electrophysiological investigation of small and large number representations. *Journal of Cognitive Neuroscience, 21*, 1039–1053.

Hyde, D. C., & Spelke, E. S. (2011). Neural signatures of number processing in human infants: Evidence for two core systems underlying nonverbal numerical cognition. *Developmental Science, 14*, 360–371.

Hyde, D. C., & Spelke, E. S. (2012). Spatiotemporal dynamics of processing nonsymbolic number: An ERP source localization study. *Human Brain Mapping, 33*, 2189–2203.

Hyde, D. C., & Wood, J. N. (2011). Spatial attention determines the nature of nonverbal numerical cognition. *Journal of Cognitive Neuroscience, 23*, 2336–2351.

Intriligator, J., & Cavanagh, P. (2001). The spatial resolution of visual attention. *Cognitive Psychology, 43*, 171–216.

Izard, V., Dehaene-Lambertz, G., & Dehaene, S. (2008). Distinct cerebral pathways for object identity and number in 3-month-old infants. *PLoS Biology, 6*, 275–285.

Izard, V., Sann, C., Spelke, E. S., & Streri, A. (2009). Newborn infants perceive abstract numbers. *Proceedings of the National Academy Sciences of the United States of America, 106*, 10382–10385.

Jevons, W. S. (1871). The power of numerical discrimination. *Nature, 3*, 281–282.

Kahneman, D., & Treisman, A. (1984). Changing views of attention and automaticity. In R. Parasuraman, D. R. Davies, & J. Beatty (Eds.), *Variants of attention* (pp. 29–61). New York: Academic Press.

Kahneman, D., Treisman, A., & Gibbs, B. J. (1992). The reviewing of object files: Object-specific integration of information. *Cognitive Psychology, 24*, 174–219.

Kaufman, E. L., Lord, M. W., Reese, T. W., & Volkmann, J. (1949). The discrimination of visual number. *American Journal of Psychology, 62*, 498–525.

Le Corre, M., & Carey, S. (2007). One, two, three, four, nothing more: An investigation of the conceptual sources of the verbal counting principles. *Cognition, 105*, 395–438.

Leslie, A. M., Gelman, R., & Gallistel, C. (2008). The generative basis of natural number concepts. *Trends in Cognitive Sciences*, *12*, 213–218.

Leslie, A. M., Xu, F., Tremoulet, P., & Scholl, B. (1998). Indexing and the object concept: Developing 'what' and 'where' systems. *Trends in Cognitive Sciences*, *2*, 10–18.

Libertus, M. E., & Brannon, E. M. (2010). Stable individual differences in number discrimination in infancy. *Developmental Science*, *13*, 900–906.

Libertus, M. E., Feigenson, L., & Halberda, J. (2011). Preschool acuity of the approximate number system correlates with school math ability. *Developmental Science*, *14*, 1292–1300.

Libertus, M. E., Woldorff, M. G., & Brannon, E. M. (2007). Electrophysiological evidence for notation independence in numerical processing. *Behavioral and Brain Functions*, *3*, 1. http://dx.doi.org/10.1186/1744-9081-3-1.

Lipton, J. S., & Spelke, E. S. (2003). Origins of number sense: Large number discrimination in human infants. *Psychological Science*, *14*, 396–401.

Lipton, J. S., & Spelke, E. S. (2004). Discrimination of large and small numerosities by human infants. *Infancy*, *5*, 271–290.

Luck, S. J. (2005). The operation of attention—millisecond by millisecond—over the first half second. In H. Ogmen, & B. G. Breitmeyer (Eds.), *The first half second: The microgenesis and temporal dynamics of unconscious and conscious visual processes* (pp. 187–206). Cambridge, MA: MIT Press.

Lyons, I. M., Ansari, D., & Beilock, S. L. (2012). Symbolic estrangement: Evidence against a strong association between numerical symbols and the quantities they represent. *Journal of Experimental Psychology: General*, *141*, 635–641.

Lyons, I. M., Price, G. R., Vaessen, A., Blomert, L., & Ansari, D. (2014). Numerical predictors of arithmetic success in grades 1-6. *Developmental Science*, *17*, 714–726.

Mandler, G., & Shebo, B. J. (1982). Subitizing: An analysis of its component processes. *Journal of Experimental Psychology: General*, *111*, 1–22.

Mazza, V., & Caramazza, A. (2011). Temporal brain dynamics of multiple object processing: The flexibility of individuation. *PLoS One*, *6*, e1743.

Mazza, V., Pagano, S., & Caramazza, A. (2012). Multiple object individuation and exact enumeration. *Journal of Cognitive Neuroscience*, *25*, 697–705.

McCrink, K. (2015). Intuitive nonsymbolic arithmetic. In D. C. Geary, D. B. Berch, & K. Mann Koepke (Eds.), *Evolutionary origins and early development of number Processing* (pp. 201–223). San Diego, CA: Academic Press/Elsevier.

Meck, W. H., & Church, R. M. (1983). A mode control model of counting and timing processes. *Journal of Experimental Psychology: Animal Behavior Processes*, *9*, 320–334.

Mou, Y., & vanMarle, K. (2014). Two core systems of numerical representation in infants. *Developmental Review*, *34*, 1–25.

Moyer, R. S., & Landauer, T. K. (1967). The time required for judgments of numerical inequality. *Nature*, *215*, 1519–1520.

Nan, Y., Knosche, T. R., & Luo, Y. J. (2006). Counting in everyday life: Discrimination and enumeration. *Neuropsychologia*, *44*, 1103–1113.

Nieder, A., & Dehaene, S. (2009). Representation of number in the brain. *Annual Review of Neuroscience*, *32*, 185–208.

Pagano, S., & Mazza, V. (2012). Individuation of multiple targets during visual enumeration: New insights from electrophysiology. *Neuropsychologia*, *50*, 754–761.

Park, J., & Brannon, E. M. (2013). Training the approximate number system improves math proficiency. *Psychological Science*, *24*, 2013–2019.

Parkman, J. M. (1971). Temporal aspects of digit and letter inequality judgments. *Journal of Experimental Psychology*, *91*, 191–205.

Piazza, M. (2010). Neurocognitive start-up tools for symbolic number representations. *Trends in Cognitive Sciences*, *14*, 542–551.

Piazza, M., Giacomini, E., Le Bihan, D., & Dehaene, S. (2003). Single-trial classification of parallel pre-attentive and serial attentive processes using fMRI. *Proceedings of the Royal Society of London: Biological Sciences*, *270*, 1237–1245.

Piazza, M., Izard, V., Pinel, P., Le Bihan, D., & Dehaene, S. (2004). Tuning curves for approximate numerosity in the human intraparietal sulcus. *Neuron*, *44*, 547–555.

Piazza, M., Mechelli, A., Butterworth, B., & Price, C. (2002). Are subitizing and counting implemented as separate or functionally overlapping processes? *NeuroImage*, *15*, 435–446.

Piazza, M., Pica, P., Izard, V., Spelke, E. S., & Dehaene, S. (2013). Education enhances the acuity of the nonverbal approximate number system. *Psychological Science*, *24*, 1037–1043.

Piazza, M., Pinel, P., Le Bihan, D., & Dehaene, S. (2007). A magnitude code common to numerosities and number symbols in human intraparietal cortex. *Neuron*, *53*, 293–305.

Pinel, P., Piazza, M., Le Bihan, D., & Dehaene, S. (2004). Distributed and overlapping representations of number, size, and luminance in parietal cortex during comparative judgements. *Neuron*, *41*, 983–993.

Posid, T., & Cordes, S. (2015). The small-large divide: A case of incompatible numerical representations in infancy. In D. C. Geary, D. B. Berch, & K. Mann Koepke (Eds.), *Evolutionary origins and early development of number processing* (pp. 253–276). San Diego, CA: Academic Press.

Pylyshyn, Z. W. (1994). Some primitive mechanisms of spatial attention. *Cognition*, *50*, 363–384.

Pylyshyn, Z. W., & Storm, R. W. (1988). Tracking multiple independent targets: Evidence for a parallel tracking mechanism. *Spatial Vision*, *3*, 179–197.

Revkin, S., Piazza, M., Izard, V., Cohen, L., & Dehaene, S. (2008). Does subitizing reflect numerical estimation? *Psychological Science*, *19*, 607–614.

Rips, L. J., Asmuth, J., & Bloomfield, A. (2006). Giving the boot to the bootstrap: How not to learn the natural numbers. Cognition, 101, B51–B60.

Rips, L. J., Bloomfield, A., & Asmuth, J. (2008). From numerical concepts to concepts of number. *Behavioral and Brain Sciences*, *31*, 623–642.

Sasanguie, D., De Smedt, B., Defever, E., & Reynvoet, B. (2012). Association between basic numerical abilities and mathematics achievement. *British Journal of Developmental Psychology*, *30*, 344–357.

Sathian, K., Simon, T. J., Peterson, S., Patel, G. A., Hoffman, J. M., & Grafton, S. T. (1999). Neural evidence linking visual object enumeration and attention. *Journal of Cognitive Neuroscience*, *11*, 36–51.

Scholl, B. J. (2001). Objects and attention: The state of the art. *Cognition*, *80*, 1–46.

Scholl, B. J. (2002). *Objects and attention*. Cambridge, MA: MIT Press.

Spelke, E. S. (2011). Natural number and natural geometry. In E. Brannon, & S. Dehaene (Eds.), *Attention and performance Vol. 24. Space, time and number in the brain: Searching for the foundations of mathematical thought* (pp. 287–317). Oxford, United Kingdom: Oxford University Press.

Spelke, E. S., & Tsivkin, S. (2001). Initial knowledge and conceptual change: Space and number. In M. Bowerman, & S. Levinson (Eds.), *Language acquisition and conceptual development* (pp. 70–93). Cambridge, UK: Cambridge University Press.

Starr, A., & Brannon, E. M. (2015). Evolutionary and developmental continuities in numerical cognition. In D. C. Geary, D. B. Berch, & K. Mann Koepke (Eds.), *Evolutionary origins and early development of number processing* (pp. 123–144). San Diego, CA: Academic Press.

Starr, A. B., Libertus, M. E., & Brannon, E. M. (2013a). Infants show ratio dependent discrimination regardless of set size. *Infancy*, *18*, 927–941.

Starr, A. B., Libertus, M. E., & Brannon, E. M. (2013b). Number sense in infancy predicts mathematical abilities in childhood. *Proceedings of the National Academy of Sciences of the United States of America*, *110*, 18116–18120.

Taves, E. H. (1941). Two mechanisms for the perception of visual numerousness. *Archives of Psychology*, *37*, 1–47.

Temple, E., & Posner, M. I. (1998). Brain mechanisms of quantity are similar in 5-year-olds and adults. *Proceedings of the National Academy of Sciences of the United States of America*, *95*, 7836–7841.

Trick, L. M., & Pylyshyn, Z. W. (1994). Why are small and large numbers enumerated differently? A limited-capacity preattentive stage in vision. *Psychological Review*, *101*, 80–102.

Van Opstal, F., Gevers, W., De Moor, W., & Verguts, T. (2008). Dissecting the symbolic distance effect: Comparison and priming effects in numerical and nonnumerical orders. *Psychonomic Bulletin & Review*, *15*, 419–425.

vanMarle, K. (2012). Communication breakdown: Object tracking and analog magnitude representations are incommensurable. *Poster presented at the XVIII Biennial International Conference on Infant Studies*, Minneapolis, MN.

vanMarle, K. (2013). Different mechanisms underlie infants' representation of small and large numbers in an ordinal choice task. *Journal of Experimental Child Psychology*, *114*, 102–110.

vanMarle, K. (2015). Foundations of the formal number concept: How preverbal mechanisms contribute to the development of cardinal knowledge. In D. C. Geary, D. B. Berch, & K. Mann Koepke (Eds.), *Evolutionary origins and early development of number processing* (pp. 175–199). San Diego, CA: Academic Press.

vanMarle, K., Chu, F. W., Li, Y., & Geary, D. C. (2014). Acuity of the approximate number system and preschoolers' quantitative development. *Developmental Science*, *17*, 492–505.

vanMarle, K., & Wynn, K. (2009). Infants' auditory enumeration: Evidence for analog magnitudes in the small number range. *Cognition*, *111*, 302–316.

vanMarle, K., & Wynn, K. (2011). Tracking and quantifying objects and non-cohesive substances. *Developmental Science*, *14*, 302–316.

vanMarle, K., Chu, F. W., Mou, Y., & Geary, D. C. (March, 2015). Attaching meaning to the number words: What mechanism underlies children's cardinal knowledge? *Paper presented at the Biennal Meeting of the Society for Research in Child Development*, Philadelphia, PA.

Warren, H. C. (1897). The reaction time of counting. *Psychological Review*, *4*, 569–591.

Whalen, J., Gallistel, C. R., & Gelman, R. (1999). Nonverbal counting in humans: The psychophysics of number representation. *Psychological Science*, *10*, 130–137.

Wilcox, T., Haslup, J. A., & Boas, D. A. (2010). Dissociation of processing of featural and spatiotemporal information in the infant cortex. *NeuroImage*, *53*, 1256–1263.

Wilcox, T., Stubbs, J., Hirshkowitz, A., & Boas, D. A. (2012). Object processing and functional organization of the infant cortex. *NeuroImage*, *62*, 1833–1840.

Wood, J., & Spelke, E. (2005). Infants' enumeration of actions: Numerical discrimination and its signature limits. *Developmental Science*, *8*, 173–181.

Wynn, K. (1990). Children's understanding of counting. *Cognition*, *36*, 155–193.

Wynn, K. (1992). Children's acquisition of the number words and the counting system. *Cognitive Psychology*, *24*, 220–251.

Wynn, K., Bloom, P., & Chiang, W. C. (2002). Enumeration of collective entities by 5-month-old infants. *Cognition*, *83*, B55–B62.

Xu, F. (2003). Numerosity discrimination in infants: Evidence for two systems of representations. *Cognition*, *89*, B15–B25.

Xu, F., & Spelke, E. S. (2000). Large number discrimination in 6-month-old infants. *Cognition*, *74*, B1–B11.

Xu, F., Spelke, E. S., & Goddard, S. (2005). Number sense in human infants. *Developmental Science*, *8*, 88–101.

Chapter 4

A Neurodevelopmental Perspective on the Role of Memory Systems in Children's Math Learning

Vinod Menon

Department of Psychiatry and Behavioral Sciences, Stanford University, Stanford, California, USA

INTRODUCTION

The emergence of memory-based problem solving is a cardinal feature of children's cognitive development. Two decades of behavioral studies with children have demonstrated that a shift from the use of procedure-based to memory-based problem solving is a hallmark of children's cognitive development (Geary, 1994; Menon, 2014; Siegler, 1996). The use of memory-based approaches to solve mathematical problems can the academic achievement of typically developing children and children with dyscalculia who do not transition to the use of memory-based strategies (Cho, Ryali, Geary, & Menon, 2011; Geary, 2011; Geary & Brown, 1991; Geary & Hoard, 2003).

During their learning and skill acquisition years, children have available to them a mix of procedural and memory-based approaches for solving problems in a given domain. Elaborate computational procedures (e.g., counting to solve arithmetic problems, phonetic decoding to sound out and read words) are used more frequently than memory-based procedures (e.g., addition fact or word retrieval) (Butterworth, 1999; Geary, 1994; Menon, 2014; Siegler, 1996). Cognitive development is characterized by the use of different overlapping strategic approaches to problem solving, with learning reflected in less efficient procedures that are gradually replaced by more efficient memory-based ones (Geary, 1994; Siegler, 1996). In contrast to the tremendous advances in our understanding of children's cognitive development at the behavioral level, knowledge of their neurobiological mechanisms is only now beginning to emerge.

Within the domain of mathematical cognition, arithmetic learning has been the focus of most investigations of efficient fact retrieval. This process is preceded by an extended period during which children use a mix of counting,

Development of Mathematical Cognition. http://dx.doi.org/10.1016/B978-0-12-801871-2.00004-6

decomposition, retrieval, and other procedures to solve addition problems (Geary, 1994; Siegler & Shrager, 1984). Despite considerable advances in our understanding of the behavioral and cognitive mechanisms characterizing these shifts (Siegler & Svetina, 2006), the role of memory systems in building representations for basic arithmetic facts, including addition, is only now being recognized. Children's arithmetic problem solving therefore provides an ideal domain for studying the brain mechanisms that underlie this cardinal feature of cognitive development.

Although the posterior parietal cortex (PPC), and its intraparietal sulcus (IPS) subdivision in particular, has been the focus of most neurobiological research on mathematical cognition and learning (Ansari, 2008; Butterworth, 1999; Dehaene, Piazza, Pinel, & Cohen, 2003), recent research is beginning to emphasize a multisystem approach (Fias, Menon, & Szucs, 2013; Menon, 2015a, Qin et al., 2014). Specifically, multiple distributed neural processes involved in number form, magnitude, and quantity representations, working memory, and declarative memory have been identified as being important for numerical problem solving and mathematical learning. This chapter focuses on the declarative memory system (Figure1), which plays a

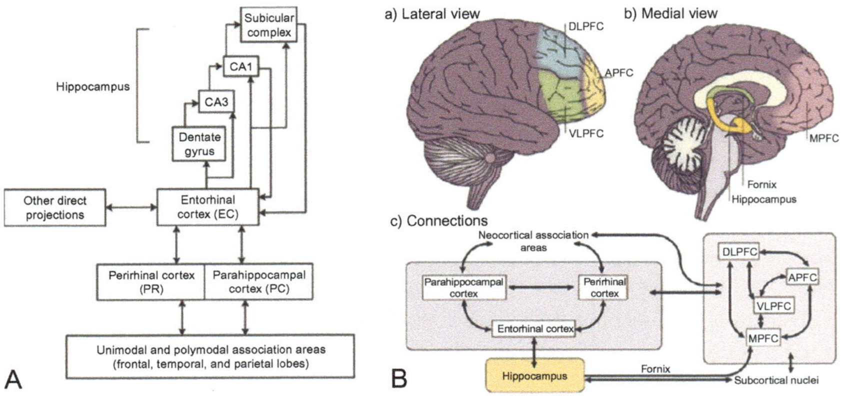

FIGURE 1 Medial temporal lobe memory system. (A) Schematic view of the medial temporal lobe declarative memory system, which is composed of the hippocampus and the perirhinal, entorhinal, and parahippocampal cortices. (B) MTL connections with the prefrontal coretex. (a,b) Major subdivisions of the prefrontal cortex (PFC) that are linked to the MTL. The dorsolateral (DLPFC, BA 46 and 9) and ventrolateral (VLPFC, BA 44, 45, and 47) are the two main PFC regions involved in mathematical cognition. (c) MTL-PFC connections. There are large cortico-cortical direct reciprocal connections between the MTL and the PFC, passing through the uncinate fascicle, anterior temporal stem, and anterior corpus callosum. The DLPFC has strong reciprocal connections with the perirhinal and entorhinal cortices. The MTL receives information from a range of unimodal and polymodal sensory association areas. This information predominantly enters through the perirhinal and parahippocampal cortices, which project back to these regions. The PFC has reciprocal connections with sensory association cortices including temporal and parietal regions and many subcortical structures. *Adapted from Squire and Dede (2015) and Simons and Spiers (2003).*

critical, but heretofore underappreciated, role in scaffolding children's mathematics learning and skill development.

The importance of the medial temporal lobe (MTL), particularly its hippocampal subdivision, in learning and memory for events in space and time is well known (Davachi, 2006; Davachi, Mitchell, & Wagner, 2003; Diana, Yonelinas, & Ranganath, 2007; Eichenbaum, Yonelinas, & Ranganath, 2007; Tulving, 1983), but its contributions to mathematics learning and cognitive development more broadly have received little attention until recently. This chapter presents a neurocognitive framework for conceptualizing the development of memory-based fact retrieval and highlights the role for which MTL systems have heretofore been largely neglected in the field. An important and novel aspect of these findings is that it cannot be assumed that brain systems engaged by children during critical stages of development are the same as those in adults with well-developed schema knowledge (McClelland, McNaughton, & O'Reilly, 1995; Tse et al., 2007; van Kesteren, Ruiter, Fernandez, & Henson, 2012).

DEVELOPMENT OF MEMORY-BASED STRATEGIES IN CHILDREN'S MATHEMATICS LEARNING

The ability to quickly and efficiently retrieve basic arithmetic facts from long-term memory is a core feature of children's early mathematical skill development. During development, children's mathematical problem-solving skills gradually become more dependent on memory-based strategies such as direct retrieval, and are less dependent on effortful procedures such as counting (Geary, Hoard, Byrd-Craven, & DeSoto, 2004). Children's cognitive development is characterized by "overlapping waves" (Siegler, 1996), wherein advances are characterized not by broad and abrupt shifts from one stage of thinking to another, but rather by changes in the distributions of strategies children use for problem solving (Figure 2). This developmental shift in strategy use is most evident during second and third grades (Ashcraft & Fierman, 1982; Carpenter & Moser, 1984; Geary, 1994; Siegler, 1996). Critically, a shift to fact retrieval serves as a foundation for the efficient solution of more complex arithmetic problems in which simple arithmetic is embedded (Geary & Widaman, 1987), and failure to undergo such a shift is a prominent deficit in children and adults with dyscalculia (Kaufmann, 2002; McCloskey, Harley, & Sokol, 1991).

Children primarily use four strategies to solve single-digit addition problems: (a) counting fingers, (b) verbal counting, (c) retrieval, and (d) decomposition (e.g., $6+7=6+(6+1)=(6+6)+1=12+1=13$) (Ashcraft, 1982; Geary, Hamson, & Hoard, 2000). Repeated use of counting results in the formation of associations between problem stems (e.g., $5+7$) and answers (e.g., 12) such that presenting the stem will eventually trigger retrieval of the correct answer (Siegler & Shrager, 1984). Such paired associations are precisely the kind of encoding processes that engage the MTL memory system (Diekelmann, Wilhelm, & Born, 2009; Frankland & Bontempi, 2005; Kumaran, Summerfield,

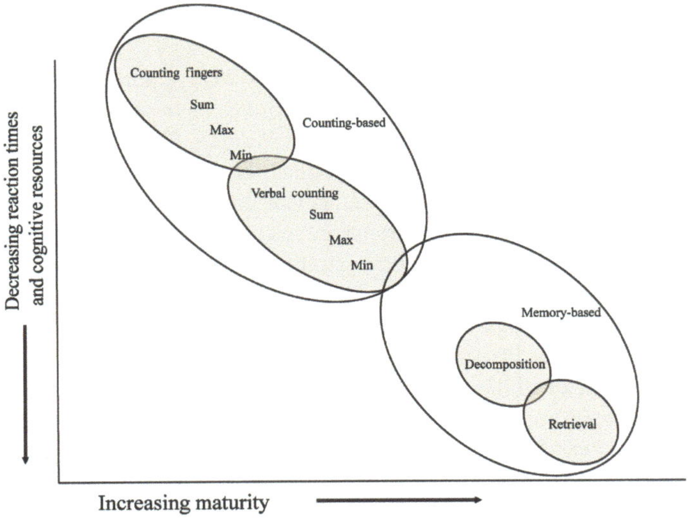

FIGURE 2 Developmental changes in children's problem-solving strategies. Decreased reliance on inefficient problem-solving strategies such as finger and verbal counting to memory-based strategies is a hallmark of children's mathematical skill development. *Adapted from Geary and Hoard (2003).*

Hassabis, & Maguire, 2009; Nadel & Moscovitch, 2001; Schacter, Addis, & Buckner, 2007; Schacter, Norman, & Koutstaal, 1998; Tse et al., 2011; Tulving, 1983, 2002).

It is also thought that the representation of addition facts in long-term memory is aided by the repeated use of counting and other procedures during problem solving (Ashcraft, 1982; Groen & Parkman, 1972; Siegler, Shipley, Simon, & Halford, 1995; Siegler & Shrager, 1984). As an example, counting up from 5 to 8 ("five, six, seven, eight") to solve the problem "5 + 3 = ?" results in an association between the answer ("eight") and the problem stem ("5 + 3"). After many such counts, children begin to retrieve the answer directly when presented with the stem (Siegler & Shrager, 1984). In this way, schema knowledge of counting principles, cardinality and ordinality, helps frame the problem space for use of counting procedures to solve arithmetic problems and supports encoding of information into long-term semantic memory.

A variety of behavioral methods have been employed to assess the mix of strategies young children use to solve arithmetic problems, ranging from verbal report (Carpenter & Moser, 1984) to mathematical modeling of reaction times (Ashcraft, 1982; Groen & Parkman, 1972). Concerns were initially raised regarding the validity of child reports (Hamann & Ashcraft, 1985), but subsequent studies showed that self-reported strategies were consistent with associated mean response time (RT) patterns and experimenter observation (Geary, 1990; Groen & Parkman, 1972; Siegler, 1987; Wu et al., 2008). Crucially, self-report, observer-report, and RT patterns show high concordance (Kappa=0.948), suggesting that by the age of 7 children are aware of, and can accurately report on the strategies they used to solve a given problem. Despite the fidelity of these measures, behavioral research leaves unclear the underlying brain systems that support the shift to fact retrieval over development and how they are related to individual differences in learning. Neurobiological investigations can provide insights into the mechanisms underlying such transitions in ways that behavioral studies alone cannot.

DECLARATIVE MEMORY AND ITS DEVELOPMENT

Declarative memory is central to our ability to remember facts and events (Davachi, 2006; Davachi et al., 2003; Diana et al., 2007; Eichenbaum et al., 2007; Tulving, 1983). It lies at the foundation of learning and memory (Diekelmann et al., 2009; Frankland & Bontempi, 2005; Kumaran et al., 2009; Nadel & Moscovitch, 2001; Schacter et al., 1998, 2007; Tse et al., 2011; Tulving, 1983, 2002). Declarative memory abilities show significant changes from childhood through adolescence and into adulthood (Ghetti & Bunge, 2012; Ghetti, DeMaster, Yonelinas, & Bunge, 2010; Kail, 1990; Ofen et al., 2007; Schneider & Pressley, 1997). Memory capacity, speed, and accuracy of retrieval as well as strategies used to retrieve information undergo a protracted developmental trajectory (Billingsley, Lou Smith, & Pat McAndrews, 2002; Cycowicz & Friedman, 2003; Cycowicz, Friedman, & Duff, 2003; Cycowicz, Friedman, Snodgrass, & Duff, 2001; Dirks & Neisser, 1977; Drummey & Newcombe, 2002; Ghetti & Bunge, 2012; Ghetti et al., 2010; Mandler & Robinson, 1978; Ofen et al., 2007; Schneider & Pressley, 1997). This increased capacity likely contributes to academic learning and knowledge acquisition in children, but the links between the two have not been adequately investigated. Even in adults, much of what we have learned about the development of declarative memory is based on retrieval of recently learned materials (e.g., minutes or hours). Thus, we know very little about how memories for facts, such as arithmetic facts, are formed, stabilized, and consolidated over longer intervals of months and years (Bauer, 2008; Ostby, Tamnes, Fjell, & Walhovd, 2011). Nevertheless, by age 7-8, the long-term memory capacity is sufficiently mature to facilitate mathematics learning and fluent retrieval of facts. As described below, this occurs in parallel with the maturation of the MTL memory system, a system important for encoding new information and forming long-term memories.

MEDIAL TEMPORAL LOBE MEMORY SYSTEM

The MTL forms the structural core of the declarative memory system (Figure 1), and research over the past two decades has clarified the specific roles of its functional subdivisions in different aspects of encoding and retrieval of novel information (Kumaran et al., 2009; Schacter et al., 1998, 2007; Tulving, 1983,

2002). The MTL and its associated functional circuits play an important role in memory encoding and retrieval in both children and adults (Ghetti et al., 2010; Menon, Boyett-Anderson, & Reiss, 2005; Ofen et al., 2007). The MTL is thought to contribute to declarative memory through binding of inputs from multiple cortical areas (Davachi, 2006; Eichenbaum, 2004; Eichenbaum et al., 2007), while its functional interactions with the prefrontal cortex (PFC) are thought to facilitate memory formation and retrieval through cognitive control processes acting on the contents of memory (Qin et al., 2007, 2009; Qin, van Marle, Hermans, & Fernandez, 2011).

Declarative memory relies on the coordinated interactions of distributed brain areas, most prominently, the hippocampus and the PFC (Figure 1) (Norman & O'Reilly, 2003; Qin, Hermans, Rijpkema, & Fernández, 2011; Simons & Spiers, 2003). Research in rodents has helped in the framing of large-scale circuit models of how newly acquired memories are consolidated into long-term memory (Diekelmann et al., 2009; Frankland & Bontempi, 2005; McGaugh, 2000). These studies suggest that newly acquired memories are strongly dependent on the hippocampus and its interactions with the PFC, and become a increasingly independent of the MTL over time (Figure 3). Such systems-level functional reorganization appears to be mediated by the lateral as well as the medial PFC (Frankland & Bontempi, 2005; Takashima et al., 2006). In adults, neuroimaging studies have recently begun to examine memory consolidation and test the translational validity of theoretical models based on studies in animals (Takashima et al., 2006, 2009; van Dongen, Takashima, Barth, & Fernandez, 2011; Wang & Morris, 2010).

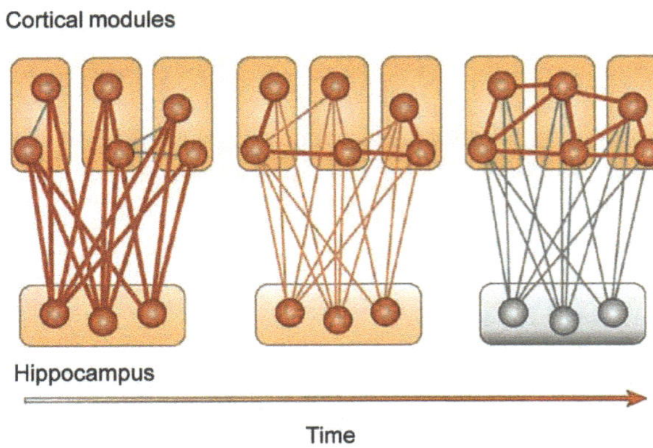

FIGURE 3 Model of time-dependent functional interactions underlying long-term memory formation and consolidation. Encoding of perceptual, motor, and cognitive information initially occurs in several specialized primary and associative cortical areas. The hippocampus integrates information from these distributed cortical modules that represent the various features of an experience, and rapidly fuses these features into a coherent memory trace. Successive reactivation of this hippocampal-cortical network leads to progressive strengthening of cortico-cortical connections. Incremental strengthening of cortico-cortical connections eventually allows new memories to become independent of the hippocampus and to be gradually integrated with preexisting cortical memories. *Adapted from Frankland and Bontempi (2005).*

This neurocognitive model provides a framework for conceptualizing how distributed neural representations for arithmetic facts develop over time, and how the declarative memory system can contribute to the formation of memories that are eventually independent of the MTL.

MEMORY PROCESSES IN THE CONTEXT OF MATHEMATICS LEARNING

The role of the declarative memory system in mathematics learning must be considered in the context of other core neurocognitive processes that support the development and mastery of numerical skills. First, the understanding of numerical quantity, knowledge of numerical magnitude, and manipulations of symbolic and nonsymbolic quantity ("number sense") may be critical building blocks from which all mathematical knowledge is constructed. These basic building blocks rely on visual and auditory association cortices that help decode the visual form and phonological features of numerical stimuli and the parietal attention system (Dehaene et al., 2003), which helps to build semantic representations of quantity (Ansari, 2008) from multiple low-level, visuo-spatial primitives such as eye gaze and pointing (Simon, Mangin, Cohen, Le Bihan, & Dehaene, 2002). Second, procedural and working memory systems anchored in the basal ganglia and fronto-parietal circuits create short-term representations that support the manipulation of multiple discrete quantities over several seconds. Third, prefrontal control processes guide and maintain attention and working memory during goal-directed decision making.

In contrast, the declarative memory system plays a distinctive role in facilitating encoding, retrieval, and long-term memory formation of mathematical information. Together, these processes contribute to associative encoding of facts, as well as generalization beyond individual problem attributes within a broader schema. Crucially, until recently, most studies of mathematics learning in both adults and children have primarily focused on the role of ventral-temporal, occipital and parietal cortices in number sense and quantity manipulation, and in so doing almost entirely neglected the role of memory systems in mathematics learning. Although the MTL is known to be critical for memory formation, its potential role in learning arithmetic facts has been largely ignored in previous brain imaging studies because adults are less likely to use this system during retrieval.

CHILDREN ENGAGE THE MTL MEMORY SYSTEM DIFFERENTLY THAN ADULTS

Over the past few years, evidence has been slowly accumulating for the differential involvement of the MTL memory system in mathematics learning, especially during key stages of skill acquisition in children. Theories of memory consolidation posit that the MTL plays an important role in the early stages of learning and retrieval, but its involvement decreases over time with a concomitant increase in reliance on neocortical memory systems. Furthermore, MTL involvement in retrieval also depends on how well schema and domain knowledge, which are poorly developed in young children, are established (Tse et al., 2007). If such a model applies to arithmetic fact learning, for example, one would predict that compared to adults, children will rely more on MTL memory systems during the initial stages of formal skill acquisition.

The first evidence for the differential engagement of the MTL memory system in arithmetic skill acquisition came from a cross-sectional study in children, adolescents, and adults who ranged in age from 8 to 19 (Rivera, Reiss, Eckert, & Menon, 2005). Participants viewed arithmetic equations in the form 'a+b=c' or 'a−b=c' and were asked to judge whether the results were correct or not. The groups were matched for accuracy, allowing the researchers to probe dissimilar trajectories of functional maturation independent of performance differences. Analysis of brain areas that showed linear changes with age revealed greater activation in the left PPC and lateral occipital temporal cortex along with lower activation in multiple PFC areas in adults (Figure 4), leading to the

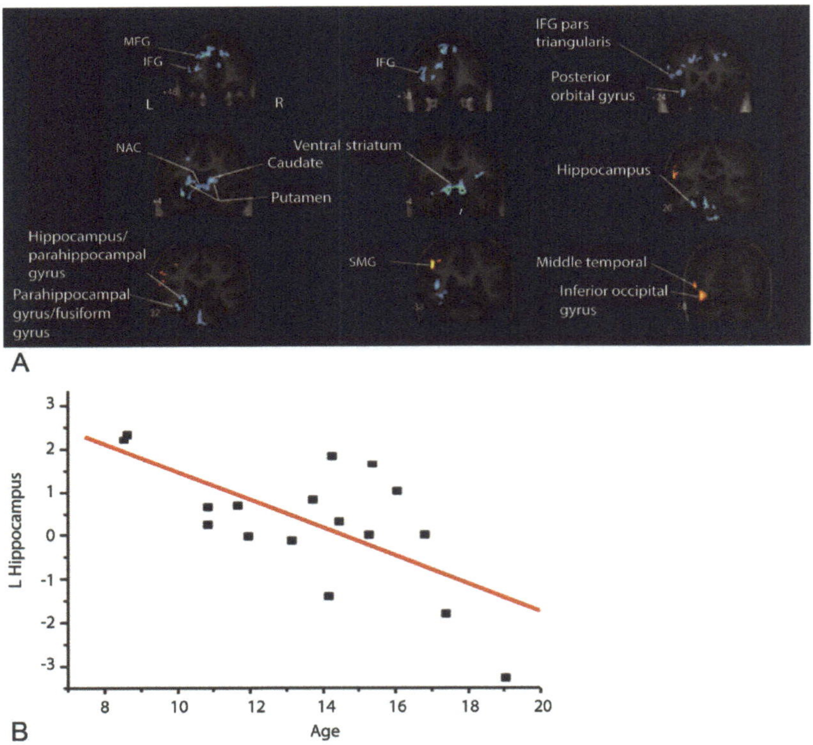

FIGURE 4 Children engage the MTL memory system differently than adults. (A) Brain areas that showed significant increases and decreases in activation with age on mental arithmetic trials. The red color scale shows areas of activation that are positively correlated with age, while the blue color scale shows areas of activation that are negatively correlated with age. (B) Age-related decreases in medial temporal lobe response. *Adapted from Rivera et al. (2005).*

proposal that increased expertise in arithmetic is accompanied by functional specialization of the PPC along with decreased reliance on attentional resources for sequencing and execution of procedures. Importantly, younger children exhibited significantly greater engagement of multiple MTL regions including the hippocampus, a subdivision of the MTL memory system that plays a major role in encoding and retrieval of novel information (Squire, Stark, & Clark, 2004). In addition, the parahippocampal gyrus—a region that mediates convergence of high-level input from the visual association cortex into the hippocampus (Suzuki & Amaral, 1994)—also showed greater activation in younger children (Figure 4). Together, these regions facilitate the associative encoding of information in short-term memory (Eichenbaum, 2000).

In the same vein, De Smedt, Holloway, and Ansari (2011) found greater hippocampal response in children as compared to adults when solving addition problems; hippocampal activation was not detected for subtraction problems that are less well rehearsed and more difficult to memorize because subtraction problems are not commutative (e.g., $5-3 \neq 3-5$). These studies provided tantalizing hints that the hippocampus may be involved in children's mathematics skill acquisition in ways that are quite different from adults.

INDIVIDUAL DIFFERENCES IN CHILDREN'S RETRIEVAL STRATEGY USE ARE ASSOCIATED WITH THE MTL

Cross-sectional developmental studies spanning wide age ranges do not adequately capture key processes involved in the acquisition of mathematics skills. To capture the mechanisms driving cognitive change, it is critical to focus on the age ranges in which the rate of change is the steepest. Examining age-related differences between children, adolescents, and adults alone is not sufficient for drawing accurate conclusions about the interaction between cognitive development and individual differences in performance; for example, comparison of 7- and 12-year-olds may be insensitive to change occurring in specific mathematical abilities between the ages of 7 and 9. Furthermore, cognitive developmental change cannot be conflated with age-related changes, and critically, it cannot be assumed that learning in the developing child's brain relies on mechanisms similar to those used by adults (Karmiloff-Smith, 2010).

As noted earlier, children's gains in problem-solving skills during the elementary school years are characterized by shifts in the mix of problem-solving approaches they use and gradual replacement of inefficient procedural strategies with direct retrieval of domain relevant facts. Cho et al. (2012) used a relatively large sample of 2nd and 3rd graders ($n=86$) to capture the full range of children's strategy mix for solving addition problems, from those who relied predominantly on counting to those who used a mix of counting and retrieval, to those who primarily used retrieval. Analysis of the associated brain imaging data focused on neurodevelopmental changes related to increased use of retrieval strategies.

Behavioral strategy use was indexed by computing the retrieval frequency in each child. During strategy assessment, participants were asked to solve each problem without the use of paper and pencil as quickly as possible and to verbally state the answer out loud. They were also asked to report how they solved each problem immediately after stating the answer. Based on each participant's self-reported strategy and the experimenter's observations, strategies for solving addition problems were classified into two major categories of interest: "counting" (such as counting fingers, verbal counting, count by numbers, or counting in mind), and "direct retrieval" (such as just know, remember, or guess responses). This approach allows us to quantify the frequency of memory-based strategy use, or "retrieval fluency," which provides a reliable assessment of putative memory-based strategy use in arithmetic problem solving (Geary et al., 2004; Wu et al., 2008). Higher retrieval fluency was associated with greater response in multiple brain regions, including the hippocampus and parahippocampal gyrus (Figure 5) subdivisions of the MTL. This study demonstrated for the first time that individual differences in children's retrieval strategy use, far from being idiosyncratic, are in fact associated with a predictable profile of MTL response. Furthermore, use of counting strategies was not associated with any consistent pattern of brain response across children, suggesting significant variability in brain response associated with counting and other inefficient strategies in young children.

DECODING BRAIN ACTIVITY PATTERNS ASSOCIATED WITH COUNTING AND RETRIEVAL STRATEGIES

Additional insights into the differential brain processes associated with distinct problem-solving strategies in 7- to 9-year old children were afforded by comparing brain activation patterns between two groups of children—those who primarily used retrieval and those who primarily used counting strategies (Cho et al., 2011, 2012). Children in the latter group were well-practiced at counting and could solve arithmetic problems as accurately and as quickly as those who used retrieval (Figure 6A). Retrieval and counting strategies were associated with different activation patterns in MTL regions important for memory encoding and retrieval, including bilateral hippocampus and parahippocampal gyrus as well as the PPC and ventrolateral PFC (Figure 6B). These results demonstrate that retrieval and counting strategies during early learning are characterized by distinct patterns of activity in a distributed network of brain regions that includes the MTL. The existence of decodable fine-scale pattern differences in fMRI signals suggests not only that the MTL is differentially engaged in relation to retrieval, but also that the underlying neural resources are accessed and used differently in each strategy. Memory-based arithmetic problem solving evokes distinct fine-scale neural representations that are independent of overall differences in task-related activation levels. This demonstrates that even when behavioral performance is equivalent, counting and retrieval are associated with

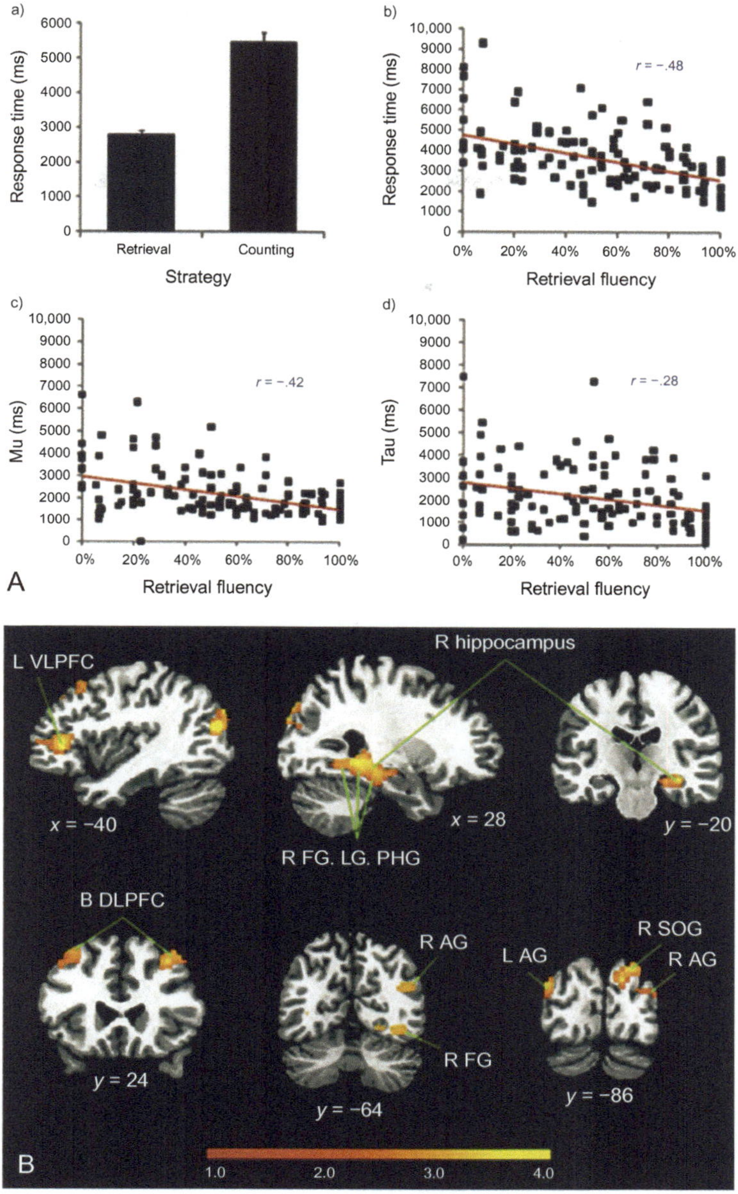

FIGURE 5 Individual differences in children's retrieval strategy use are associated with the MTL. (A) Behavioral results from the in-scanner addition task. (a) Children were significantly faster in their RTs for the control task compared with the addition task. (b) Children were significantly more accurate on the control task compared with the addition task. (c,d) Children with higher retrieval fluency were faster and more accurate on the addition task. (B) Brain responses associated with greater retrieval fluency. Activity levels were positively correlated with retrieval fluency in the right hippocampus and parahippocampal gyrus (PHG), fusiform gyrus (FG), and lingual gyrus (LG), left ventrolateral PFC, bilateral dorsolateral PFC, bilateral posterior angular gyrus (AG), right superior parietal lobule (SPL), and superior occipital gyrus (SOG). No brain areas showed negative correlations with retrieval fluency. L = left; R = right.

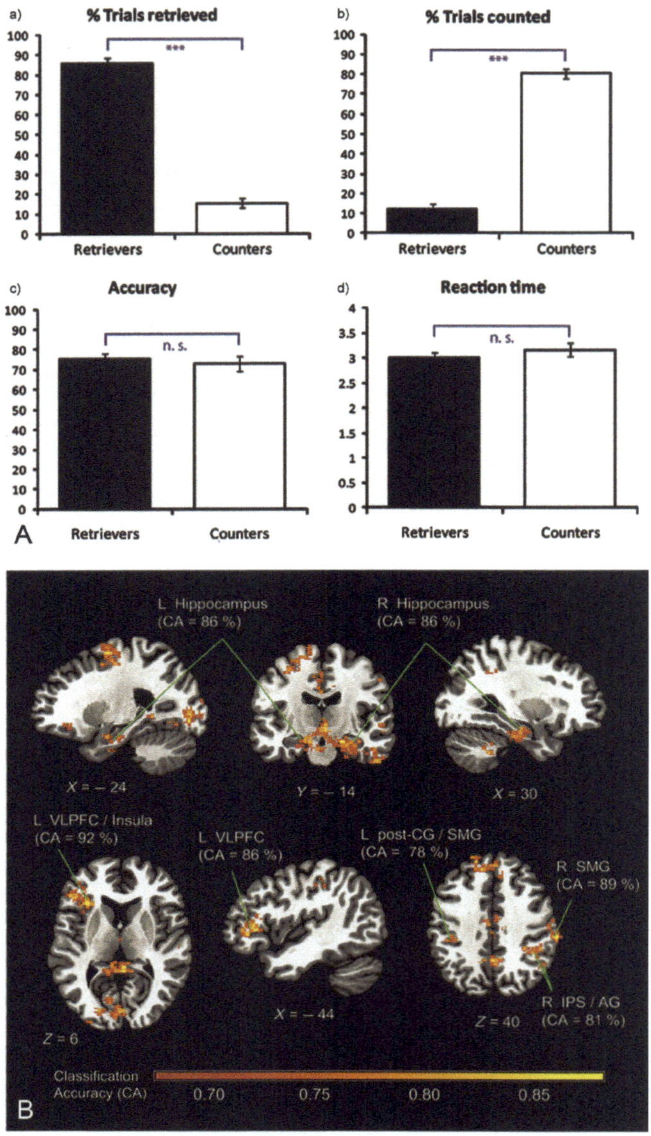

FIGURE 6 Differential brain responses associated with retrieval and countring strategy use. (A) (a,b) Differential strategy use in Retrievers and Counters measured from the strategy assessment session outside the scanner. The two groups differed significantly on percentage of trials in which retrieval or counting were used. (c,d) Performance in Retrievers and Counters on the Addition task performed inside the scanner. The two groups were well matched on performance, with no differences in either accuracy or reaction time (RT). *** $p < .001$; ns, not significant. (B) Multivariate pattern analysis (MPA) revealed significant differences in spatial activation patterns between Retrievers and Counters in three brain areas: (1) bilateral medial temporal lobe—hippocampi and parahippocampal gyri; (2) left prefrontal cortex—ventrolateral prefrontal cortex (VLPFC) and adjoining anterior insula; (3) bilateral posterior parietal cortex—left and right supramarginal gyrus (SMG), right intra-parietal sulcus (IPS) and angular gyrus (AG). Peak classification accuracies (CA) in each brain region are shown in parentheses. *Adapted from Cho et al. (2011).*

differential patterns of activity in core brain regions that support memory encoding and formation (Tse et al., 2007; Wang & Morris, 2010).

HIPPOCAMPAL-PREFRONTAL CORTEX CIRCUITS AND THEIR ROLE IN CHILDREN'S MATHEMATICS LEARNING

The dynamic coordination between the MTL and PFC plays an important role in memory formation (Norman & O'Reilly, 2003; Qin, Hermans, Rijpkema, & Fernández, 2011; Simons & Spiers, 2003), and connectivity analyses are beginning to shed light on how the MTL and PFC interact to support memory formation for arithmetic facts. Both the ventrolateral and dorsolateral PFC are associated with increased retrieval use (Cho et al., 2012). The ventrolateral PFC is known to play a prominent role in cognitive control over memory retrieval processes, both in adults (Badre & Wagner, 2007; Koechlin, Ody, & Kouneiher, 2003; Miller, 2000) and children (Adleman et al., 2002; Bunge & Wright, 2007; Houde, Rossi, Lubin, & Joliot, 2010; Kwon, Reiss, & Menon, 2002). In children, these control processes are thought to be important for accurate retrieval of relevant facts and inhibition of irrelevant information (DeStefano & LeFevre, 2004; Kaufmann, 2002; Kaufmann, Lochy, Drexler, & Semenza, 2004; Logie, Gilhooly, & Wynn, 1994).

Analysis of task-related MTL connectivity has identified distributed functional circuits associated with retrieval fluency. In particular, right hippocampus connectivity with bilateral ventrolateral and dorsolateral PFC were strongly correlated with retrieval fluency (Figure 7A) (Cho et al., 2012). Additionally, dynamic causal modeling of fMRI data provides further insights into the temporal profile of interactions between MTL and PFC regions involved in mediating retrieval fluency. This analysis revealed strong bidirectional interactions between the hippocampus and the left ventrolateral and dorsolateral PFC (Figure 7B). Crucially, causal influences from the left ventrolateral PFC to the hippocampus appear to serve as the main "top-down" component, while causal influences from the hippocampus to the left dorsolateral PFC serve as the main "bottom-up" component of this retrieval network. While still preliminary, these analyses highlight the differential contribution of hippocampal-prefrontal circuits to the early development of retrieval fluency in arithmetic problem solving, and provide a novel framework for studying dynamic developmental processes involving the MTL and PFC that accompany the maturation of cognitive skills. Further research is needed to investigate how these processes contribute to concomitant improvements in cognitive control over retrieval, including successful inhibition of irrelevant information, such as incorrect answers, intermediate steps, and operand intrusions (Barrouillet & Lepine, 2005; Passolunghi & Siegel, 2004).

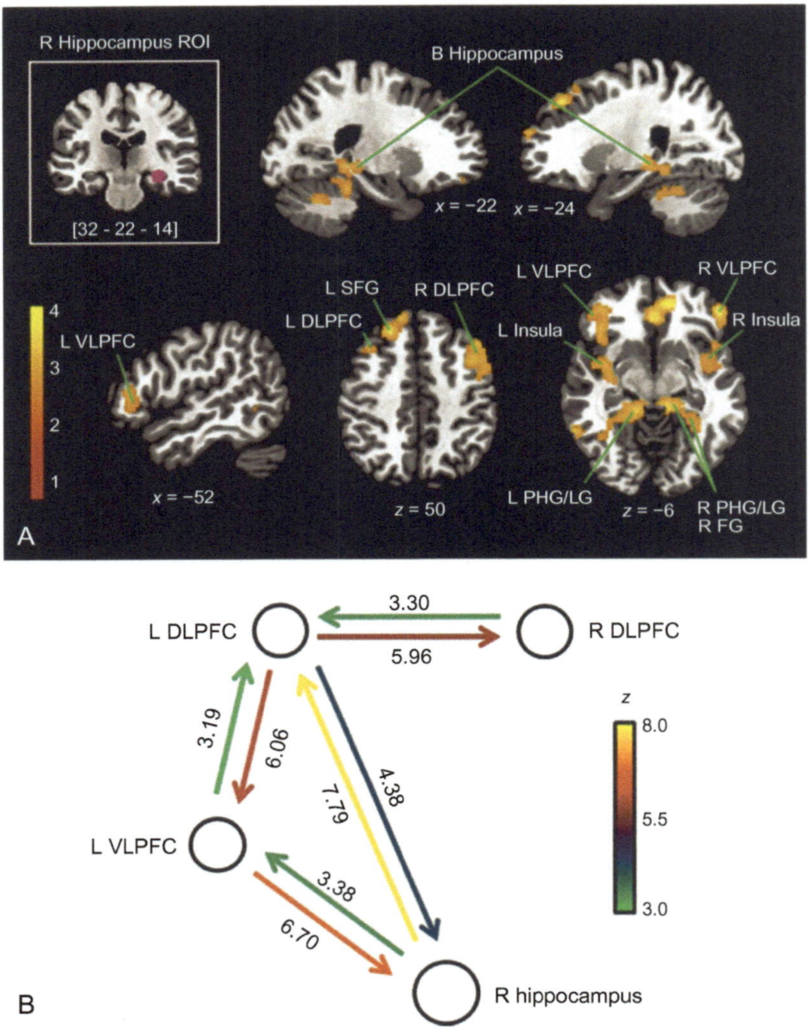

FIGURE 7 Hippocampal-prefrontal cortex circuits in children's math learning. (A) Functional connectivity of right hippocampus (top-left inset), showing greater effective connectivity during addition problem solving, compared with a control task. Greater connectivity was observed in the bilateral hippocampus, bilateral VLPFC, bilateral DLPFC, left SFG, bilateral insula, bilateral LG, bilateral PHG, and right FG. (B) Dynamic causal interactions in the hippocampal-prefrontal retrieval network. Analysis of causal interactions between left VLPFC, left and right DLPFC, and right hippocampus. Both the left VLPFC and the left DLPFC showed highly significant direct causal influences with the right hippocampus (all $ps < .00001$). Causal links depicted were all significant using $p < .01$, after Bonferroni correction.

LONGITUDINAL CHANGES IN MTL RESPONSE, REPRESENTATIONS AND CONNECTIVITY ASSOCIATED WITH MEMORY-BASED RETRIEVAL

A major limitation of past research is the use of cross-sectional designs, where inferences are made about development on the basis of data acquired from cohorts that differ in age. Cross-sectional studies cannot provide a clear picture of how individuals change over time, why certain cognitive abilities grow faster in some individuals than in others, and what factors predict heterogeneous growth rates (Kraemer, Yesavage, Taylor, & Kupfer, 2000). Developmental researchers emphasize the importance of longitudinal studies for examining individual variability in mathematical skill development (Bjorklund, 1999; Stern, 1999), as this can provide a more comprehensive picture of the maturation of brain systems with learning. Longitudinal studies are beginning to provide crucial new information about the neurobiological mechanisms that underlie the transition to memory-based problem solving and long-term memory formation on an individual level.

Qin et al. (2014) investigated the transition from procedure-based to memory-based problem-solving strategies using longitudinal fMRI data from 7- to 9-year-old children. The transition from use of counting to memory-based retrieval strategies over a 1.2-year interval was mediated by increased hippocampal activation and decreased prefrontal-parietal engagement (Figure 8). Following an initial increase in hippocampal engagement during middle childhood, this hippocampal dependency decreased during adolescence and adulthood despite further improvements in memory-based problem solving (Figure 9). This pattern of initial increase and subsequent decrease in activation provides novel support for models of long-term memory consolidation which posit that the hippocampus plays a time-limited role in the early phase of knowledge acquisition (McClelland et al., 1995; Tse et al., 2007). Consistent with this pattern of developmental change, previous studies of adults have reported no reliable hippocampal engagement in basic arithmetic tasks. Thus, the hippocampal system is critical to children's early learning of arithmetic facts (Cho et al., 2011, 2012; De Smedt et al., 2011), while retrieval is largely dependent on the neocortex in adults (Dehaene et al., 2003; Menon, 2015a). Through further schooling and experience with mathematics, fact retrieval becomes increasingly independent of the hippocampal memory system during adolescence and adulthood.

Furthermore, longitudinal improvements in retrieval-strategy use were best predicted by increased functional connectivity in hippocampal-neocortical circuits (Qin et al., 2014). Increased hippocampal functional coupling with prefrontal and parietal cortices was positively correlated with individual gains in use of memory-based strategies (Figure 10). Hippocampal-neocortical functional reorganization therefore plays an important role in children's shift from effortful counting to more efficient memory-based problem solving. Consistent

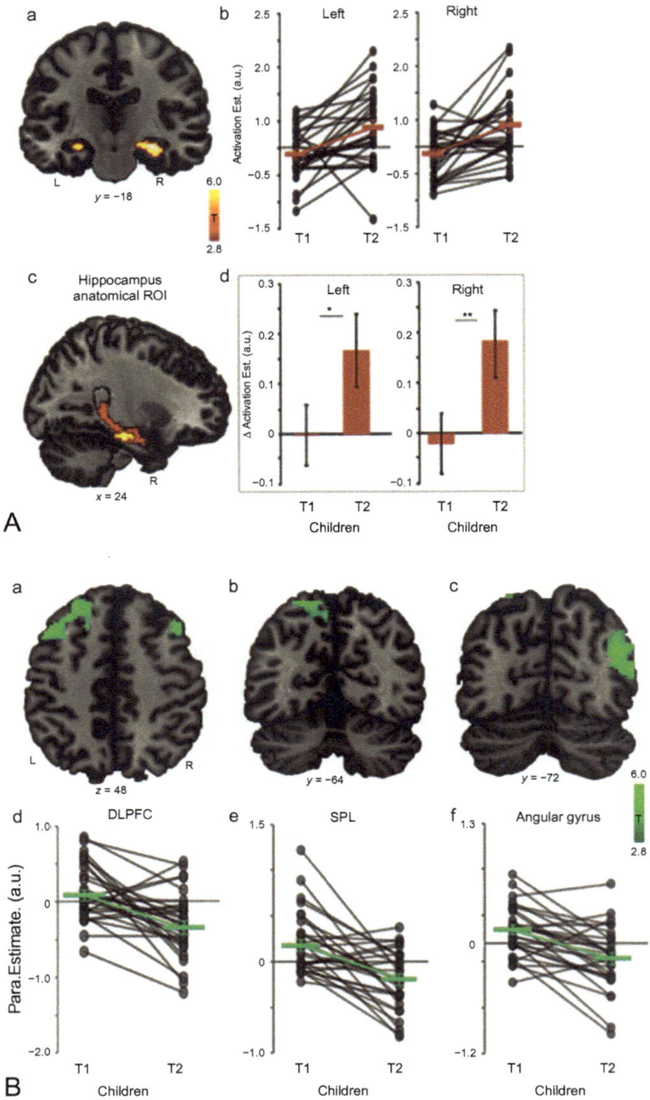

FIGURE 8 Longitudinal changes in MTL associated with memory-based retrieval. (A) Longitudinal changes in hippocampal engagement in children. (a) Left and right hippocampus clusters which showed increased engagement during addition problem solving between time 1 (T1) and time 2 (T2). (b) Each line represents individual developmental trajectories of hippocampal engagement over time. Bold red lines represent group means at T1 and T2. (c) Sagittal view of anatomically defined region of interest (ROI) encompassing the entire right hippocampus (coded in red) and significant functional clusters from 2A (coded in hot orange). (d) Bar graphs depict increased engagement from T1 to T2 for the anatomically defined entire left and right hippocampus. * $p<0.05$; ** $p<0.01$. (B) Longitudinal changes in task-related activation in the prefrontal cortex and the parietal cortex during addition problem solving (a–c) Bilateral dorsolateral prefrontal cortex (DLPFC), left superior parietal lobule (SPL), right parietal-occipital regions (i.e., angular gyrus, AG) that show significant decreases in task-related activation from Time-2 to Time-1. (d–f) Each line represents each individual's developmental trajectories of the DLPFC, SPL, AG engagement over time. Bold green lines represent the mean across all individuals for Time-1 and Time-2 separately.

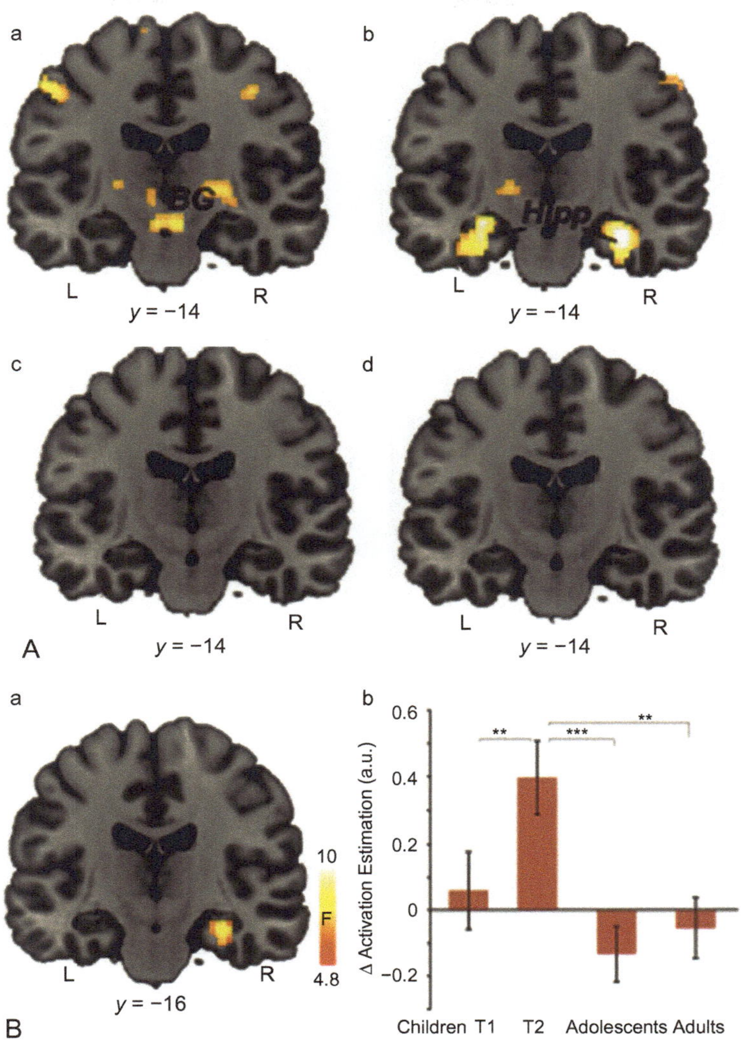

FIGURE 9 Developmental changes in medial temporal lobe engagement from childhood through adolescence into adulthood. (A) Developmental changes in medial temporal lobe engagement from childhood through adolescence into adulthood. Coronal view of activation patterns in the medial temporal lobe during addition (vs. control) problem solving in children at Time-1 (a), Time-2 (b), adolescents (c), and adults (d). The most prominent activation of the medial temporal lobe was observed in children at Time-2, not in children at Time-1, adolescents or adults. Hipp, hippocampus; BG, basal ganglia. (B) Longitudinal changes in hippocampal engagement during childhood, and further development through adolescence into adulthood. (a) Right hippocampus showing main effect of group across children, adolescents, and adults (omnibus *F*-contrast). (b) Bar graphs depict developmental changes in the functionally defined hippocampus cluster.

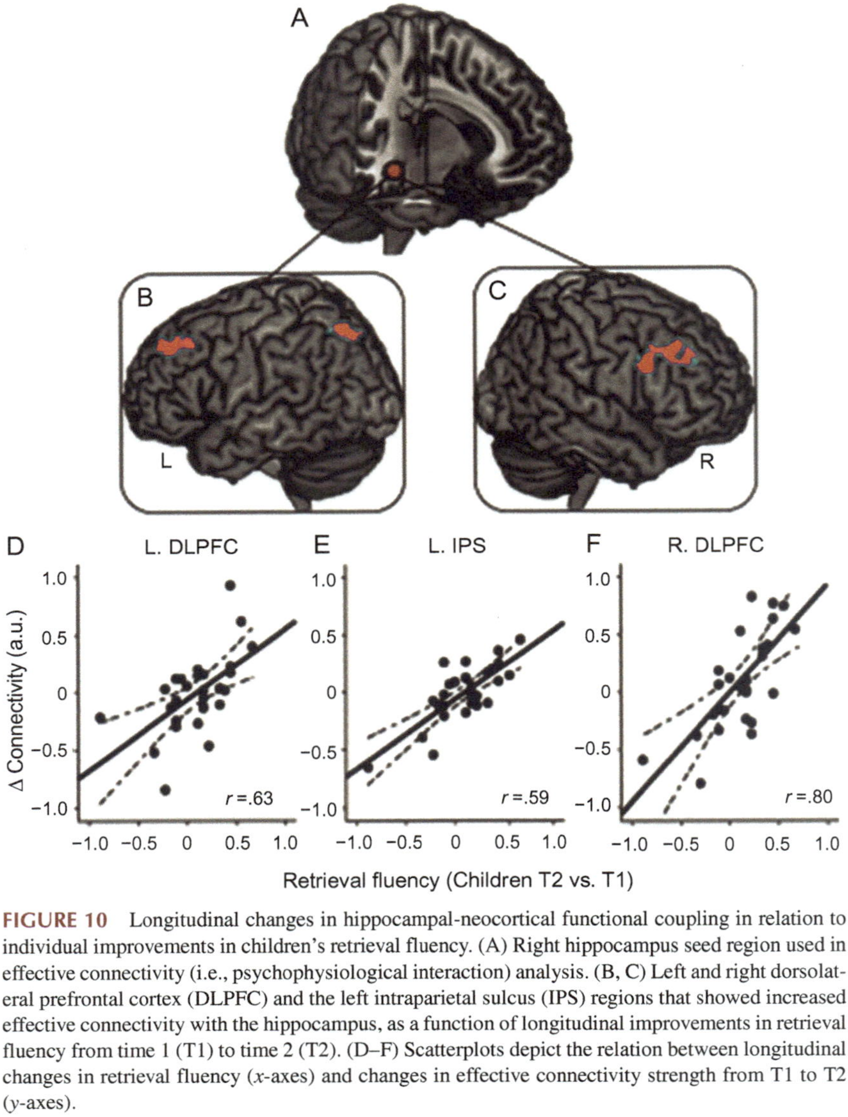

FIGURE 10 Longitudinal changes in hippocampal-neocortical functional coupling in relation to individual improvements in children's retrieval fluency. (A) Right hippocampus seed region used in effective connectivity (i.e., psychophysiological interaction) analysis. (B, C) Left and right dorsolateral prefrontal cortex (DLPFC) and the left intraparietal sulcus (IPS) regions that showed increased effective connectivity with the hippocampus, as a function of longitudinal improvements in retrieval fluency from time 1 (T1) to time 2 (T2). (D–F) Scatterplots depict the relation between longitudinal changes in retrieval fluency (*x*-axes) and changes in effective connectivity strength from T1 to T2 (*y*-axes).

with this view, increased intrinsic connectivity of the hippocampus with the neocortex predicted performance gains associated with 8 weeks of mathematics tutoring in children ages 7–9 (Supekar et al., 2013) (Figure 11).

More broadly, hippocampal-neocortical interactions have been implicated in human episodic memory (Eichenbaum, 2004; Squire & Zola-Morgan, 1991) and early stages of conceptual learning in adults (Kumaran et al., 2009; McClelland et al., 1995), but there has been limited evidence for its role in

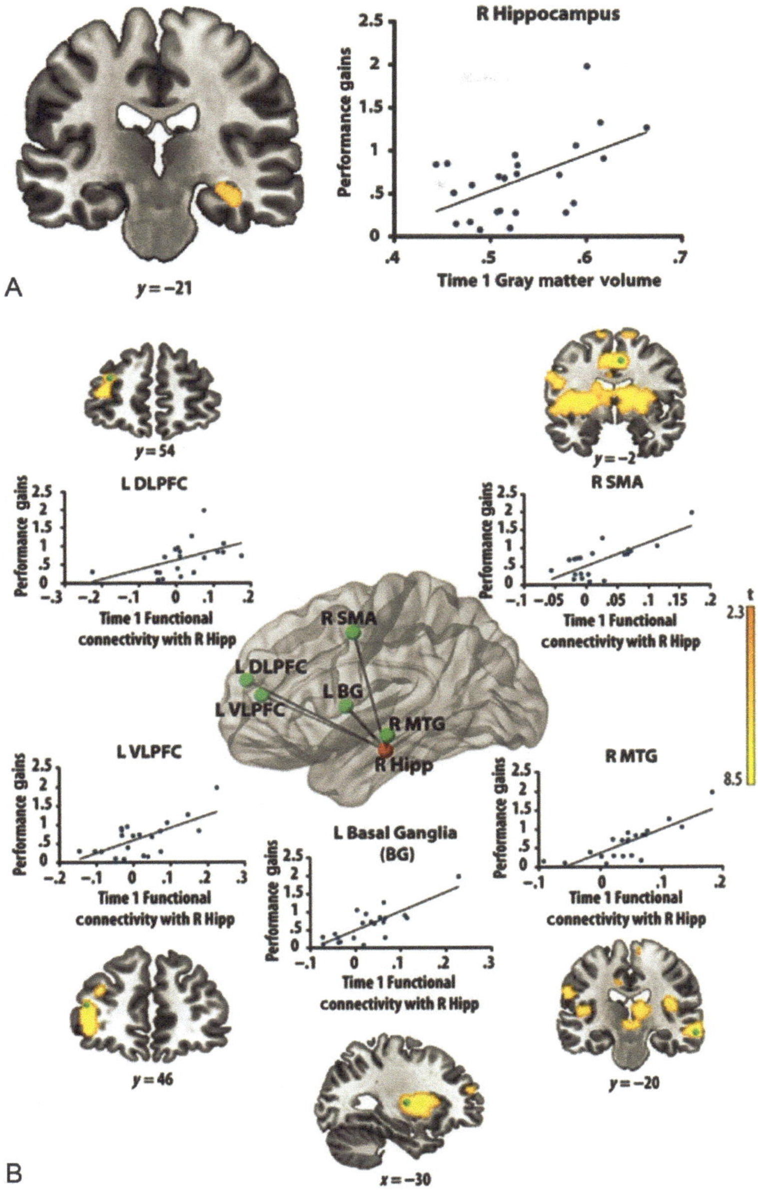

FIGURE 11 MTL structure and connectivity predict children's math learning. (A) Gray matter volume in hippocampus correlates with improvement in arithmetic performance in response to 8 weeks of one-to-one math tutoring. (B) Intrinsic functional connectivity of the hippocampus correlates with improvement in arithmetic performance in response to 8 weeks of one-to-one math tutoring. Performance gains were predicted by hippocampal connectivity with the left dorsolateral prefrontal cortex (L DLPFC), left ventrolateral prefrontal cortex (L VLPFC), right supplementary motor area (R SMA), left basal ganglia (L BG), and right middle temporal gyrus (R MTG). Composite 3D view of connectivity network is shown in the central panel with the right hippocampus seed ROI highlighted in red and voxels showing peak connectivity with the hippocampus highlighted in green. Surrounding panels show brain areas correlated with performance gains with tutoring. Scatterplots in each panel are based on voxels showing peak connectivity.

the development of children's memory-based problem-solving skills. Findings from longitudinal and training studies in children thus establish a tighter link between hippocampal-neocortical functional reorganization and the emergence of memory-based problem solving during childhood.

Beyond childhood, retrieval strategy use continues to improve through adolescence into adulthood, and is associated with decreased activation as well as more stable inter-problem representations in the hippocampus. Specifically, although mean hippocampal activation decreases during adolescence, inter-problem pattern stability in the hippocampus, prefrontal and temporal cortices increases with the refinement of memory-based problem solving from childhood to adolescence and adulthood. Although retrieval of arithmetic facts no longer requires the same level of engagement of the hippocampus-dependent memory system, the manner in which it is engaged gets refined over time and gradually becomes more stable across individual problems. This pattern of results is consistent with recent adult studies in memory encoding and retrieval tasks that found that greater pattern similarity across stimuli in the MTL predicts better behavioral performance through strengthening of discrete information distributed over neocortical regions (Kriegeskorte & Kievit, 2013; Xue et al., 2010). Such developmental changes likely reflect fine-tuned integration and reconfiguration of local functional circuits through experience-dependent plasticity (Casey, Tottenham, Liston, & Durston, 2005; Qin, Young, Supekar, Uddin, & Menon, 2012; Toga, Thompson, & Sowell, 2006). In sum, emerging evidence suggests that children's mathematics learning is characterized by reorganization and refinement of neural activity patterns in the MTL and its associated neocortical circuits, and this process plays a dominant role in the transition to memory-based problem solving.

WHY ADULTS MAY NOT RELY ON MTL MEMORY SYSTEMS FOR MATHEMATICS PERFORMANCE AND LEARNING

The pattern of results from the cross-sectional and longitudinal studies we have reviewed in this chapter raises the question of whether adults might engage the MTL for learning complex arithmetic problems. Findings to date from several short-term training studies involving multi-digit subtraction and multiplication problems do not provide support for such a role (Delazer et al., 2005; Ischebeck, Zamarian, Egger, Schocke, & Delazer, 2007; Zamarian, Ischebeck, & Delazer, 2009). For example, Ischebeck et al. (2007)) trained adult participants to solve complex arithmetic problems (e.g., 13 x 8 using repeated problems during a fMRI session). Accuracy and RT improved over time for the trained problems—training effects were rapid and became significant after approximately eight repetitions of a problem and remained stable over the course of the experiment. The brain basis of these improvements was investigated by contrasting responses to repeated versus novel problems in each adult. Increased proficiency with recently learned arithmetic facts was associated with reduced responses in the IPS and ventrolateral PFC, and increased activation in the angular gyrus (Delazer et al., 2005; Ischebeck et al., 2006, 2007). Critically, there was no evidence for the involvement of the MTL memory system and the route to learning appears to be mediated entirely by neocortical systems.

Finally, a related question is whether adults engage the MTL differentially for problems in which they use retrieval strategies. Grabner and colleagues used trial-by-trial strategy self-reports to identify brain regions underlying retrieval and calculation strategies in arithmetic problem solving (Grabner et al., 2009). They found that self-reported retrieval use was associated with greater response in the angular gyrus, whereas self-reported procedural strategies were associated with responses in IPS and superior parietal lobule regions of the PPC as well as the PFC, occipital cortex, and basal ganglia. No differential responses were found in the MTL, similar to findings from the developmental study by Qin et al. (2014).

Taken together, these findings suggest that children and adults likely engage the MTL system differently during arithmetic problem solving and mathematics learning, and that even as memory-based retrieval continues to improve beyond childhood, MTL responses decrease over time. Further research is needed to investigate whether adolescents and adults engage the MTL during learning of more complex mathematical materials, for which they may lack appropriate schema and domain knowledge.

CONCLUSIONS

The development of efficient problem-solving strategies is one of the most important milestones of children's learning. Emerging data suggest that it is no longer appropriate to focus solely on parietal circuits as the basis of arithmetic skill development (Fias et al., 2013; Menon, 2015a; Qin et al., 2014; see also Chapter 8). A number of scaffolding systems are engaged during development, and new studies are beginning to investigate brain systems that are otherwise missed in studies involving adults. In this context, the present review has underscored the crucial, and heretofore unappreciated, role played by the MTL memory system in the development of children's mathematical skills.

This review has highlighted recent discoveries regarding an essential building block of children's cognitive development—the transition from procedure-based to memory-based problem-solving strategies. Recent research suggests that the hippocampal system is pivotal to this strategic transition. Functional reorganization and refinement of neural activity within the MTL as well as its associated neocortical regions plays a crucial role in the development of memory-based problem solving in young children. Findings to date highlight the dynamic role of the MTL in the maturation of memory-based problem-solving strategies, characterized by greater engagement in childhood, followed by a decreased role in adolescence and adulthood. Hippocampal-neocortical reorganization facilitates fluent retrieval and long-term neocortical memory consolidation in children, eventually resulting in retrieval processes that are independent of the MTL. At the same time, problem representations in the MTL become more stable with development, suggesting alternate mechanisms by which information can be retrieved in adults. In this manner, the hippocampal system fosters the gradual establishment of long-lasting knowledge represented in the neocortex, through its role in rapid learning and integration of new information into existing knowledge schemas (McClelland et al., 1995; Tse et al., 2007; van Kesteren et al., 2012). In this view, the hippocampus plays a critical, but time-limited, role in the early phase of knowledge acquisition, and this hippocampal dependence is reduced following reconfiguration of neocortical connections and stabilization of newly acquired knowledge (McClelland et al., 1995; Squire & Zola-Morgan, 1991).

More broadly, this pattern of changes provides new insights into integration and differentiation of functional circuits over development, mediated by transient engagement, refinement of local circuits, functional specialization, strengthening of specific prefrontal connections, and experience-dependent plasticity—reflecting general organizing principles of neurocognitive development (Bourgeois & Rakic, 1993; Fair et al., 2007; Rakic, Bourgeois, & Goldman-Rakic, 1994; Supekar, Musen, & Menon, 2009; Toga et al., 2006; Uddin, Supekar, & Menon, 2010).

This review has also highlighted several important differences between brain systems engaged in children and adults. In the early stages of arithmetic skill acquisition, children show a different profile of brain responses associated with mathematics learning and skill acquisition than that found in adults. A truly developmental approach requires focusing on a narrow window of schooling in order to understand graduated change over time (Karmiloff-Smith, 2010) and studies comparing children with adults are likely to be insensitive to major developmental changes that occur during early elementary schooling. Critically, brain systems and circuits in the developing child's brain are clearly not the same as those seen in more mature adult brains sculpted by years of learning.

ACKNOWLEDGMENTS

It is a pleasure to thank Teresa Iuculano and Christian Battista for valuable feedback on the chapter and Tricia Ngoon for assistance with the figures. Supported by grants from the NIH (HD047520, HD059205).

REFERENCES

Adleman, N. E., Menon, V., Blasey, C. M., White, C. D., Warsofsky, I. S., Glover, G. H., et al. (2002). A developmental fMRI study of the Stroop color-word task. *NeuroImage*, *16*(1), 61–75. http://dx.doi.org/10.1006/nimg.2001.1046, S1053811901910468 [pii].

Ansari, D. (2008). Effects of development and enculturation on number representation in the brain. *Nature Reviews Neuroscience*, *9*(4), 278–291. http://dx.doi.org/10.1038/nrn2334.

Ashcraft, M. H. (1982). The development of mental arithmetic—A chronometric approach. *Developmental Review*, *2*(3), 213–236.
Ashcraft, M. H., & Fierman, B. A. (1982). Mental addition in third, fourth, and sixth graders. *Journal of Experimental Child Psychology*, *33*, 18.
Badre, D., & Wagner, A. D. (2007). Left ventrolateral prefrontal cortex and the cognitive control of memory. *Neuropsychologia*, *45*(13), 2883–2901. http://dx.doi.org/10.1016/j.neuropsychologia.2007.06.015 S0028-3932(07)00221-7 [pii].
Barrouillet, P., & Lepine, R. (2005). Working memory and children's use of retrieval to solve addition problems. *Journal of Experimental Child Psychology*, *91*(3), 183–204. http://dx.doi.org/10.1016/j.jecp.2005.03.002.
Bauer, P. J. (2008). Toward a neuro-developmental account of the development of declarative memory. *Developmental Psychobiology*, *50*(1), 19–31. http://dx.doi.org/10.1002/dev.20265.
Billingsley, R. L., Lou Smith, M., & Pat McAndrews, M. (2002). Developmental patterns in priming and familiarity in explicit recollection. *Journal of Experimental Child Psychology*, *82*(3), 251–277, S0022-0965(02)00007-3 [pii].
Bjorklund, D. F. (1999). What individual differences can teach us about developmental function, and vice versa. In W. Schneider (Ed.), *Individual development from 3 to 12: Findings from the Munich longitudinal study* (pp. 29–37). Cambridge: Cambridge University Press.
Bourgeois, J. P., & Rakic, P. (1993). Changes of synaptic density in the primary visual cortex of the macaque monkey from fetal to adult stage. *The Journal of Neuroscience*, *13*(7), 2801–2820.
Bunge, S. A., & Wright, S. B. (2007). Neurodevelopmental changes in working memory and cognitive control. *Current Opinion in Neurobiology*, *17*(2), 243–250. http://dx.doi.org/10.1016/j.conb.2007.02.005, S0959-4388(07)00029-3 [pii].
Butterworth, B. (1999). *The mathematical brain*. London: Macmillan.
Carpenter, T. P., & Moser, J. M. (1984). The acquisition of addition and subtraction concepts in grades one through three. *Journal for Research in Mathematics Education*, *15*(3), 179–202. http://dx.doi.org/10.2307/748348.
Casey, B. J., Tottenham, N., Liston, C., & Durston, S. (2005). Imaging the developing brain: What have we learned about cognitive development? *Trends in Cognitive Sciences*, *9*(3), 104–110. http://dx.doi.org/10.1016/j.tics.2005.01.011, S1364-6613(05)00030-6 [pii].
Cho, S., Metcalfe, A. W., Young, C. B., Ryali, S., Geary, D. C., & Menon, V. (2012). Hippocampal-prefrontal engagement and dynamic causal interactions in the maturation of children's fact retrieval. *Journal of Cognitive Neuroscience*, *24*(9), 1849–1866. http://dx.doi.org/10.1162/jocn_a_00246.
Cho, S., Ryali, S., Geary, D. C., & Menon, V. (2011). How does a child solve 7 + 8? Decoding brain activity patterns associated with counting and retrieval strategies. *Developmental Science*, *14*(5), 989–1001. http://dx.doi.org/10.1111/j.1467-7687.2011.01055.x.
Cycowicz, Y. M., & Friedman, D. (2003). Source memory for the color of pictures: Event-related brain potentials (ERPs) reveal sensory-specific retrieval-related activity. *Psychophysiology*, *40*(3), 455–464.
Cycowicz, Y. M., Friedman, D., & Duff, M. (2003). Pictures and their colors: What do children remember? *Journal of Cognitive Neuroscience*, *15*(5), 759–768. http://dx.doi.org/10.1162/089892903322307465.
Cycowicz, Y. M., Friedman, D., Snodgrass, J. G., & Duff, M. (2001). Recognition and source memory for pictures in children and adults. *Neuropsychologia*, *39*(3), 255–267. S0028-3932(00)00108-1 [pii].
Davachi, L. (2006). Item, context and relational episodic encoding in humans. *Current Opinion in Neurobiology*, *16*(6), 693–700.

Davachi, L., Mitchell, J. P., & Wagner, A. D. (2003). Multiple routes to memory: Distinct medial temporal lobe processes build item and source memories. *Proceedings of the National Academy of Sciences of the United States of America*, *100*(4), 2157–2162. http://dx.doi.org/10.1073/pnas.0337195100 0337195100 [pii].

De Smedt, B., Holloway, I. D., & Ansari, D. (2011). Effects of problem size and arithmetic operation on brain activation during calculation in children with varying levels of arithmetical fluency. *NeuroImage*, *57*(3), 771–781. http://dx.doi.org/10.1016/j.neuroimage.2010.12.037.

Dehaene, S., Piazza, M., Pinel, P., & Cohen, L. (2003). Three parietal circuits for number processing. *Cognitive Neuropsychology*, *20*(3-6), 487–506.

Delazer, M., Ischebeck, A., Domahs, F., Zamarian, L., Koppelstaetter, F., Siedentopf, C. M., et al. (2005). Learning by strategies and learning by drill—Evidence from an fMRI study. *NeuroImage*, *25*(3), 838–849.

DeStefano, D., & LeFevre, J. A. (2004). The role of working memory in mental arithmetic. *European Journal of Cognitive Psychology*, *16*(3), 353–386. http://dx.doi.org/10.1080/09541440244000328|issn 0954-1446.

Diana, R. A., Yonelinas, A. P., & Ranganath, C. (2007). Imaging recollection and familiarity in the medial temporal lobe: A three-component model. *Trends in Cognitive Sciences*, *11*(9), 379–386.

Diekelmann, S., Wilhelm, I., & Born, J. (2009). The whats and whens of sleep-dependent memory consolidation. *Sleep Medicine Reviews*, *13*(5), 309–321. http://dx.doi.org/10.1016/j.smrv.2008.08.002, S1087-0792(08)00102-0 [pii].

Dirks, J., & Neisser, U. (1977). Memory for objects in real scenes: The development of recognition and recall. *Journal of Experimental Child Psychology*, *23*, 315–328.

Drummey, A. B., & Newcombe, N. S. (2002). Developmental changes in source memory. *Developmental Science*, *5*, 502–513.

Eichenbaum, H. (2000). Hippocampus: Mapping or memory? *Current Biology*, *10*, R785–R787.

Eichenbaum, H. (2004). Hippocampus: Cognitive processes and neural representations that underlie declarative memory. *Neuron*, *44*(1), 109–120. http://dx.doi.org/10.1016/j.neuron.2004.08.028, S089662730400529X [pii].

Eichenbaum, H., Yonelinas, A. P., & Ranganath, C. (2007). The medial temporal lobe and recognition memory. *Annual Review of Neuroscience*, *30*, 123–152.

Fair, D. A., Dosenbach, N. U., Church, J. A., Cohen, A. L., Brahmbhatt, S., Miezin, F. M., et al. (2007). Development of distinct control networks through segregation and integration. *Proceedings of the National Academy of Sciences of the United States of America*, *104*(33), 13507–13512.

Fias, W., Menon, V., & Szucs, D. (2013). Multiple components of developmental dyscalculia. *Trends in Neuroscience and Education*, *2*(2), 43–47.

Frankland, P. W., & Bontempi, B. (2005). The organization of recent and remote memories. *Nature Reviews Neuroscience*, *6*(2), 119–130. http://dx.doi.org/10.1038/nrn1607, nrn1607 [pii].

Geary, D. C. (1990). A componential analysis of an early learning deficit in mathematics. *Journal of Experimental Child Psychology*, *49*(3), 363–383.

Geary, D. C. (1994). *Children's mathematical development: Research and practical applications.* Washington, DC: American Psychological Association.

Geary, D. C. (2011). Cognitive predictors of achievement growth in mathematics: A 5-year longitudinal study. *Developmental Psychology*, *47*(6), 1539–1552. http://dx.doi.org/10.1037/a0025510.

Geary, D. C., & Brown, S. C. (1991). Cognitive addition: Strategy choice and speed-of-processing differences in gifted, normal, and mathematically disabled children. *Developmental Psychology*, *27*(3), 398–406.

Geary, D. C., Hamson, C. O., & Hoard, M. K. (2000). Numerical and arithmetical cognition: A longitudinal study of process and concept deficits in children with learning disability. *Journal of Experimental Child Psychology*, *77*(3), 236–263.

Geary, D. C., & Hoard, M. K. (2003). Learning disabilities in basic mathematics: Deficits in memory and cognition. In J. M. Royer (Ed.), *Mathematical cognition* (pp. 93–115). Greenwich, CT: Information Age Publishing.

Geary, D. C., Hoard, M. K., Byrd-Craven, J., & DeSoto, M. C. (2004). Strategy choices in simple and complex addition: Contributions of working memory and counting knowledge for children with mathematical disability. *Journal of Experimental Child Psychology*, *88*(2), 121–151.

Geary, D. C., & Widaman, K. F. (1987). Individual differences in cognitive arithmetic. *Journal of Experimental Psychology General*, *116*(2), 154–171.

Ghetti, S., & Bunge, S. A. (2012). Neural changes underlying the development of episodic memory during middle childhood. *Developmental Cognitive Neuroscience*, http://dx.doi.org/10.1016/j.dcn.2012.05.002, S1878-9293(12)00057-6 [pii].

Ghetti, S., DeMaster, D. M., Yonelinas, A. P., & Bunge, S. A. (2010). Developmental differences in medial temporal lobe function during memory encoding. *The Journal of Neuroscience*, *30*(28), 9548–9556. http://dx.doi.org/10.1523/JNEUROSCI.3500-09.2010, 30/28/9548 [pii].

Grabner, R. H., Ansari, D., Koschutnig, K., Reishofer, G., Ebner, F., & Neuper, C. (2009). To retrieve or to calculate? Left angular gyrus mediates the retrieval of arithmetic facts during problem solving. *Neuropsychologia*, *47*(2), 604–608. http://dx.doi.org/10.1016/j.neuropsychologia.2008.10.013.

Groen, G. J., & Parkman, J. M. (1972). Chronometric analysis of simple addition. *Psychological Review*, *79*(4), 329–343.

Hamann, M. S., & Ashcraft, M. H. (1985). Simple and complex mental addition across development. *Journal of Experimental Child Psychology*, *40*, 49–72.

Houde, O., Rossi, S., Lubin, A., & Joliot, M. (2010). Mapping numerical processing, reading, and executive functions in the developing brain: An fMRI meta-analysis of 52 studies including 842 children. *Developmental Science*, *13*(6), 876–885.

Ischebeck, A., Zamarian, L., Egger, K., Schocke, M., & Delazer, M. (2007). Imaging early practice effects in arithmetic. *NeuroImage*, *36*(3), 993–1003.

Ischebeck, A., Zamarian, L., Siedentopf, C., Koppelstatter, F., Benke, T., Felber, S., et al. (2006). How specifically do we learn? Imaging the learning of multiplication and subtraction. *NeuroImage*, *30*(4), 1365–1375. http://dx.doi.org/10.1016/j.neuroimage.2005.11.016.

Kail, R. V. (1990). *The development of memory in children*. New York: Freeman.

Karmiloff-Smith, A. (2010). Neuroimaging of the developing brain: Taking "developing" seriously. *Human Brain Mapping*, *31*(6), 934–941. http://dx.doi.org/10.1002/hbm.21074.

Kaufmann, L. (2002). More evidence for the role of the central executive in retrieving arithmetic facts—A case study of severe developmental dyscalculia. *Journal of Clinical and Experimental Neuropsychology*, *24*(3), 302–310.

Kaufmann, L., Lochy, A., Drexler, A., & Semenza, C. (2004). Deficient arithmetic fact retrieval—Storage or access problem? A case study. *Neuropsychologia*, *42*(4), 482–496, S0028393203002343 [pii].

Koechlin, E., Ody, C., & Kouneiher, F. (2003). The architecture of cognitive control in the human prefrontal cortex. *Science*, *302*(5648), 1181–1185. http://dx.doi.org/10.1126/science.1088545 302/5648/1181 [pii].

Kraemer, H. C., Yesavage, J. A., Taylor, J. L., & Kupfer, D. (2000). How can we learn about developmental processes from cross-sectional studies, or can we? *The American Journal of Psychiatry*, *157*(2), 163–171.

Kriegeskorte, N., & Kievit, R. A. (2013). Representational geometry: Integrating cognition, computation, and the brain. *Trends in Cognitive Sciences*, *17*(8), 401–412. http://dx.doi.org/10.1016/j.tics.2013.06.007.

Kumaran, D., Summerfield, J. J., Hassabis, D., & Maguire, E. A. (2009). Tracking the emergence of conceptual knowledge during human decision making. *Neuron*, *63*(6), 889–901. http://dx.doi.org/10.1016/j.neuron.2009.07.030, S08966273(09)00618-7 [pii].

Kwon, H., Reiss, A. L., & Menon, V. (2002). Neural basis of protracted developmental changes in visuo-spatial working memory. *Proceedings of the National Academy of Sciences of the United States of America*, *99*(20), 13336–13341.

Logie, R. H., Gilhooly, K. J., & Wynn, V. (1994). Counting on working memory in arithmetic problem solving. *Memory & Cognition*, *22*(4), 395–410.

Mandler, J. M., & Robinson, C. A. (1978). Developmental changes in picture recognition. *Journal of Experimental Child Psychology*, *26*(1), 122–136, 0022-0965(78)90114-5 [pii].

McClelland, J. L., McNaughton, B. L., & O'Reilly, R. C. (1995). Why there are complementary learning systems in the hippocampus and neocortex: Insights from the successes and failures of connectionist models of learning and memory. *Psychological Review*, *102*(3), 419–457.

McCloskey, M., Harley, W., & Sokol, S. M. (1991). Models of arithmetic fact retrieval: An evaluation in light of findings from normal and brain-damaged subjects. *Journal of Experimental Psychology Learning, memory, and cognition*, *17*(3), 377–397.

McGaugh, J. L. (2000). Memory—A century of consolidation. *Science*, *287*(5451), 248–251.

Menon, V. (2014). Arithmetic in child and adult brain. In R. Cohen Kadosh & A. Dowker (Eds.), *Handbook of Mathematical Cognition*. Oxford: Oxford University Press.

Menon, V. (2015a). *Arithmetic in the child and adult brain*. Oxford: Oxford University.

Menon, V., Boyett-Anderson, J. M., & Reiss, A. L. (2005). Maturation of medial temporal lobe response and connectivity during memory encoding. *Brain Research Cognitive Brain Research*, *25*(1), 379–385. http://dx.doi.org/10.1016/j.cogbrainres.2005.07.007, S0926-6410(05)00193-X [pii].

Miller, E. K. (2000). The prefrontal cortex and cognitive control. *Nature Reviews Neuroscience*, *1*(1), 59–65. http://dx.doi.org/10.1038/35036228.

Nadel, L., & Moscovitch, M. (2001). The hippocampal complex and long-term memory revisited. *Trends in Cognitive Sciences*, *5*(6), 228–230, S13646613(00)01664-8 [pii].

Norman, K. A., & O'Reilly, R. C. (2003). Modeling hippocampal and neocortical contributions to recognition memory: A complementary-learning-systems approach. *Psychological Review*, *110*(4), 611–646. http://dx.doi.org/10.1037/0033-295X.110.4.611, 2003-08987-001 [pii].

Ofen, N., Kao, Y. C., Sokol-Hessner, P., Kim, H., Whitfield-Gabrieli, S., & Gabrieli, J. D. (2007). Development of the declarative memory system in the human brain. *Nature Neuroscience*, *10*(9), 1198–1205. http://dx.doi.org/10.1038/nn1950, nn1950 [pii].

Ostby, Y., Tamnes, C. K., Fjell, A. M., & Walhovd, K. B. (2011). Dissociating memory processes in the developing brain: The role of hippocampal volume and cortical thickness in recall after minutes versus days. *Cerebral Cortex*, *22*(2), 381–390. http://dx.doi.org/10.1093/cercor/bhr116, bhr116 [pii].

Passolunghi, M. C., & Siegel, L. S. (2004). Working memory and access to numerical information in children with disability in mathematics. *Journal of Experimental Child Psychology*, *88*(4), 348–367. http://dx.doi.org/10.1016/j.jecp.2004.04.002, S0022096504000669 [pii].

Qin, S., Cho, S., Chen, T., Rosenberg-Lee, M., Geary, D. C., & Menon, V. (2014). Hippocampal-neocortical functional reorganization underlies children's cognitive development. *Nature Neuroscience*, *17*(9), 1263–1269. http://dx.doi.org/10.1038/nn.3788.

Qin, S., Hermans, E. J., Rijpkema, M., & Fernández, G. (2011). *Adaptive memory: Imaging medial temporal and prefrontal memory systems*. Nijmegen, The Netherlands: Radboud Universiteit Nijmegen, PhD.

Qin, S., Piekema, C., Petersson, K. M., Han, B., Luo, J., & Fernandez, G. (2007). Probing the transformation of discontinuous associations into episodic memory: An event-related fMRI study. *NeuroImage, 38*(1), 212–222. http://dx.doi.org/10.1016/j.neuroimage.2007.07.020, S1053-8119(07)00655-6 [pii].

Qin, S., Rijpkema, M., Tendolkar, I., Piekema, C., Hermans, E. J., Binder, M., et al. (2009). Dissecting medial temporal lobe contributions to item and associative memory formation. *NeuroImage, 46*(3), 874–881. http://dx.doi.org/10.1016/j.neuroimage.2009.02.039, S1053-8119(09)00195-5 [pii].

Qin, S., van Marle, H. J., Hermans, E. J., & Fernandez, G. (2011). Subjective sense of memory strength and the objective amount of information accurately remembered are related to distinct neural correlates at encoding. *The Journal of Neuroscience, 31*(24), 8920–8927. http://dx.doi.org/10.1523/JNEUROSCI.2587-10.2011, 31/24/8920 [pii].

Qin, S., Young, C. B., Supekar, K., Uddin, L. Q., & Menon, V. (2012). Immature integration and segregation of emotion-related brain circuitry in young children. *Proceedings of the National Academy of Sciences of the United States of America, 109*(20), 7941–7946. http://dx.doi.org/10.1073/pnas.1120408109, 1120408109 [pii].

Rakic, P., Bourgeois, J. P., & Goldman-Rakic, P. S. (1994). Synaptic development of the cerebral cortex: Implications for learning, memory, and mental illness. *Progress in Brain Research, 102*, 227–243. http://dx.doi.org/10.1016/S0079-6123(08)60543-9, S0079-6123(08)60543-9 [pii].

Rivera, S. M., Reiss, A. L., Eckert, M. A., & Menon, V. (2005). Developmental changes in mental arithmetic: Evidence for increased functional specialization in the left inferior parietal cortex. *Cerebral Cortex, 15*(11), 1779–1790. http://dx.doi.org/10.1093/cercor/bhi055, bhi055 [pii].

Schacter, D. L., Addis, D. R., & Buckner, R. L. (2007). Remembering the past to imagine the future: the prospective brain. *Nature Reviews Neuroscience, 8*(9), 657–661.

Schacter, D. L., Norman, K. A., & Koutstaal, W. (1998). The cognitive neuroscience of constructive memory. *Annual Review of Psychology, 49*, 289–318.

Schneider, W. E., & Pressley, M. J. (1997). *Memory development between 2 and 20* (second ed.). New York: Springer.

Siegler, R. S. (1987). The perils of averaging data over strategies: An example from children's addition. *Journal of Experimental Psychology General, 116*, 250–264.

Siegler, R. S. (1996). *Emerging minds: The process of change in children's thinking* (1st ed.). New York: Oxford University Press.

Siegler, R. S., Shipley, C., Simon, T. J., & Halford, G. S. (1995). Variation, selection, and cognitive change. In Anonymous (Eds.), *Developing cognitive competence: New approaches to process modeling* (pp. 31–76). L. Erlbaum: Hillsdale, New Jersey.

Siegler, R. S., & Shrager, J. (1984). Strategy choice in addition and subtraction: How do children know what to do? In C. Sophian (Ed.), *Origins of cognitive skills* (pp. 229–293). Hillsdale, NJ: L. Erlbaum.

Siegler, R. S., & Svetina, M. (2006). What leads children to adopt new strategies? A microgenetic/cross-sectional study of class inclusion. *Child Development, 77*(4), 997–1015. http://dx.doi.org/10.1111/j.1467-8624.2006.00915.x.

Simon, O., Mangin, J. F., Cohen, L., Le Bihan, D., & Dehaene, S. (2002). Topographical layout of hand, eye, calculation, and language-related areas in the human parietal lobe. *Neuron, 33*(3), 475–487.

Simons, J. S., & Spiers, H. J. (2003). Prefrontal and medial temporal lobe interactions in long-term memory. *Nature Reviews Neuroscience*, *4*(8), 637–648.

Squire, L. R., Stark, C. E., & Clark, R. E. (2004). The medial temporal lobe. *Annual Review of Neuroscience*, *27*, 279–306.

Squire, L. R., & Zola-Morgan, S. (1991). The medial temporal lobe memory system. *Science*, *253*(5026), 1380–1386.

Stern, E. (1999). Development of mathematical competencies. In W. Schneider (Ed.), *Individual development from 3 to 12: Findings from the Munich longitudinal study* (pp. 154–170). Cambridge: Cambridge University Press.

Supekar, K., Musen, M., & Menon, V. (2009). Development of large-scale functional brain networks in children. *PLoS Biology*, *7*(7), http://dx.doi.org/10.1371/journal.pbio.1000157, e1000157.

Supekar, K., Swigart, A. G., Tenison, C., Jolles, D. D., Rosenberg-Lee, M., Fuchs, L., et al. (2013). Neural predictors of individual differences in response to math tutoring in primary-grade school children. *Proceedings of the National Academy of Sciences of the United States of America*, *110*(20), 8230–8235. http://dx.doi.org/10.1073/pnas.1222154110.

Suzuki, W. A., & Amaral, D. G. (1994). Perirhinal and parahippocampal cortices of the macaque monkey: Cortical afferents. *The Journal of Comparative Neurology*, *350*(4), 497–533.

Takashima, A., Nieuwenhuis, I. L., Jensen, O., Talamini, L. M., Rijpkema, M., & Fernandez, G. (2009). Shift from hippocampal to neocortical centered retrieval network with consolidation. *The Journal of Neuroscience*, *29*(32), 10087–10093. http://dx.doi.org/10.1523/JNEUROSCI.0799-09.2009, 29/32/10087 [pii].

Takashima, A., Petersson, K. M., Rutters, F., Tendolkar, I., Jensen, O., Zwarts, M. J., et al. (2006). Declarative memory consolidation in humans: A prospective functional magnetic resonance imaging study. *Proceedings of the National Academy of Sciences of the United States of America*, *103*(3), 756–761. http://dx.doi.org/10.1073/pnas.0507774103, 0507774103 [pii].

Toga, A. W., Thompson, P. M., & Sowell, E. R. (2006). Mapping brain maturation. *Trends in Neurosciences*, *29*(3), 148–159. http://dx.doi.org/10.1016/j.tins.2006.01.007, S01662236(06)00024-5 [pii].

Tse, D., Langston, R. F., Kakeyama, M., Bethus, I., Spooner, P. A., Wood, E. R., et al. (2007). Schemas and memory consolidation. *Science*, *316*(5821), 76–82. http://dx.doi.org/10.1126/science.1135935.

Tse, D., Takeuchi, T., Kakeyama, M., Kajii, Y., Okuno, H., Tohyama, C., et al. (2011). Schema-dependent gene activation and memory encoding in neocortex. *Science*, *333*(6044), 891–895. http://dx.doi.org/10.1126/science.1205274, science.1205274 [pii].

Tulving, E. (1983). *Elements of episodic memory*. New York: Oxford University Press.

Tulving, E. (2002). Episodic memory: From mind to brain. *Annual Review of Psychology*, *53*, 1–25. http://dx.doi.org/10.1146/annurev.psych.53.100901.135114 53/1/1 [pii].

Uddin, L. Q., Supekar, K., & Menon, V. (2010). Typical and atypical development of functional human brain networks: Insights from resting-state FMRI. *Frontiers in Systems Neuroscience*, *4*, 21. http://dx.doi.org/10.3389/fnsys.2010.00021.

van Dongen, E. V., Takashima, A., Barth, M., & Fernandez, G. (2011). Functional connectivity during light sleep is correlated with memory performance for face-location associations. *NeuroImage*, *57*(1), S1053–S8119. http://dx.doi.org/10.1016/j.neuroimage.2011.04.019, S1053-8119(11)00397-1 [pii].

van Kesteren, M. T., Ruiter, D. J., Fernandez, G., & Henson, R. N. (2012). How schema and novelty augment memory formation. *Trends in Neurosciences*, *35*(4), 211–219. http://dx.doi.org/10.1016/j.tins.2012.02.001.

Wang, S., & Morris, R. (2010). Hippocampal-neocortical interactions in memory formation, consolidation, and reconsolidation. *Annual Review of Psychology*, *61*(49-79), C41–C44. http://dx.doi.org/10.1146/annurev.psych.093008.100523.

Wu, S., Meyer, M., Maeda, U., Salimpoor, V., Tomiyama, S., Geary, D., et al. (2008). Standardized assessment of strategy use and working memory in early mental arithmetic performance. *Developmental Neuropsychology*, *33*(3), 365–393.

Xue, G., Dong, Q., Chen, C., Lu, Z., Mumford, J. A., & Poldrack, R. A. (2010). Greater neural pattern similarity across repetitions is associated with better memory. *Science*, *330*(6000), 97–101. http://dx.doi.org/10.1126/science.1193125, science.1193125 [pii].

Zamarian, L., Ischebeck, A., & Delazer, M. (2009). Neuroscience of learning arithmetic—Evidence from brain imaging studies. *Neuroscience and Biobehavioral Reviews*, *33*(6), 909–925. http://dx.doi.org/10.1016/j.neubiorev.2009.03.005.

Chapter 5

Finger Representation and Finger-Based Strategies in the Acquisition of Number Meaning and Arithmetic

Ilaria Berteletti[1], James R. Booth[2]
[1]*Department of Psychology, University of Illinois, Champaign, Illinois, USA*
[2]*Department of Communication Sciences and Disorders, The University of Texas at Austin, Austin, Texas, USA*

INTRODUCTION

The use of symbolic representations to keep track of discrete amounts is estimated to go back to the Paleolithic era, where tally marks were carved on animal bones (i.e., tally sticks). Some suggest that the Lebombo bone was used as alunar calendar and is dated at 35,000 BCE (see Bangura, 2012). However, tally sticks were not the only means to represent numerical quantities. People in several cultures throughout history have appealed to fingers and body parts to represent discrete quantities (Ifrah, 2000). There is, however, an important difference between fingers and body part counting. Body part counting often starts on fingers but moves to other parts such as the wrist, elbow, shoulder, neck, and so on (Ifrah, 2000). These parts are sequentially listed in a fixed order and in a one-to-one correspondence, but the numerical quantity that they represent is not in the body part per se: expressing a quantity by naming a body part is analogous to the use of number words. Fingers also represent items in a one-to-one correspondence by being raised sequentially; however, they are analogous to tally marks in that they transparently represent the numerical quantity of the counted set. Although cultural differences exist in the use of fingers to represent quantities (Bender & Beller, 2012; Pika, Nicoladis, & Marentette, 2009), the numerical quantity conveyed may be reconstructed by one-to-one correspondence even in the absence of language (Figure 1).

Probably because of their transparent representation of numerical quantities, many cultures use fingers as tools to support the teaching of numbers

Development of Mathematical Cognition. http://dx.doi.org/10.1016/B978-0-12-801871-2.00005-8

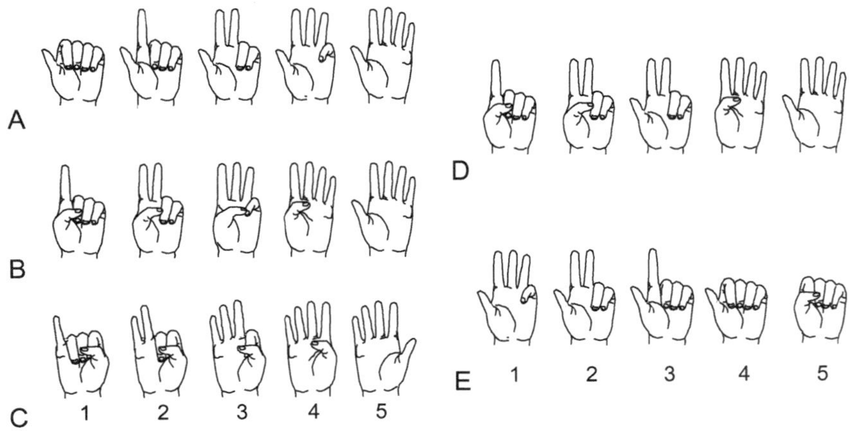

FIGURE 1 Representation of five different finger-counting systems from (A) large parts of Europe, (B) China and the United States, (C) Iran, (D) American Sign Language, and (E) Pekai-Alue in Papua New Guinea. These systems differ in terms of starting finger and if fingers are extended or folded to represent counted items. *From Bender and Beller (2012).*

(Ashcraft, 1982; Barrouillet, Mignon, & Thevenot, 2008; Butterworth, 1999; Groen & Parkman, 1972; see also: Ginsburg, 1977; Svenson, Hedenborg, & Lingman, 1976; Woods, Resnick, & Grocn, 1975). In this chapter, we present evidence suggesting that finger representation and finger strategies play a functional role in number and arithmetic processing, and that the use of finger-based strategies during childhood may play a role in creating strong associations between finger and numerical representations (Di Luca & Pesenti, 2011; Fischer & Brugger, 2011). We also suggest that finger-based strategies can play a different role depending on the arithmetic operation. The role of finger-counting strategies in the development of spatial-numerical associations will also be discussed. Finally, after reviewing the three prominent views that account for the common neural substrates between finger representation and numerical processing, we conclude with potential future directions.

FINGERS IN NUMERICAL AND ARITHMETIC PROCESSING

The Role of Fingers and Finger Representation in Number Processing

Prior to acquiring proficiency in speech production, toddlers as well as their parents use fingers and gestures to communicate numerical and mathematical concepts (Lee, Kotsopoulos, Tumber, & Makosz, 2014). Children are repeatedly primed throughout early childhood to use fingers and gestures in various numerical contexts: the role of gestures has shown to be central in preschoolers' and kindergartners' understanding of early numerical concepts and counting by reducing working memory load (Alibali & DiRusso, 1999; Graham, 1999).

Ultimately, relying on and raising fingers to count becomes natural and, based on handedness, a strong digit mapping direction (i.e., right-to-left for right-handed children) already emerges by age 5 (Sato & Lalain, 2008). In fact, the use of fingers as a strategy to manipulate numerical quantities precedes the use of verbal and written codes (Butterworth, 1999). Because fingers are a tangible support, they are thought to help children understand important numerical concepts such as cardinality (Butterworth, 1999) and one-to-one correspondence by sequentially raising each finger or pointing to objects for each subsequent number word recited (Gallistel & Gelman, 1992; Gelman & Gallistel, 1978).

This extensive reliance on fingers during childhood has been shown to create a relation between fingers and numerical representations that endures into adulthood. A number of studies with adults have found that hand configurations are automatically associated with numerical representations (Badets & Pesenti, 2010; Badets, Pesenti, & Olivier, 2010; Di Luca, Granà, Semenza, Seron, & Pesenti, 2006; Di Luca & Pesenti, 2008). In a study using an Arabic digit-to-finger mapping task, Di Luca et al. (2006) found that participants were faster and made fewer errors when the mapping matched personal finger-counting habits than with different mappings (e.g., left-to-right orientation). This result was irrespective of the physical hand position (i.e., hand palm facing up or down), suggesting that personal finger-counting habits influence the way numerical information is mentally represented and processed. In a subsequent study, they showed that number naming was faster when participants were presented with canonical finger configurations (i.e., those compatible with their personal finger-counting habits) as compared to noncanonical ones, and that only the former provided automatic access to number semantics (Di Luca & Pesenti, 2008). According to the authors, this result demonstrates that personal finger configurations are stored in long-term memory as a consequence of the repeated use of fingers when counting during childhood. Finally, these authors have also found that canonical but not noncanonical finger configurations activate a place-coding numerical representation (i.e., where activation spreads symmetrically to smaller and larger magnitudes) rather than a summation-coding representation (i.e., where activation accumulates across all smaller numerical magnitudes up to the target). These findings highlight the qualitative difference between the two and the possible increased value derived by the repetitive use of fingers (Di Luca, Lefèvre, & Pesenti, 2010).

Interestingly, the relation between fingers and numerical meaning goes beyond hand configuration. In a series of studies, Andres and colleagues demonstrated a bidirectional relation between finger movements and numerical magnitude (Andres, Davare, Pesenti, Olivier, & Seron, 2004; Andres, Ostry, Nicol, & Paus, 2008; Badets, Andres, Di Luca, & Pesenti, 2007; Badets & Pesenti, 2010). Adult participants were required to adjust the finger aperture in a precision grasping task where blocks of the same size were tagged with small (e.g., 2) or large (e.g., 8) digits on their visible side (Andres et al., 2008). Movements for grasping blocks tagged with larger numbers were wider than those performed for blocks of the same physical size but tagged with smaller numbers. This effect suggests that the number magnitude processing during finger movements interferes with object size estimation. Because number magnitude interference was not observed in a comparison task of object size estimation, the authors conclude that interference is specific to action. Moreover, Badets et al. (2010) have also shown that interference was bidirectional as perceiving opening or closing grip movements influenced numerical processing.

All these studies support the hypothesis that relying on fingers and gestures during childhood creates persistent long-term associations between fingers and numerical representations. These associations influence both numerical processing as well as finger actions that involve magnitude estimation.

Neural Substrates for Hand and Number Processing

Several clinical findings support the relation between finger and hand representations with numerical processing and arithmetic skill (Butterworth, 1999; Gerstmann, 1940; Gracia-Bafalluy & Noël, 2008; Noël, 2005). In Gerstmann syndrome patients suffer a deficit in representing fingers (i.e., finger agnosia), a disability in performing calculations (i.e., acalculia) along with right-left disorientation and agraphia (i.e., a deficit in writing) as a consequence of a lesion in the inferior parietal lobule near the angular gyrus. The co-occurrence of deficits might not be the disruption of a unique representation common to all tasks as suggested by some (Butterworth, 1999; Gerstmann, 1940), but rather the consequence of a lesion in a common vascular territory affecting processes in adjacent regions encompassing areas involved in spatial attention and number processing (Dehaene, Piazza, Pinel, & Cohen, 2003). In the latter case, finger-based strategies during childhood may play a role in creating a strong association between adjacent representations and support the acquisition of proficient counting (Di Luca & Pesenti, 2011; Fischer & Brugger, 2011).

Following Gerstmann's studies, Strauss and Werner (1938) were the first to report a significant positive relation between a deficit in finger representation and arithmetic skill in a group of children with low math achievement but normal IQ. This relation has been found to be particularly relevant in solving word problems requiring active manipulation of small magnitudes from 1 to 10 (Costa et al., 2011). Moreover, finger gnosia (i.e., the ability to discriminate one's fingers) has been found to correlate with and selectively predict future numerical and arithmetic ability beyond standardized scores (Fayol, Barrouillet, & Marinthe, 1998; Noël, 2005; Reeve & Humberstone, 2011). Gracia-Bafalluy and Noël (2008) have shown that training finger discrimination in children with poor finger gnosis improved performance on numerical tasks such as finger counting and subitizing (i.e., quickly determining the number of objects in sets of up to 4 items). These authors suggest a functional link between hand representation and number skill (but see Fischer, 2010 for statistical considerations).

Finally, another study found that the development of finger gnosis, spatial abilities, and nonsymbolic numerical abilities were correlated independently of chronological age (Chinello, Cattani, Bonfiglioli, Dehaene, & Piazza, 2013). The hypothesis suggested by the authors is that finger counting could be the source of this correlation. Finger gnosis would improve with finger counting by increasing individual finger representations and their relative position in space. Moreover finger counting would improve spatial abilities by tracking multiple items in parallel concurrently with the corresponding mental representation of numerical quantities. Indeed, in children attending kindergarten, it was found that strategies for solving addition problems, which included heavy reliance on finger counting, were significantly correlated with spatial but not verbal subtests of the Wechsler Preschool and Primary Scale of Intelligence (WPPSI; Geary & Burlingham-Dubree, 1989).

Evidence for shared processes between hand representation and numerical processing has also been found in electrophysiological and neuroimaging studies. Using transcranial magnetic stimulation (TMS: see Sarkar & Cohen Kadosh, this volume)—a noninvasive technique that consists of generating a brief magnetic field next to the scalp stimulating the neuronal elements in the underlying cortex—studies have found evidence for overlapping finger and numerical representations (Andres, Seron, & Olivier, 2007; Sato, Cattaneo, Rizzolatti, & Gallese, 2007). Sato et al. (2007) stimulated the finger motor area and recorded finger motor-evoked potentials while participants were performing a visual Arabic digits parity judgment task. In line with their hypothesis, they found a close relationship between number and finger cortical representations by showing a specific increase in the corticospinal excitability of the right-hand muscles for small numbers. The authors interpret this result in terms of an automatic activation of fingers as a consequence of an embodied finger strategy used during childhood to represent numbers. Furthermore, in the only functional magnetic resonance imaging (fMRI) study that independently localized the hand motor network by asking participants to perform a finger-moving task, it was found that perceiving numbers as digits or written words showed enhanced activations in these regions (Tschentscher, Hauk, Fischer, & Pulvermüller, 2012). The hand motor network was also modulated by individual finger-counting habits (i.e., left- vs. right-hand starters) suggesting that activation of motor circuits during number processing in adults is the consequence of motor activations induced by finger-counting strategies when learning to count during childhood (Butterworth, 1999; Di Luca & Pesenti, 2011).

These results suggest that it could precisely be the counting strategy and the one-to-one correspondence that relates finger representations to numbers (Andres et al., 2007; Imbo, Vandierendonck, & Fias, 2011; Kansaku et al., 2007). In a study where adults were asked to perform single-digit addition problems by counting on (i.e., 7+3=7 ... 8 ... 9 ... 10), passive hand movement disrupted performance more as compared to strategies requiring retrieval or intermediate steps (i.e., 8+3=8+2+1=10+1=11; Imbo et al., 2011). Additionally, in an fMRI study, Kansaku et al. (2007) showed that the left ventral premotor cortex was specifically activated when counting large sequences (beyond 20 elements) and, by applying TMS over this area, participants' counting ability was disrupted. Finally, increased excitability using TMS in the motor cortex has also been found in a task where participants had to keep track of visually presented elements (Andres et al., 2007) either using number words or letters. The authors suggested that the motor circuit has a functional role in assisting the counting process by keeping track of the elements and is not uniquely associated with reciting the numerical sequence. These results support the involvement of hand motor circuits in adults' arithmetic abilities as a result of the use of finger-counting during childhood.

Activation in finger representation areas in adults performing numerical tasks suggests that long-term associations were created during childhood while using finger-based strategies in numerical contexts. Because activation has also been found in hand motor circuits, finger-related activations may be more than a mere reminiscence of long-term associations but functionally support processes that are involved in counting such as keeping track of items, in particular when quantities exceed the first 10 digits that map on fingers. Accordingly, the observed relation between finger gnosis and later arithmetic skill goes beyond the ability to simply recognize fingers but should indicate a more comprehensive and functional ability to use fingers to support item tracking in a numerical context.

Finger-Based Strategies and Finger Representation in Arithmetic

Beyond counting and helping in understanding numerical concepts, fingers are also important as a physical support in mental calculations: children keep track of the counted items on their fingers which alleviates working memory load when procedures are not automatized (Alibali & DiRusso, 1999; Geary, 2005). Children go through several developmental steps when learning arithmetic on their fingers (Ashcraft, 1982; Barrouillet et al., 2008; Ginsburg, 1977; Groen & Parkman, 1972; Svenson et al., 1976; Woods et al., 1975). For example, when solving addition problems they start by representing the operands separately on their hands and then count all the fingers raised (i.e., count all strategy). With practice and an understanding of the cardinal value of numerals, they learn to start from the first operand and count only the fingers corresponding to the second operand. Finally, when a sound understanding of the relative cardinal values of the two operands occurs, children are able to select the largest operand to start from and count only the fingers corresponding to the smaller operand (i.e., count min; Ginsburg, 1977; Groen & Parkman, 1972; Siegler & Shrager, 1984). A recent study found that counting habits (i.e., right vs. left hand start) in both children (age 5–12 years) and adult participants significantly influence RTs for nonautomatized, addition problems (i.e., single-digit problems, such as $4+9$ for children and two-digit problems, such as 31 + 17, for adults), where right-hand starters outperformed left-hand starters (Newman & Soylu, 2014). These results indicate that the quality of finger-based counting influences procedural strategies for solving addition problems and that this influence is stronger for situations involving greater working memory load (e.g., two-digit addition problems) even in math-skilled adult participants.

Domahs, Krinzinger, and Willmes (2008) have also found traces of finger-based counting in arithmetic calculation. They analyzed elementary school children's responses to simple and complex addition (i.e., single-digit problems with results smaller or larger than 10) and subtraction (i.e., with single- or double-digit minuend) problems and observed an above-chance proportion of errors consistent with representational failure in keeping track of "full hands"—what they termed split-five errors (i.e., the answer deviates exactly by ±5 from the correct result; e.g., 32 + 17 = 44 or 54). The observed subbase-five structure is evidence of how finger representation influences mental arithmetic in children and suggests that early external finger-based representations become internalized during primary school. This internalization of finger-based representation has been shown to persist beyond childhood because subbase-five effects have also been observed in adult participants performing mental addition tasks (Klein, Moeller, Willmes, Nuerk, & Domahs, 2011). Interestingly, in German languages (e.g., English, Swedish) it was found that colloquial speech has an overrepresentation of subbased-five phrases for estimating time (see you in 5, 10 min) and quantities (i.e., "about 20–25 students in the class"; Eriksson, Bailey, & Geary, 2010). Consequently, it would be interesting to investigate if such prevalence is stronger when finger-based strategies are promoted. This would suggest that the internalized full-hand configuration has a broader influence in everyday estimation of numerical ranges.

Other studies of adults suggest automatic activation of finger-based representations and their corresponding hand configuration during mental calculation (Badets et al., 2010). Participants were faster and more accurate in judging the outcome of a simple addition if it was presented as a correct canonical finger configuration compared to when it was presented as a noncanonical configuration or when compared to sets of rods positioned similarly to canonical and noncanonical finger configurations. This facilitation effect, according to the authors, is due to an automatic anticipation of the correct hand configuration associated with the addition problem. Finally, adult participants performing a simple sequential tapping task with their right hand while solving single-digit operations experienced a reaction time cost for solving problems (Michaux, Masson, Pesenti, & Andres, 2013). Importantly, this interference effect was found only for those operations that during childhood rely the most on finger based strategies: addition and subtraction problems.

These studies demonstrate how finger-based strategies used in childhood are internalized with practice and how the quality of such internalized representations may shape arithmetic performance and even colloquial speech

(Eriksson et al., 2010). These results open new questions on the relation between use of such strategies and the development of the numerical representation. Indeed, no study has investigated the role of finger-based strategies in supporting the mapping of number words onto the internal numerical representation. We do not know whether fingers are mediating the association between number words and quantities in early childhood or whether this process might be impaired in math difficulty.

Children with a math learning disability have been found to rely more on back-up strategies such as finger counting (Geary, Brown, & Samaranayake, 1991; Jordan, Hanich, & Kaplan, 2003; Jordan & Montani, 1997). They tend to make more errors when relying on them (Geary, 1990; Gersten, Jordan, & Flojo, 2005), and their poor understanding of the principles governing efficient counting impairs the transition to more sophisticated finger-counting strategies (Geary, Bow-Thomas, & Yao, 1992). Importantly, to successfully use the most sophisticated finger-counting strategy (i.e., count min), children need to understand the relative cardinal value of two numbers to determine the larger one. However, what remains to be understood is which aspects of finger-counting strategies are dysfunctional: applying the counting procedures properly or internalizing the cardinality associated with finger configurations. Moreover, recent evidence suggests that math learning disability is a heterogeneous disorder (Bartelet, Ansar, Vaessen, & Blomert, 2014; Geary, 2010; Von Aster, 2000). Therefore, although not all subtypes might show delays in finger-counting strategies, different subtypes of math learning disability might relate to different aspects of finger-counting strategies.

A second possibility is that math difficulty could be the consequence of a poor understanding of the procedures involved, such as putting together two quantities to add, yielding poor implementation of any strategy regardless of whether a fast access to a refined numerical representation is necessary. Finally, a third possibility is a deficit in creating strong finger-number associations. Indeed, in one study it was reported that a deficit in finger gnosia in children with mathematical difficulty was distinct from a deficit in analog numerical representation (Costa et al., 2011). Although children with mathematical difficulty did not show working memory impairments, they were unable to use fingers to represent numbers efficiently.

Because differences in finger gnosia have been described in young children (Noël, 2005), and because children commonly use finger-based strategies before formal education, future investigations in this direction have the potential to determine early markers of math difficulty (Geary et al., 1992). A greater capacity to differentiate among children with math learning disability based on the presence of finger-based strategy impairments, and which aspects of finger counting are impaired, would enable the design of specific intervention strategies to either reinforce finger representation or implement external support strategies to compensate for their inability to rely on fingers.

Neural Substrates for Finger-Related Activation During Arithmetic Problem Solving

Supporting the hypothesis of a progressive internalization of finger-based strategies, a number of brain imaging studies of adults have reported incidental motor and premotor activations while performing arithmetic tasks, and these have been interpreted in terms of concurrent finger-related brain activations indicative of finger-based strategies used during childhood. Pesenti, Thioux, Seron, and De Volder (2000) found activation in premotor areas during single-digit addition problems. Finding greater activation in the left precentral gyrus for simple addition problems compared to a digit comparison task, they argued that these fronto-lateral activations when combined with the parietal activations reflect the involvement of a finger-movement network underlying finger counting. Hanakawa et al. (2002) compared nonmotor mental tasks including repeated addition (e.g., 5 + 3 + 4 + 7...) with silent verbal rehearsal of seven-digit numbers. They found the anterior and dorsal part of the lateral premotor cortex, considered to be central in planning motor actions, to be significantly more activated for the mental addition than for the verbal rehearsal of digits. Simon and colleagues (Simon et al., 2004; Simon, Mangin, Cohen, Le Bihan, & Dehaene, 2002), in a simple subtraction task, found a predominantly left lateralized network comprising the horizontal segment of the intraparietal sulcus (IPS), the premotor and motor cortices and anteriorly the middle frontal gyrus. Activation in prefrontal and precentral areas were also found specifically for two-digit addition and subtraction problems in adults (Knops & Willmes, 2014).

Task difficulty (e.g., comparing multiplication problems with smaller and larger operands) has also been shown to increase activation in the premotor and frontal cortices in addition to a network subtending direct retrieval of answers (Jost, Khader, Burke, Bien, & Rösler, 2009). According to the authors, such activations reflect increased difficulty during fact retrieval processes and support the coordination of multiple processing steps for solving more complex arithmetic problems. Therefore, when retrieval fails, premotor activations might reflect the execution of intermediary calculations (e.g., $6 \times 9 = [5 \times 9] + 9$) that rely on finger-based strategies.

Overall, in a meta-analysis examining brain activation for the different operations, Arsalidou and Taylor (2011) showed that across fMRI studies, the prefrontal cortex including the precentral gyrus, and thus the motor cortex, is significantly active when people solve addition, subtraction, multiplication, and division problems. Furthermore, a strong co-occurrence of activation in the prefrontal cortex (i.e., middle frontal gyrus [MFG] and premotor cortex) and the IPS has also been highlighted in a review on the neurobiological underpinnings of mathematical cognition (Ashkenazi, Black, Abrams, Hoeft, & Menon, 2013). An early strong connection between these areas could reflect a natural predisposition to rely on fingers. In a recent review on white matter tracts, Matejko and Ansari (2015) suggest that the corticospinal tract and the superior corona radiata are two tracts that have often been found to be related to numerical and mathematical processing. If the superior corona radiata is usually found in higher-level cognitive processes, the corticospinal tract is typically associated with motor functions. This could indicate a relation between finger-based strategies and numerical representations. Future studies should investigate the development of these white matter tracts and whether strong relations in these tracts are present before the learning of numerical concepts. The early presence of connections between numerical and finger representations could suggest a predisposition to use fingers in representing numerical quantities. Moreover, these white matter tracts should show greater correlations with operations that have been learned through finger-based strategies (e.g., subtraction problems) compared with verbally retrieved operations (e.g., multiplication problems).

Activation in the IPS is considered to be reflective of numerical quantity processing (Dehaene et al., 2003; Piazza, Izard, Pinel, Le Bihan, & Dehaene, 2004), whereas activation in premotor cortex is argued to reflect the motor planning and action related to finger-counting and finger-based strategies. However, it should be noted that the parietal cortex, and specifically the IPS, plays an important role in tasks involving the hand, such as pointing and grasping (De Lange, Hagoort, & Toni, 2005; Simon et al., 2004, 2002; Vingerhoets, 2014). In particular, tasks such as mental imagery of hand and finger movements, movement preparation, and object manipulation have been shown to rely on a parieto-prefrontal network (Binkofski et al., 1999; De Lange et al., 2005; Haaland, Elsinger, Mayer, Durgerian, & Rao, 2004). According to De Lange and colleagues, the parietal cortex also plays an important role in combining somatosensory and visuomotor information during hand movements.

To our knowledge only three studies have directly tested the involvement of finger/hand representation during arithmetic processing. An fMRI study showed that simple arithmetic and finger discrimination tasks induce common activations in the horizontal IPS and posterior superior parietal lobule (Andres, Michaux, & Pesenti, 2012). The relation was found to be stronger for subtraction compared to multiplication problems and for the left compared to the right hemisphere. Indeed, subtraction problems rely more on quantity processing compared to multiplication problems, which rely more on direct retrieval of answers from long-term memory (Fayol & Thevenot, 2012). In this case, the authors argue that the common activation found in IPS could be part of a broader network related to the mental imagery of finger-based strategies.

A second study focused on the neural networks involved in different numerical tasks (i.e., symbolic and nonsymbolic number comparison, symbolic and nonsymbolic addition and counting tasks) and their relationship to activations underlying finger representations (i.e., guided finger movement task) and eye movements (saccades) in children (Krinzinger et al., 2011). The contribution of the finger-related network, including the ventral precentral sulcus, the supplementary motor area, and the dorso-lateral prefrontal cortex, was the strongest for nonsymbolic and symbolic addition tasks regardless of presentation format. Finally, in our recent fMRI study we independently localized finger-related somatosensory and motor areas (i.e., pre- and post-central gyri) and analyzed activation induced by single-digit subtraction and multiplication problems in school-age children (Berteletti & Booth, 2015). Results showed activation of finger motor areas for subtraction problems and increased activation in somatosensory areas for larger (e.g., 9-3) versus smaller (e.g., 4-3) problems. Finally, a significant brain-behavior relation was found between activation in the somatosensory area and performance to subtractions problems, providing the first neurological evidence for a functional role of finger representation in proficient arithmetic problem solving.

In summary, neuroimaging studies show finger-related activations during arithmetic problem solving in particular for problems that rely more on counting procedures and quantity manipulation such as subtraction and addition problems as well as harder problems. Activation in premotor and motor areas in adults and children have therefore been interpreted as supporting the planning of finger-based strategies, whereas activation in the finger somatosensory area has been found to be related specifically to performance on subtraction problems in children. Finally, although the IPS is considered to be the locus for the numerical representation, it also plays an important role in hand and finger movements. Results indicate that common activations for arithmetic processing and finger discrimination could be accounted for by processes involved in mental imagery of finger counting in particular for those operations that require greater quantity manipulation. These results raise the question of whether finger-based strategies can influence operation-specific networks.

FINGER-BASED STRATEGIES AND OPERATION-SPECIFIC PROCESSES

Theories on the solving of and representation of answers to single-digit arithmetic problems date back to the early 1970s (Groen & Parkman, 1972). In these first models, answers to problems are assumed to be retrieved from representational networks in long-term memory. Children initially solve problems through procedural strategies such as counting, and with practice they eventually consolidate the problem with the corresponding answer in a long-term memory network (Siegler & Shrager, 1984). According to these models, a proficient solver retrieves the answer directly from memory in operation-specific tables (Ashcraft, 1982, 1992). However, recent clinical, behavioral, and neuroimaging data cast doubts on the use of the retrieval strategy for all operations, suggesting that different arithmetic operations may be processed differently (Barrouillet et al., 2008; Dehaene, 1992; Dehaene & Cohen, 1995; Fayol & Thevenot, 2012; LeFevre et al., 1996; Michaux et al., 2013) and may rely on partially distinct neural networks (Berteletti & Booth, 2015; Dehaene et al., 2003; Demir, Prado, & Booth, 2014; Prado et al., 2011; Prado, Mutreja, & Booth, 2014; see Arsalidou & Taylor, 2011 for a meta-analysis). In this section we discuss studies showing such dissociation and argue that finger-based strategies might shape these different processes during childhood.

A Model Supporting Operation-Specific Processes

In the Triple Code model, Dehaene argues for a distinction between numerical representations responsible for different operations (Dehaene, 1992; Dehaene & Cohen, 1995). In this model, the auditory verbal word frame is responsible for verbal counting and retrieving arithmetic facts stored in long-term memory. In particular, the model postulates that only answers to addition and multiplication problems are verbally stored and retrieved. The visual-Arabic number form is responsible for parity (odd-even) judgments and operations in which most answers are not stored verbally, such as multidigit subtraction problems. Increasing behavioral, neuropsychological, and neuroimaging data support such a distinction (Barrouillet & Thevenot, 2013; Dehaene, 1992; Dehaene et al., 2003; Lemer, Dehaene, Spelke, & Cohen, 2003; McCloskey, 1992; Prado et al., 2014, 2011). The neuropsychological literature has reported several cases where operation-specific deficits were found after brain lesions in the parietal lobe, in particular in the left hemisphere. The ability to retrieve answers to simple multiplication problems can be selectively impaired despite an intact ability to solve simple subtraction problems (Cohen, Dehaene, Chochon, Lehéricy, & Naccache, 2000; Dagenbach & McCloskey, 1992; McNeil & Warrington, 1994; Pesenti, Seron, & van der Linden, 1994) as well as the converse (Cohen & Dehaene, 2000; Dagenbach & McCloskey, 1992; Dehaene & Cohen, 1997; Delazer & Benke, 1997; Lee, 2000; Pesenti et al., 1994; Van Harskamp & Cipolotti, 2001).

Moreover, Dehaene and Cohen (1997) reported a double dissociation where one patient, after a right inferior parietal lesion, had spared multiplication performances but failed in subtraction problems and numerical magnitude judgments, whereas the second patient showed the reverse pattern following a left subcortical lesion. Finally, a recent study presented evidence of a dissociation between subtraction and multiplication problems in the right parietal cortex from intraoperative cortical electrostimulation (Yu et al., 2011). A patient undergoing a neuro-surgical intervention was stimulated at three right parietal sights underpinning quantity processing while solving simple subtraction and multiplication problems. Stimulation in these loci affected performance of subtraction problems only, indicating that the parietal lobe is differentially engaged in the two operations, at least in adults. According to Dehaene et al. (2003), the observed clinical dissociations all fit with the assumption of two distinct processes: one dedicated to quantity manipulation supporting subtraction and numerical quantity comparisons and the other devoted to verbal storage and retrieval of multiplication problems as well as counting.

Behavioral Evidence for Operation-Specific Processes

Behavioral studies have also demonstrated that a different mix of processes support problem solving for the different operations. A number of studies have investigated what strategies adults and children use to solve arithmetic problems either by analyzing response times or the self-reported strategy (Barrouillet & Fayol, 1998; Barrouillet & Lépine, 2005; Barrouillet & Thevenot, 2013; Campbell & Xue, 2001; Geary & Wiley, 1991; Geary, Frensch, & Wiley, 1993; Groen & Parkman, 1972; Ilg & Ames, 1951; LeFevre, DeStefano, Penner-Wilger, & Daley, 2006; LeFevre, Sadesky, & Bisanz, 1996; Seyler, Kirk, & Ashcraft, 2003; Siegler, 1987; Starkey & Gelman, 1982). Overall, results indicated that solutions to multiplication as well as smaller addition and subtraction problems were retrieved, whereas larger problems were more often solved by computation (LeFevre, Sadesky, & Bisanz, 1996; LeFevre, Bisanz, et al., 1996); however, the amount of automatized retrieval and procedural execution differed across ages (Geary et al., 1993; Geary & Wiley, 1991). Additionally, a study with children showed that solutions to subtraction problems are less likely retrieved than solved through computation, and when retrieval occurs it is restricted to specific problems involving 1 (Barrouillet & Thevenot, 2003).

However, these studies generally make one of two assumptions: either that participants are able to reliably report the strategy they use or that no difference in reaction times across operations indicates similar underlying processing. These two assumptions were questioned in a recent study where priming the operation sign 150 ms before the operands of the problem induced facilitation effects differently depending on the operation (Fayol & Thevenot, 2012). In this study, adult participants showed faster reaction times in the primed condition only for addition and subtraction problems compared to multiplication problems, which did not show a facilitation effect. The authors argue that the only possible explanation for such a priming effect is that an operation-specific procedure is preactivated by the arithmetic sign thus facilitating its implementation, but only for those operations that require computation. For multiplication problems, no such prime effect occurred, suggesting that the operands are necessary to retrieve the result from long-term memory supporting the retrieval strategy. Importantly, they observed that the facilitation effect was present even for very small problems such as $2+3$, the answers to which are commonly considered to be retrieved from long-term memory (Barrouillet & Thevenot, 2013; LeFevre et al., 2006; LeFevre, Sadesky, & Bisanz, 1996). Their results demonstrate that procedural knowledge can be activated and implemented quickly by individuals and that procedural processing times can be indistinguishable from retrieval strategy times.

Two other studies found evidence for dissociation between subtraction and multiplication problems in children and adults with developmental dyslexia (i.e., a learning disability affecting phonological processing). In these studies, multiplication problems were more impaired relative to subtraction problems, supporting the hypothesis that only multiplication problems rely on direct retrieval of answers (Boets & De Smedt, 2010; De Smedt & Boets, 2010). Finally, a study using a dual-task paradigm showed that accuracy on multiplication problems decreased when participants were in a phonological suppression condition, whereas performance on subtraction problems was disrupted in a visuospatial suppression condition (Lee & Kang, 2002), indicating that different working memory components are necessary for solving problems requiring these two operations—at least in adults (see DeStefano & LeFevre, 2004 for a review on working memory involvement in mental arithmetic).

According to these behavioral studies, proficiency in solving multiplication problems would result in accessing the answer from long-term verbal memory, whereas proficiency in subtraction and addition problems would result in efficient execution of procedures, even for very small and frequent problems.

Operation-Specific Neural Networks

Various neuroimaging studies have also provided evidence in both direct and indirect support of distinct neural substrates for different operations (Fehr, Code, & Herrmann, 2007; Prado et al., 2011, 2014; Zhou et al., 2007; for a meta-analysis see Arsalidou & Taylor, 2011). In a study investigating arithmetic complexity in the four operations, Fehr et al. (2007) concluded that "different operators are processed utilizing shared, but predominantly different, brain networks" (p. 99). Furthermore, different processing pathways have been reported depending on the operation. In a study with adult acute stroke patients, lesions in different white matter tracts were differentially related to performance in single-digit arithmetic problems, corroborating the finding of (at least partially) operation-specific networks (Mihulowicz, Willmes, Karnath, & Klein, 2014).

Zhou et al. (2007) compared fMRI activations of Chinese undergraduates elicited by single-digit multiplication and addition problems. Although both operations elicited a partially similar network, additions elicited a visuospatial processing network, including the right IPS, to a greater extent than multiplications. Conversely, multiplications relied more on a left lateralized network encompassing the superior temporal gyrus (STG), indicating a greater verbal involvement. These findings were confirmed in a study that explicitly tested the differential involvement of verbal and spatial-numerical processing in single-digit subtraction and multiplication problems. Prado et al. (2011) independently localized the verbal and the spatial-numerical processing networks. They found that multiplication problems elicited greater activations in the left middle temporal gyrus (MTG) and the inferior frontal gyrus (IFG), areas dedicated to verbal long-term storage and retrieval. Subtraction problems conversely relied to a greater extent on the right IPS, an area dedicated to numerical magnitude comparison. Furthermore, in a developmental study using the same localizers and the same tasks, Prado et al. (2014) reported a developmental dissociation in the neural correlates underlying the two operations in children between 2nd and 7th grade. Activity in the fronto-temporal network, subserving verbal processing, increased with grade only for multiplication problems, whereas grade-related increase of activity was found in the right parietal cortex only for subtraction problems. These results again highlight how proficiency in single-digit problems may be achieved through different neural networks supporting different types of problem-solving strategies. Finally, a larger involvement of the STG and MTG was also found during multiplication compared to subtraction problems by Andres, Pelgrims, Michaux, Olivier, and Pesenti (2011) in an fMRI-guided TMS study.

These findings indicate substantial neural differences in processing strategies depending on the operation. The differences, persistent in adulthood, indicate that multiplication problems are preferentially solved through a verbal retrieval strategy in a verbal network (i.e., IFG, MTG, and STG) whereas addition and subtraction problems are solved through procedural strategies involving brain areas encompassing spatial manipulation and transformation of numerical quantities (i.e., IPS and SPL). Importantly, the Fayol and Thevenot (2012) study highlights how procedural strategies become increasingly automatized for easier and over-practiced problems up to the point that RTs are indistinguishable from retrieval RTs.

Operation-Specific Processes as a Consequence of Operation-Dependent Teaching Methods

Different teaching methods could account for operation-specific networks found in children and adults. In the first years of schooling, children have to learn how to solve arithmetic operations and use a variety of strategies before becoming proficient (Siegler, 1991; Siegler & Jenkins, 1989). Studies have found strong evidence for counting-based strategies in preschoolers and elementary school pupils (Ashcraft, 1982). Children use progressively more refined strategies depending on mastery and on operation: count all, count min/max, transformation (e.g., $6+8=6+6=12+2$) and retrieval (Barrouillet et al., 2008; Ginsburg, 1977; Groen & Parkman, 1972; Svenson et al., 1976; Woods et al., 1975). Moreover, during school, a different emphasis is placed on retrieval depending on the operation and commutativity. Indeed, children are encouraged to retrieve multiplication problems by using a rote learning approach (Campbell & Xue, 2001). Conversely, subtraction problems are taught by means of procedures without emphasizing memorization (Barrouillet et al., 2008; Campbell & Xue, 2001; Dehaene, 1992; Thevenot & Barrouillet, 2006). Specifically, children solve addition and subtraction problems through various finger-based calculation strategies, which is not the case for multiplication problems (Baroody, 1987; Siegler & Shipley, 1987; Siegler & Shrager, 1984).

Practical reasons might explain why different methods are preferentially used depending on the operation. Indeed, the additive and subtractive concepts are very intuitively represented on fingers. Raising or lowering fingers clearly translate the transformation required by the operation sign. Moreover, although only 25 single-digit additive or subtractive problems (i.e., addition problems with results up to 10 and subtraction problems with a minuend up to 10) may be fully represented on the two hands, any other problem can be solved on fingers by using the count on strategy with the only requirement to keep track of decades. For multiplication problems, using fingers to solve even easy operations such as 3×6 is very impractical as a child would have to keep track of the number of times a given number is added and keep track of the decade. Some less commonly used finger-based algorithms exist to solve part of the multiplication table problems (e.g., 6×6 to 9×9); however, these require multiple steps and the multiplicative transformation is nontransparent (Kolpas, 2002). Because finger-based strategies in multiplication are impractical and nontransparent, a rote verbal strategy appears advantageous and is likely the reason why it is intensively used only for multiplication problems.

In summary, extensive evidence supports the hypothesis that different strategies are employed for different arithmetic operations. However, in the debate about the strategies used to learn and solve simple arithmetical facts, little attention has been given to the role of fingers and hand representation in shaping the underlying neural network for subtraction and addition problems. Indeed, fingers are the first and most accessible support used by children to count and compare numerical quantities (Butterworth 1999; Sato & Lalain, 2008). It is therefore reasonable that hand representation influences arithmetical facts, learning, and strategy use.

Operation-Dependent Finger-Related Activations

Given the literature on the involvement of hand-related processes and neural networks in arithmetic problem solving (Arsalidou & Taylor, 2011; Ashkenazi et al., 2013; Hanakawa et al., 2002; Knops & Willmes, 2014; Pesenti et al., 2000; Simon et al., 2004, 2002) and the different strategies employed by both children and adults depending on the operation (Fayol & Thevenot, 2012; Roussel, Fayol, & Barrouillet, 2002), a differential relation between hand-related activation and type of operation can be hypothesized. On the one hand, multiplication has generally been taught by rote memorization and has been shown to rely on retrieval strategies in a verbal neural network (Boets & De Smedt, 2010; De Smedt & Boets, 2010; Lee & Kang, 2002; and see De Smedt, this volume, for a thorough review of the role of phonological processing in arithmetic fact retrieval). On the other hand, addition and subtraction problems are generally taught by means of counting-based procedures and rely to a greater extent on a spatial-numerical network (Dehaene et al., 2003; Jost et al., 2009; Lee & Kang, 2002; Newman & Soylu, 2013). Therefore, we expect finger-related activation to play a greater role in the acquisition and processing of addition and subtraction problems.

To our knowledge, only one behavioral study with adult participants (Michaux et al., 2013) and one neuroimaging study with children (Berteletti & Booth, 2015) have addressed this issue. In the adult study, the authors assessed whether finger movements interfered more with addition and subtraction or with multiplication problem solving. Because the former benefit the most from finger-based calculation strategies in childhood, the authors predicted that adult participants' performance for solving addition and subtraction problems would decrease more compared to multiplication problems if participants were required to perform a finger-tapping dual task. A control condition, where participants were asked to move their foot, along with a rest condition allowed controlling for general working memory load and proficiency in each operation. Results showed that only finger movements induced interference, as reflected in slower RTs, for solving addition and subtraction but not for multiplication problems. These findings demonstrate the specific role of finger-related processes in solving addition and subtraction problems, suggesting that finger movements and mental arithmetic are functionally related even in adult participants. According to the authors, "solving an arithmetic (i.e., addition or subtraction) problem can then be viewed as implicitly eliciting mental activation of finger-based representations and/or procedures acquired during childhood" (p. 202).

In our study with children (Berteletti & Booth, 2015), we specifically investigated whether different operations elicit finger-related activations differently and whether this activation is related to problem-solving efficiency in children. Moreover, we also investigated the specific contribution of finger representation (i.e., the somatosensory activation) as compared to finger movement (i.e., motor activation). Our aim was to understand skill-based effects in finger-related activation in both somatosensory and motor cortex during single-digit subtraction and multiplication problems. Finger somatosensory and motor areas were independently localized, and brain activation revealed that only subtraction problems significantly activated finger motor areas and larger subtraction problems yielded greater somatosensory activation compared to smaller problems. The two operations did not differ in accuracy, both yielding results that were 90% correct. These results are consistent with the earlier described enhanced involvement of finger-based strategies for subtraction problems compared to multiplication problems (Michaux et al., 2013) and for problems that require the manipulation of larger quantities (Jost et al., 2009). Our results provide the first neurological evidence for the importance of fine-grained finger representation in arithmetic skill specifically for problems that rely most on finger-based strategies. A question that remains open is whether activation for hand representation persists even in participants that have had longer exposure to facts, as our group of children had only started mastering arithmetic facts.

Finger Counting, Cultural Influence, and Spatial-Numerical Relations

Thus far we have seen that finger-counting strategies play a role in both numerical and arithmetic processing. However, as Bender and Beller (2012) point out, hands are naturally accessible and cognitively instrumental but crucially may be used differently in different cultures. Indeed, some cultures do not even possess number words beyond 5 (Pica, Lemer, Izard, & Dehaene, 2004). Therefore, some studies have started investigating cultural differences in finger-counting strategies (Bender & Beller, 2011, 2012; Fuson & Kwon, 1992; Lindemann, Alipour, & Fischer, 2011; Pika et al., 2009; Tschentscher et al., 2012). Strategy differences have in fact been observed between cultures as well as between linguistic groups, supporting the possibility that counting strategies are the result of cultural conventions (Lindemann et al., 2011; Pika et al., 2009). In a large-scale survey, researchers investigated both culture and reading direction as factors for finger-counting habits. Overall, Western individuals (reading from left to right) showed a marked preference for starting to count with the thumb of the left hand, whereas Iranians (reading from right to left) preferred starting with the little finger of the right hand. Interestingly, marked differences were also found within Western cultures where individuals from the United States and the United Kingdom were more consistent in declaring they started with the left hand (>80%) compared with about 50% of Italians or Belgians (Lindemann et al., 2011). Variations on the starting finger have also been found: Germans preferred the thumb compared to native English speakers who started counting on their index finger (Pika et al., 2009). Still, there is considerable variation in how fingers are used for number representation and problem solving within cultures and linguistic groups.

The low intragroup consistency for finger-counting direction could also account for apparently discordant findings on the relation between the starting hand for counting and general hand preference for manual activities (Sato & Lalaìn, 2008).

It appears that when cultural conventions are loose, hand preference may influence counting strategy and starting hand for counting. Furthermore, another finding that could account for observed differences is that intra-individual differences in number-finger relations exist and depend on the numerical principle to be represented (Wasner, Moeller, Fischer, & Nuerk, 2015). Except for number 10, the finger gestures for expressing the cardinality or ordinality of a set were highly discordant. The authors argue that finger-counting habits depend on the perceptual and proprioceptive situation and task demands strongly influence the produced gesture. For example, when enumerating "three" concepts (i.e., ordinality) an individual could start by raising the thumb followed by the index and middle fingers whereas the same individual could use the index, middle. and ring fingers to order three drinks in a noisy restaurant (i.e., cardinality). Therefore, even small differences in task instructions between studies could yield very different finger strategies.

The relation between finger-counting direction and number-space representations has also been investigated. Indeed, numbers are thought to be represented on an oriented mental number line with small numbers on the left and larger numbers on the right (Dehaene, 1992). This association has been repeatedly found in many different contexts (Fias & Fischer, 2005; Gevers & Lammertyn, 2005; Hubbard, Piazza, Pinel, & Dehaene, 2005; Wood, Willmes, Nuerk, & Fischer, 2008 for a meta-analysis). In particular, when participants are required to make a parity judgment, response times are faster when small numbers are responded with the left hand and large numbers with the right hand. This effect is known as the spatial-numerical association of response codes (SNARC), and it was initially argued that reading direction influences the orientation of the mental number line (Dehaene, Bossini, & Giraux, 1993). However, some studies have suggested that finger-counting strategy may be a better determinant of the spatial-numerical association.

A seminal study by Di Luca et al. (2006) asked Italian adults to respond to single-digit numbers by using one finger for each digit. They systematically varied the finger-digit mapping to either correspond to individual's counting-mapping (i.e., starting to count on the thumb of the right hand) or to correspond to the expected SNARC mapping (i.e., where numbers were assigned from left to right with number 1 being the small finger of the left hand). They also created several control conditions among which the starting hand was switched preserving the finger-mapping, or where the SNARC mapping was reversed. Comparing conditions indicated that the personal finger-mapping showed better performance than the SNARC-compatible condition, suggesting that finger-counting is a better determinant of the association between hand response and numerical magnitude.

This result was corroborated in a study explicitly comparing two groups of adults who differed in their starting hand for counting when performing a classic single-digit parity judgment (Fisher, 2008): the typical SNARC effect was found only in the group of left-hand starters. These results therefore question the initial hypothesis that reading direction is the basis for the SNARC effect and suggest that finger-counting habits influence numerical cognition.

Furthermore, Fischer and Brugger (2011) argue that reading direction may not fully account for the spatial-numerical association since: Hebrew participants, who read from right-to-left, show a left-to-right SNARC (Shaki, Fischer, & Petrusic, 2009); the effects of reading on spatial associations can be easily overturned (Fischer, Mills, & Shaki, 2010; Fischer, Shaki, & Cruise, 2009; Shaki & Fischer, 2008); pre-reading children explore objects more efficiently if they are enumerated from left-to-right (Opfer & Furlong, 2011; Opfer, Thompson, & Furlong, 2010); and finally spatial-numerical associations have also been found in animals (Rugani, Kelly, Szelest, Regolin, & Vallortigara, 2010). However, even finger-based representations have been found to be limited in explaining the spatial-numerical association. Brozzoli et al. (2008) demonstrated that numerical representations rely either on body-based (i.e., determined by individual's finger-counting direction) or space-based (i.e., determined by the internal mental number line) hand configurations, and that when the two are in competition, the latter dominates.

In summary, specific finger-counting strategies appear to be culturally defined and to play at least a partial role in shaping an individual's numerical representation. Therefore, given the Bender and Beller (2012) typology for understanding different counting strategies based on different properties, it would be important to explore the developmental implications of such differences. These properties include: dimensionality (i.e., a distinct symbols/gestures vs. a recursive principle using a finite set of symbols/gestures for composing numbers), base and subbase values (e.g., base 10, 20, and subbase 5), extent (i.e., the total extent of the integers that may be represented), regularity (i.e., whether the system contains exceptions such as the words eleven and twelve which are not built using the decade-unit rule governing the successive decades), and sign count (i.e., total number of distinct signs). Indeed, some characteristics, such as dimensionality, not only might help novices learn the system or children learn the counting strategy faster, but may also have implications for number processing and the internal numerical representation as well as increased efficiency in supporting arithmetic problem solving. In fact, some finger-counting strategies could foster stronger operation-related effects compared to others. Therefore, the distinction between finger-related neural networks that depend on the operation should be modulated by the strength of the operation-related effects due to different finger-counting strategies.

FUTURE DIRECTIONS

Two main views account for the relation between finger and numerical representations: the functionalist view (Butterworth, 1999) and the redeployment view (Penner-Wilger & Anderson, 2013). The functionalist view postulates that finger and numerical representations become linked through development because of the repeated reliance on finger-counting and finger-based strategies to represent numerical quantities and solve arithmetic problems (Butterworth, 1999; Di Luca & Pesenti, 2011; Michaux, Pesenti, Badets, Di Luca, & Andres, 2010). This view also postulates that children with better finger-related abilities (gnosia or fine motor control) should show higher arithmetic skill and that precise motor execution of finger gestures to represent numbers is necessary for the development of efficient numerical representation. The alternative view is based on the redeployment hypothesis which suggests that the same brain area that computes a predetermined low-level processing may be re-used for many different higher-level cognitive processes (Geary, 2005; Marcus, 2004; Penner-Wilger & Anderson, 2008, 2013). Because this low-level computation is common across several cognitive processes, an efficient system will likely re-use the same brain area. Importantly, this view differs from that of neuronal recycling (Dehaene & Cohen, 2007) in the time course of the mechanism: the former occurs over evolutionary time whereas the latter during ontogenetic development.

In favor of the functionalist view, studies on the influence of finger-counting systems have found that individuals using explicit subbase-five systems, as shown in Figure 2 for Germans and German Deaf Signers, showed greater subbase-five effects compared to individuals using non-subbase-five systems, as in Chinese (Domahs et al., 2010). Because the two former groups were slower than Chinese in comparing Arabic digits that crossed the five boundary

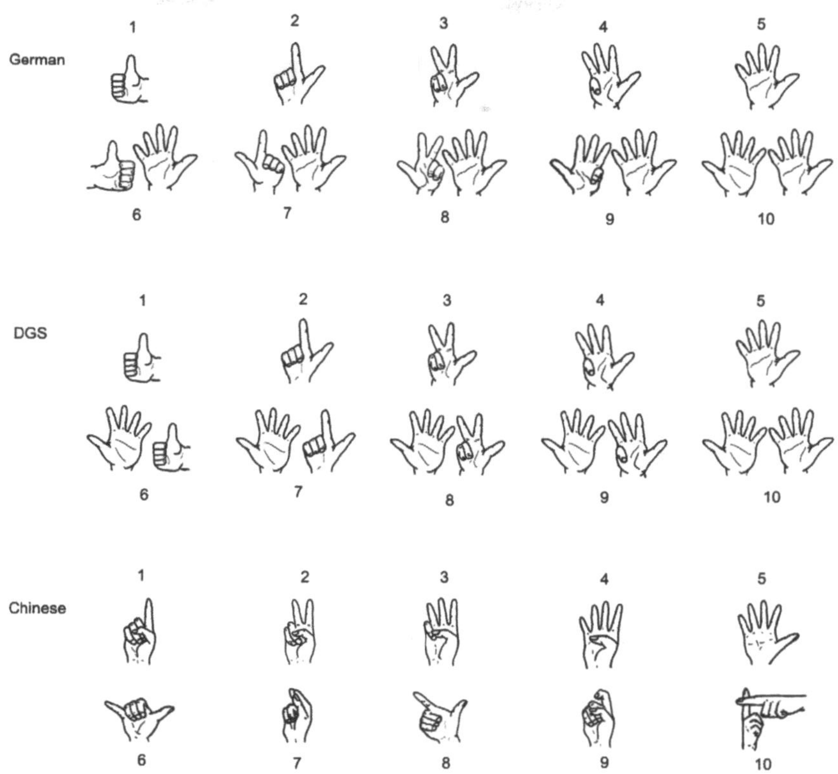

FIGURE 2 Finger-counting systems used by Germans, German Deaf Signers (DGS), and Chinese. German Deaf Signers differ from non-deaf German individuals by changing the hand that represents the subbase five for numbers above 5. Chinese, however, represent all numbers on the same hand with the exception of number 10. *From Domahs, Moeller, Huber, Willmes, and Nuerk (2010).*

(i.e., 4 vs. 6, Figure 2), and because reaction times could not be predicted by numerical magnitude alone, the authors concluded that finger-counting strategies can influence symbolic numerical processing in a functional way and that such a process is not exclusively abstract. Moreover, differences in counting strategies also influence brain neural networks. Tschentscher et al. (2012) found that the contralateral motor area of the hand subtending small number counting was activated when small Arabic digits were presented despite the absence of overt finger movement in the MRI scanner. The authors therefore conclude that an intrinsic functional link exists between finger-counting and number processing.

One of the assumptions of the redeployment view is that individuals who do not rely on finger-counting strategies should nevertheless show interference when finger and numerical tasks compete for the same neuronal resources. However, two studies speak against the necessity of relying on fingers to be proficient counters and to create a functional, internal numerical representation.

These studies challenge the assumption that finger counting is a universal mechanism for numerical cognition by showing that individuals who do not rely on finger-counting strategies are not affected by finger-related interference. In these studies, the authors (Crollen, Mahe, Collignon, & Seron, 2011; Crollen et al., 2014) investigate the role of vision in the development of finger counting comparing performances of congenitally blind individuals to those of sighted individuals.

In the first study with children, the authors found that blind children spontaneously used finger-counting strategies significantly less than sighted control children in counting and numerical quantity tasks (i.e., only 3 out of 16 blind children used fingers consistently). Importantly, the two groups did not differ in terms of basic finger discrimination abilities and exhibited similar performances on enumeration tasks. Blind children showed greater difficulty than controls only on those tasks that loaded working memory heavily, in particular, the phonological loop. These results speak against the functionalist view of a necessary involvement of finger-counting strategies to develop numerical representations. Additionally, to specifically test the redeployment hypothesis the same authors tested adult participants in an interfering hand condition (Crollen et al., 2014). Indeed, this theory suggests that number representation is located in areas that originally were dedicated to finger representation (Anderson, 2010; Michaux et al., 2010; Penner-Wilger & Anderson, 2008, 2013). Therefore, even if blind individuals might not explicitly show a reliance on finger-based strategies, according to the redeployment theory they should still show impaired performance due to hand movement interference when counting. Contrary to this theory, blind participants who did not use finger counting were not impaired, highlighting the important role of vision in the development of number-finger associations. The authors suggest that vision is important in fostering the development of finger-based strategies more than the proporioceptive dimension related to fingers.

Of course, because the brain has shown great reorganizational capacities in congenitally blind individuals (Kujala, Alho, & Näätänen, 2000), it can be argued that their cortical mapping might be different and might allow individuals to overcome the predicted interference. It is fair to assume that perhaps interference in congenitally blind individuals occurs only with greater working memory demands, thus impelling processes to rely more on finger-based strategies (Jost et al., 2009). In their studies, Crollen and colleagues tested only enumeration and counting tasks. Therefore, to discount the redeployment view, it would be necessary to assess whether congenitally blind individuals may also be able to perform more complex tasks without finger movement interference. Two pieces of data indicate this possibility (Crollen et al., 2011). First, in the congenitally blind group, some individuals were spontaneously relying on finger-counting strategies, suggesting the possibility that the subsets might differ on other cognitive skills leading one group to use fingers. Second, in their study on children, they showed that performance in blind children decreased for those tasks involving working memory. These results suggest that there could be qualitative differences among congenitally blind individuals in their performance in arithmetic depending on their use of fingers, in line with the functionalist view, and that there might be differences in arithmetic performance between blind and sighted individuals, in accord with both the redeployment and the functionalist views. It would have been informative if Crollen and colleagues tested arithmetic performance in their groups.

Future studies should also more systematically test the relation between the quality of finger-counting strategies and performance in numerical processing and arithmetic problem solving. Based on the Bender and Beller (2012) framework for classifying and understanding the different finger strategies, it would be interesting to specifically investigate the influence of different counting strategies on arithmetic performance, on spatial-numerical associations, and on their interaction. If differences in performance were modulated by the functional properties of the finger-counting strategies, results would be more consistent with the functionalist view. Indeed, the authors supporting the redeployment view argue that current results have not shown a relation between finger-related skills (i.e., agility) and numerical representation, which should be expected based on the functionalist view. The absence of a modulation is in fact assumed in the redeployment view (a relation should only be expected between finger gnosia and numerical representation): finger representation and numerical processing share only low-level common processes. However, our study (Berteletti & Booth, 2015) is the first to demonstrate a relation between the amount of brain activation found in the finger representation (finger gnosia) area and performance in arithmetic. Further studies are necessary to unravel whether this relation causes impaired finger-counting strategies (functionalist view) or if this area being shared by the two representations influences performance on both (redeployment view).

Finally, these accounts also fail to integrate a number of findings that support the role of language in fostering the use of efficient finger-counting strategies. In children 2 to 5 years of age who are still learning to associate number words with their numerical quantity, it has been shown that performance on tasks requiring the reproduction or enumeration of sets of items in a finger condition are not superior to a number words-only condition (Nicoladis, Pika, & Marentette, 2010). These results are inconsistent with an advantage for finger strategies in young children. Moreover, it has been shown that home signers are inaccurate in the use of fingers to represent numbers beyond three and are not able to reproduce a set of elements accurately (Spaepen, Coppola, Spelke, Carey, & Goldin-Meadow, 2011). Similarly, the Mondurukù people from the Amazonian forest who do not possess number words for numbers beyond 5 (Pica et al., 2004) also lack strong finger-counting strategies (see also Butterworth, Reeve, & Reynolds, 2011). However, these authors also report a movie of an older man who claims being able to count beyond 5 (www.unicog.org/docs/MovieDocumentationCountingInMunduruku.htm). In his counting attempt, the man uses his fingers and then toes with hesitation and with loose number-finger associations. For example, for the number 5 he uses both the word corresponding to "all fingers" and "one hand." The authors also report that even though counting on fingers was slow and effortful, his performance appeared to be better than when the same man was asked to count seeds without the support of fingers.

CONCLUSIONS

In this chapter, we have presented evidence suggesting that finger-based strategies and finger representation play a substantial role in the development of numerical processing, arithmetic, and numerical representation. We show that fingers and numerical processing rely on common neural substrates and that the relation between finger representation and arithmetical skills differs depending on the operation. We also present evidence suggesting that finger-counting strategies are culturally transmitted and that visual and verbal abilities may play a role in fostering such strategies. The quality of finger-counting strategies as well as the quality of finger representation have been shown to influence numerical processing, arithmetic performance, and the spatial-numerical association. It remains to be determined if fingers play a mediating role in the mapping between number words and the internal numerical representation, and if qualitative differences in finger strategies also influence arithmetic performance. It would be particularly informative to use the Bender and Beller framework for assessing whether training on finger-counting strategies that differ on total extent of the integers represented, total distinct hand-finger configurations, or the base and subbase system would yield better arithmetic performance, and whether these would also relate differently to performance depending on the operation. Finally, it would be important, specifically for educational purposes, first to understand if an interaction exists between the influence of individual characteristics and finger-based strategies on performance, and second to differentiate the numerical and arithmetic tasks that might rely more on finger-based strategies.

REFERENCES

Alibali, M. W., & DiRusso, A. (1999). The function of gesture in learning to count: More than keeping track. *Cognitive Development*, *14*, 37–56.

Anderson, M. L. (2010). Neural reuse: A fundamental organizational principle of the brain. *The Behavioral and Brain Sciences*, *33*(4), 245–266, discussion 266–313.

Andres, M., Davare, C. A. M., Pesenti, M., Olivier, E., & Seron, X. (2004). Number magnitude and grip aperture interaction. *NeuroReport*, *15*(18), 73–77.

Andres, M., Michaux, N., & Pesenti, M. (2012). Common substrate for mental arithmetic and finger representation in the parietal cortex. *NeuroImage*, *62*(3), 1520–1528.

Andres, M., Ostry, D. J., Nicol, F., & Paus, T. (2008). Time course of number magnitude interference during grasping. *Cortex*, *44*(4), 414–419.

Andres, M., Pelgrims, B., Michaux, N., Olivier, E., & Pesenti, M. (2011). Role of distinct parietal areas in arithmetic: An fMRI-guided TMS study. *NeuroImage*, *54*(4), 3048–3056.

Andres, M., Seron, X., & Olivier, E. (2007). Contribution of hand motor circuits to counting. *Journal of Cognitive Neuroscience*, *19*(4), 563–576.

Arsalidou, M., & Taylor, M. J. (2011). Is 2+2=4? Meta-analyses of brain areas needed for numbers and calculations. *NeuroImage*, *54*(3), 2382–2393.

Ashcraft, M. H. (1982). The development of mental arithmetic: A chronometric approach. *Developmental Review*, *2*, 213–236.

Ashcraft, M. H. (1992). Cognitive arithmetic: A review of data and theory. *Cognition*, *44*, 75–106.

Ashkenazi, S., Black, J. M., Abrams, D.a, Hoeft, F., & Menon, V. (2013). Neurobiological underpinnings of math and reading learning disabilities. *Journal of Learning Disabilities*, *46*(6), 549–569.

Badets, A., Andres, M., Di Luca, S., & Pesenti, M. (2007). Number magnitude potentiates action judgements. *Experimental Brain Research*, *180*(3), 525–534.

Badets, A., & Pesenti, M. (2010). Creating number semantics through finger movement perception. *Cognition*, *115*(1), 46–53.

Badets, A., Pesenti, M., & Olivier, E. (2010). Response-effect compatibility of finger-numeral configurations in arithmetical context. *Quarterly Journal of Experimental Psychology (2006)*, *63*(1), 16–22.

Bangura, A. K. (2012). African Mathematics: From bones to computers. Lanham, MD: University Press of America.

Baroody, A. J. (1987). *Children's mathematical thinking*. New York, NY: Teachers College Press.

Barrouillet, P., & Fayol, M. (1998). From algorithmic computing to direct retrieval: Evidence from number and alphabetic arithmetic in children and adults. *Memory & Cognition*, *26*(2), 355–368.

Barrouillet, P., Mignon, M., & Thevenot, C. (2008). Strategies in subtraction problem solving in children. *Journal of Experimental Child Psychology*, *99*(4), 233–251.

Barrouillet, P., & Lépine, R. (2005). Working memory and children's use of retrieval to solve addition problems. *Journal of Experimental Child Psychology*, *91*(3), 183–204.

Barrouillet, P., Mignon, M., & Thevenot, C. (2008). Strategies in subtraction problem solving in children. *Journal of Experimental Child Psychology*, *99*(4), 233–251.

Barrouillet, P., & Thevenot, C. (2013). On the problem-size effect in small additions: Can we really discard any counting-based account? *Cognition*, *128*(1), 35–44.

Bartelet, D., Ansar, D., Vaessen, A., & Blomert, L. (2014). Cognitive subtypes of mathematics learning difficulties in primary education. *Research in Developmental Disabilities*, *35*(3), 657–670.

Bender, A., & Beller, S. (2011). Fingers as a tool for counting—Naturally fixed or culturally flexible? *Frontiers in Psychology*, *2*, 256.

Bender, A., & Beller, S. (2012). Nature and culture of finger counting: Diversity and representational effects of an embodied cognitive tool. *Cognition*, *124*(2), 156–182.

Berteletti, I., & Booth, J. R. (2015). Perceiving fingers in single-digit arithmetic problems. *Frontiers in Psychology*, *6*, 226. http://dx.doi.org/10.3389/fpsyg.2015.00226.

Binkofski, F., Buccino, G., Posse, S., Seitz, R. J., Rizzolatti, G., & Freund, H.-J. (1999). A fronto-parietal circuit for object manipulation in man: Evidence from an fMRI-study. *European Journal of Neuroscience*, *11*, 3276–3286.

Boets, B., & De Smedt, B. (2010). Single-digit arithmetic in children with dyslexia. *Dyslexia*, *16*, 183–191.

Brozzoli, C., Ishihara, M., Göbel, S. M., Salemme, R., Rossetti, Y., & Farnè, A. (2008). Touch perception reveals the dominance of spatial over digital representation of numbers. *Proceedings of the National Academy of Sciences of the United States of America*, *105*(14), 5644–5648.

Butterworth, B. (1999). What counts: How every brain is hardwired for math. New York, NY: Free Press.

Butterworth, B., Reeve, R., & Reynolds, F. (2011). Using mental representations of space when words are unavailable: Studies of enumeration and arithmetic in indigenous Australia. *Journal of Cross-Cultural Psychology*, *42*(4), 630–638.

Campbell, J. I. D., & Xue, Q. (2001). Cognitive arithmetic across cultures. *Journal of Experimental Psychology: General*, *130*(2), 299–315.

Chinello, A., Cattani, V., Bonfiglioli, C., Dehaene, S., & Piazza, M. (2013). Objects, numbers, fingers, space: Clustering of ventral and dorsal functions in young children and adults. *Developmental Science*, *16*(3), 377–393.

Cohen, L., & Dehaene, S. (2000). Calculating without reading: Unsuspected residual abilities in pure alexia. *Cognitive Neuropsychology*, *17*(6), 563–583.

Cohen, L., Dehaene, S., Chochon, F., Lehéricy, S., & Naccache, L. (2000). Language and calculation within the parietal lobe: A combined cognitive, anatomical and fMRI study. *Neuropsychologia*, *38*(10), 1426–1440.

Costa, A. J., Silva, J. B. L., Chagas, P. P., Krinzinger, H., Lonneman, J., Willmes, K., et al. (2011). A hand full of numbers: A role for offloading in arithmetics learning? *Frontiers in Psychology*, *2*, 368.

Crollen, V., Mahe, R., Collignon, O., & Seron, X. (2011). The role of vision in the development of finger-number interactions: Finger-counting and finger-montring in blind children. *Journal of Experimental Child Psychology*, *109*(4), 525–539.

Crollen, V., NoëL, M.-P., Seron, X., Mahau, P., Lepore, F., & Collignon, O. (2014). Visual experience influences the interactions between fingers and numbers. *Cognition*, *133*(1), 91–96.

Dagenbach, D., & McCloskey, M. (1992). The organization of arithmetic facts in memory: Evidence from a brain-damaged patient. *Brain and Cognition*, *20*(2), 345–366.

De Lange, F. P., Hagoort, P., & Toni, I. (2005). Neural topography and content of movement representations. *Journal of Cognitive Neuroscience*, *17*(1), 97–112.

De Smedt, B., & Boets, B. (2010). Phonological processing and arithmetic fact retrieval: Evidence from developmental dyslexia. *Neuropsychologia*, *48*(14), 3973–3981.

Dehaene, S. (1992). Varieties of numerical abilities. *Cognition*, *44*, 1–42.

Dehaene, S., Bossini, S., & Giraux, P. (1993). The mental representation of parity and number magnitude. *Journal of Experimental Psychology: General*, *122*(3), 371–396.

Dehaene, S., & Cohen, L. (1995). Towards an anatomical and functional model of number processing. *Mathematical Cognition*, *1*, 83–120.

Dehaene, S., & Cohen, L. (1997). Cerebral pathways for calculation: Double dissociation between rote verbal and quantitative knowledge of arithmetic. *Cortex*, *33*(2), 219–250.

Dehaene, S., & Cohen, L. (2007). Cultural recycling of cortical maps. *Neuron*, *56*(2), 384–398.

Dehaene, S., Piazza, M., Pinel, P., & Cohen, L. (2003). Three parietal circuits for number processing. *Cognitive Neuropsychology*, *20*(3), 487–506.

Delazer, M., & Benke, T. (1997). Arithmetic facts without meaning. *Cortex*, *33*, 697–710.

Demir, O. E., Prado, J., & Booth, J. R. (2014). The differential role of verbal and spatial working memory in the neural basis of arithmetic. *Developmental Neuropsychology*, *39*(6), 440–458.

DeStefano, D., & LeFevre, J. (2004). The role of working memory in mental arithmetic. *European Journal of Cognitive Psychology*, *16*(3), 353–386.

Di Luca, S., Granà, A., Semenza, C., Seron, X., & Pesenti, M. (2006). Finger-digit compatibility in Arabic numeral processing. *Quarterly Journal of Experimental Psychology*, *59*(9), 1648–1663.

Di Luca, S., Lefèvre, N., & Pesenti, M. (2010). Place and summation coding for canonical and non-canonical finger numeral representations. *Cognition*, *117*(1), 95–100.

Di Luca, S., & Pesenti, M. (2008). Masked priming effect with canonical finger numeral configurations. *Experimental Brain Research*, *185*(1), 27–39.

Di Luca, S., & Pesenti, M. (2011). Finger numeral representations: More than just another symbolic code. *Frontiers in Psychology*, *2*, 272.

Domahs, F., Krinzinger, H., & Willmes, K. (2008). Mind the gap between both hands: Evidence for internal finger-based number representations in children's mental calculation. *Cortex*, *44*(4), 359–367.

Domahs, F., Moeller, K., Huber, S., Willmes, K., & Nuerk, H.-C. (2010). Embodied numerosity: Implicit hand-based representations influence symbolic number processing across cultures. *Cognition*, *116*(2), 251–266.

Eriksson, K., Bailey, D., & Geary, D. (2010). The grammar of approximating number pairs. *Memory & Cognition*, *38*(3), 333–343.

Fayol, M., Barrouillet, P., & Marinthe, C. (1998). Predicting arithmetical achievement from neuro-psychological performance: A longitudinal study. *Cognition*, *68*(2), B63–B70.

Fayol, M., & Thevenot, C. (2012). The use of procedural knowledge in simple addition and subtraction problems. *Cognition*, *123*(3), 392–403.

Fehr, T., Code, C., & Herrmann, M. (2007). Common brain regions underlying different arithmetic operations as revealed by conjunct fMRI-BOLD activation. *Brain Research*, *1172*, 93–102.

Fias, W., & Fischer, M. H. (2005). Spatial representation of numbers. In J. Campbell (Ed.), *Handbook of Mathematical Cognition*. New York and Hove: Psychology Press.

Fischer, M. H. (2008). Finger counting habits modulate spatial-numerical associations. *Cortex*, *44*(4), 386–392.

Fischer, J.-P. (2010). Numerical performance increased by finger training: A fallacy due to regression toward the mean? *Cortex*, *46*(2), 272–273.

Fischer, M. H., & Brugger, P. (2011). When digits help digits: Spatial-numerical associations point to finger counting as prime example of embodied cognition. *Frontiers in Psychology*, *2*, 260.

Fischer, M. H., Mills, R.a., & Shaki, S. (2010). How to cook a SNARC: Number placement in text rapidly changes spatial-numerical associations. *Brain and Cognition*, *72*(3), 333–336.

Fischer, M. H., Shaki, S., & Cruise, A. (2009). It takes just one word to quash a SNARC. *Experimental Psychology*, *56*(5), 361–366.

Fuson, K., & Kwon, Y. (1992). Koeran children's single-digit addition and subtraction: Numbers structured by ten. *Journal of Research in Mathematics Education*, *23I*(2), 148–165.

Gallistel, C. R., & Gelman, R. (1992). Preverbal and verbal counting and computation. *Cognition*, *44*, 43–74.

Geary, D. C. (1990). A componential analysis of an early learning deficit in mathematics. *Journal of Experimental Child Psychology*, *49*(3), 363–383.

Geary, D. C. (2005). The origin of mind: Evolution of brain, cognition, and general intelligence. Washington, DC: American Psychological Association.

Geary, D. C. (2010). Mathematical disabilities: Reflections on cognitive, neuropsychological, and genetic components. *Learning and Individual Differences*, *20*, 130–133.

Geary, D. C., Bow-Thomas, C., & Yao, Y. (1992). Counting knowledge and skill in cognitive addition: A comparison of normal and mathematically disabled children. *Journal of Experimental Child Psychology*, *54*, 372–391.

Geary, D. C., Brown, S. C., & Samaranayake, V. A. (1991). Cognitive addition: A short longitudinal study of strategy choice and speed-of-processing differences in normal and mathematically disabled children. *Developmental Psychology*, *27*(5), 787–797.

Geary, D. C., & Burlingham-Dubree, M. (1989). External validation of the strategy choice model for addition. *Journal of Experimental Child Psychology*, *47*(2), 175–192.

Geary, D. C., Frensch, P. A., & Wiley, J. D. (1993). Simple and complex mental subtraction: Strategy choice and speed-of-processing differences in younger and older adults. *Psychology and Aging*, *8*(2), 242–256.

Geary, D. C., & Wiley, J. D. (1991). Cognitive addition: Strategy choice and speed-of-processing difference in young and elderly adults. *Psychology and Aging*, *6*(3), 474–483.

Gersten, R., Jordan, N. C., & Flojo, J. R. (2005). Early identification and interventions for students with mathematics difficulties. *Journal of Learning Disabilities*, *38*(4), 293–304.

Gerstmann, J. (1940). Syndrome of finger agnosia, disorientation from right and left, agraphia and acalculia. *Neurology and Psychiatry*, *44*(2), 398–408.

Gelman, R., & Gallistel, C. R. (1978). *The young child's understanding of number: A window on early cognitive development*. Cambridge, MA: Harvard University Press.

Gevers, W., & Lammertyn, J. (2005). The hunt for SNARC. *Psychology Science*, *47*(1), 10–21.

Ginsburg, H. (1977). Children's arithmetic: The learning process. New York, NY: Van Nostrand.

Gracia-Bafalluy, M., & Noël, M.-P. (2008). Does finger training increase young children's numerical performance? *Cortex*, *44*(4), 368–375.

Graham, T.a. (1999). The role of gesture in children's learning to count. *Journal of Experimental Child Psychology*, *74*(4), 333–355.

Groen, G. J., & Parkman, J. M. (1972). A chronometric analysis of simple addition. *Psychological Review*, *79*(4), 329–343.

Haaland, K. Y., Elsinger, C. L., Mayer, A. R., Durgerian, S., & Rao, S. M. (2004). Motor sequence complexity and performing hand produce differential patterns of hemispheric lateralization. *Journal of Cognitive Neuroscience*, *16*(4), 621–636.

Hanakawa, T., Honda, M., Sawamoto, N., Okada, T., Yonekura, Y., Fukuyama, H., et al. (2002). The role of rostral Brodmann area 6 in mental-operation tasks: An integrative neuroimaging approach. *Cerebral Cortex*, *12*(11), 1157–1170.

Hubbard, E. M., Piazza, M., Pinel, P., & Dehaene, S. (2005). Interactions between number and space in parietal cortex. *Nature Reviews Neuroscience*, *6*(6), 435–448.

Ifrah, G. (2000). The universal history of numbers: From prehistory to the invention of the computer (D. Vellos, E. F. Harding, S. Wood, & I. Monk, Trans.). New York, NY: John Wiley and Sons.

Ilg, F., & Ames, L. B. (1951). Developmental Trends in Arithmetic. *The Pedagogical Seminary and Journal of Genetic Psychology*, *79*(1), 3–28.

Imbo, I., Vandierendonck, A., & Fias, W. (2011). Passive hand movements disrupt adults' counting strategies. *Frontiers in Psychology*, *2*, 201.

Jordan, N. C., Hanich, L. B., & Kaplan, D. (2003). Arithmetic fact mastery in young children: A longitudinal investigation. *Journal of Experimental Child Psychology*, *85*(2), 103–119.

Jordan, N. C., & Montani, T. O. (1997). Cognitive arithmetic and problem solving: A comparison of children with specific and general mathematics difficutlies. *Journal of Learning Disabilities*, *30*(6), 624–634.

Jost, K., Khader, P., Burke, M., Bien, S., & Rösler, F. (2009). Dissociating the solution processes of small, large, and zero multiplications by means of fMRI. *NeuroImage*, *46*(1), 308–318.

Kansaku, K., Carver, B., Johnson, A., Matsuda, K., Sadato, N., & Hallett, M. (2007). The role of the human ventral premotor cortex in counting successive stimuli. *Experimental Brain Research*, *178*(3), 339–350.

Klein, E., Moeller, K., Willmes, K., Nuerk, H.-C., & Domahs, F. (2011). The influence of implicit hand-based representations on mental arithmetic. *Frontiers in Psychology*, *2*, 197.

Knops, A., & Willmes, K. (2014). Numerical ordering and symbolic arithmetic share frontal and parietal circuits in the right hemisphere. *NeuroImage*, *84*, 786–795.

Kolpas, S. J. (2002). Let your fingers do the multiplying. *The Mathematics Teacher*, *95*(4), 246–251.

Krinzinger, H., Koten, J. W., Horoufchin, H., Kohn, N., Arndt, D., Sahr, K., et al. (2011). The role of finger representations and saccades for number processing: An FMRI study in children. *Frontiers in Psychology*, *2*, 373.

Kujala, T., Alho, K., & Näätänen, R. (2000). Cross-modal reorganization of human. *Trends in Neurosciences*, *23*(3), 115–120.

Lee, K. M. (2000). Cortical areas differentially involved in multiplication and subtraction: A functional magnetic resonance imaging study and correlation with a case of selective acalculia. *Annals of Neurology*, *48*(4), 657–661.

Lee, K.-M., & Kang, S.-Y. (2002). Arithmetic operation and working memory: Differential suppression in dual tasks. *Cognition*, *83*(3), B63–B68.

Lee, J., Kotsopoulos, D., Tumber, A., & Makosz, S. (2014). Gesturing about number sense. *Journal of Early Childhood Research*, 1–17. http://dx.doi.org/10.1177/1476718X13510914.

LeFevre, J., Bisanz, J., Daley, K. E., Buffone, L., Greenham, S. L., & Sadesky, G. S. (1996). Multiple routes to solution of single-digit multiplication problems. *Journal of Experimental Psychology: General*, *125*(3), 284–306.

LeFevre, J., DeStefano, D., Penner-Wilger, M., & Daley, K. E. (2006). Selection of procedures in mental subtraction. *Canadian Journal of Experimental Psychology*, *60*(3), 209–220.

LeFevre, J., Sadesky, G. S., & Bisanz, J. (1996). Selection of procedures in mental addition: Reassessing the problem size effect in adults. *Journal of Experimental Psychology: Learning, Memory, and Cognition*, *22*(1), 216–230.

Lemer, C., Dehaene, S., Spelke, E. S., & Cohen, L. (2003). Approximate quantities and exact number words: Dissociable systems. *Neuropsychologia*, *41*(14), 1942–1958.

Lindemann, O., Alipour, A., & Fischer, M. H. (2011). Finger counting habits in middle eastern and western individuals: An online survey. *Journal of Cross-Cultural Psychology*, *42*(4), 566–578.

Marcus, G. F. (2004). The birth of the mind: How a tiny number of genes creates the complexities of human thought. New York: Basic Books.

Matejko, A. A., & Ansari, D. (2015). Drawing connections between white matter and numerical and mathematical cognition: A literature review. *Neuroscience and Behavioral Reviews*, *48*, 35–52.

McCloskey, M. (1992). Cognitive mechanisms in numerical processing: Evidence from acquired dyscalculia. *Cognition*, *44*(1–2), 107–157.

McNeil, J. E., & Warrington, E. K. (1994). A dissociation between addition and subtraction with written calculation. *Neuropsychologia*, *32*(6), 717–728.

Michaux, N., Masson, N., Pesenti, M., & Andres, M. (2013). Selective interference of finger movements on basic addition and subtraction problem solving. *Experimental Psychology*, *60*(3), 197–205.

Michaux, N., Pesenti, M., Badets, A., Di Luca, S., & Andres, M. (2010). Let us redeploy attention to sensorimotor experience. *Behavioral and Brain Sciences*, *33*(4), 283–284.

Mihulowicz, U., Willmes, K., Karnath, H.-O., & Klein, E. (2014). Single-digit arithmetic processing—Anatomical evidence from statistical voxel-based lesion analysis. *Frontiers in Human Neuroscience*, *8*, 1–10.

Newman, S. D., & Soylu, F. (2014). The impact of finger counting habits on arithmetic in adults and children. *Psychological Research*, *78*, 549–556. http://dx.doi.org/10.1007/s00426-013-0505-9.

Nicoladis, E., Pika, S., & Marentette, P. (2010). Are number gestures easier than number words for preschoolers? *Cognitive Development*, *25*(3), 247–261.

Noël, M.-P. (2005). Finger gnosia: A predictor of numerical abilities in children? *Child Neuropsychology: A Journal on Normal and Abnormal Development in Childhood and Adolescence*, *11*(5), 413–430.

Opfer, J. E., & Furlong, E. E. (2011). How numbers bias preschoolers' spatial search. *Journal of Cross-Cultural Psychology*, *42*(4), 682–695.

Opfer, J. E., Thompson, C. A., & Furlong, E. E. (2010). Early development of spatial-numeric associations: Evidence from spatial and quantitative performance of preschoolers. *Developmental Science*, *13*(5), 761–771.

Penner-Wilger, M., & Anderson, M. L. (2008). An alternative view of the relation between finger gnosis and math ability: Redeployment of finger representations for the representation of number. In: *Proceedings of the 30th Annual Meeting of the Cognitive Science Society. Austin, TX* (pp. 1647–1652).

Penner-Wilger, M., & Anderson, M. L. (2013). The relation between finger gnosis and mathematical ability: Why redeployment of neural circuits best explains the finding. *Frontiers in Psychology*, *4*, 877.

Pesenti, M., Seron, X., & Van Der Linden, M. (1994). Selective impairment as evidence for mental organization of arithmetical facts - Bb, a case of preserved subtraction. *Cortex*, *30*(4), 661–671.

Pesenti, M., Thioux, M., Seron, X., & De Volder, A. (2000). Neuroanatomical substrates of arabic number processing, numerical comparison, and simple addition: A PET study. *Journal of Cognitive Neuroscience*, *12*(3), 461–479.

Piazza, M., Izard, V., Pinel, P., Le Bihan, D., & Dehaene, S. (2004). Tuning curves for approximate numerosity in the human intraparietal sulcus. *Neuron*, *44*(3), 547–555.

Pica, P., Lemer, C., Izard, V., & Dehaene, S. (2004). Exact and approximate arithmetic in an Amazonian indigene group. *Science*, *306*(5695), 499–503.

Pika, S., Nicoladis, E., & Marentette, P. (2009). How to order a beer: Cultural differences in the use of conventional gestures for numbers. *Journal of Cross-Cultural Psychology*, *40*(1), 70–80.

Prado, J., Mutreja, R., & Booth, J. R. (2014). Developmental dissociation in the neural responses to simple multiplication and subtraction problems. *Developmental Science*, *17*(4), 537–552.

Prado, J., Mutreja, R., Zhang, H., Mehta, R., Desroches, A. S., Minas, J. E., et al. (2011). Distinct representations of subtraction and multiplication in the nueral system for numerosity and language. *Human Brain Mapping*, *32*(11), 1932–1947.

Reeve, R., & Humberstone, J. (2011). Five- to 7-year-olds' finger gnosia and calculation abilities. *Frontiers in Psychology*, *2*, 359.

Roussel, J.-L., Fayol, M., & Barrouillet, P. (2002). Procedural vs. direct retrieval strategies in arithmetic: A comparison between additive and multiplicative problem solving. *European Journal of Cognitive Psychology*, *14*(1), 61–104.

Rugani, R., Kelly, D. M., Szelest, I., Regolin, L., & Vallortigara, G. (2010). Is it only humans that count from left to right? *Biology Letters*, *6*(3), 290–292.

Sato, M., Cattaneo, L., Rizzolatti, G., & Gallese, V. (2007). Numbers within our hands: Modulation of corticospinal excitability of hand muscles during numerical judgment. *Journal of Cognitive Neuroscience*, *19*(4), 684–693.

Sato, M., & Lalain, M. (2008). On the relationship between handedness and hand-digit mapping in finger counting. *Cortex*, *44*(4), 393–399.

Seyler, D. J., Kirk, E. P., & Ashcraft, M. H. (2003). Elementary subtraction. *Journal of Experimental Psychology Learning, Memory, and Cognition*, *29*(6), 1339–1352.

Shaki, S., & Fischer, M. H. (2008). Reading space into numbers: A cross-linguistic comparison of the SNARC effect. *Cognition*, *108*(2), 590–599.

Shaki, S., Fischer, M. H., & Petrusic, W. M. (2009). Reading habits for both words and numbers contribute to the SNARC effect. *Psychonomic Bulletin & Review*, *16*(2), 328–331.

Siegler, R. S. (1987). The perils of averaging data over strategies: An example from children's addition. *Journal of Experimental Psychology: General*, *116*(3), 250–264.

Siegler, R. S., & Shipley, C. (1987). The role of learning in children's strategy choices. In L. S. Liben (Ed.), *Developmeni and learning: Conflict or congruence*. Hillsdale, NJ: Erlbaum.

Siegler, R. S. (1991). Strategy choice and strategy discovery. *Learning and Instruction*, *1*(1), 89–102.

Siegler, R. S., & Jenkins, E. (1989). How children discover new strategies. Hillsdale, NJ: Erlbaum.

Siegler, R. S., & Shrager, J. (1984). Strategy choices in addition: How do children know what to do? In C. Sophian (Ed.), Origins of cognitive skills (pp. 229–293). Hillsdale, NJ: Erlbaum.

Simon, O., Kherif, F., Flandin, G., Poline, J.-B., Rivière, D., Mangin, J.-F., et al. (2004). Automatized clustering and functional geometry of human parietofrontal networks for language, space, and number. *NeuroImage*, *23*(3), 1192–1202.

Simon, O., Mangin, J.-F., Cohen, L., Le Bihan, D., & Dehaene, S. (2002). Topographical layout of hand, eye, calculation, and language-related areas in the human parietal lobe. *Neuron*, *33*(3), 475–487.

Spaepen, E., Coppola, M., Spelke, E. S., Carey, S., & Goldin-Meadow, S. (2011). Number without a language model. *Proceedings of the National Academy of Sciences of the United States of America*, *108*(8), 3163–3168.

Strauss, A., & Werner, H. (1938). Deficiency in the finger schema in relation to arithmetic disability (finger agnosia and acalculia). *The American Journal of Orthopsychiatry*, *8*(4), 719–725.

Starkey, P., & Gelman, R. (1982). The development of addition and subtraction abilities prior to formal schooling in arithmetic. In T. P. Carpenter, J. M. Moser, & T. A. Romberg (Eds.), *Addition and subtraction: A cognitive perspective*. Hillsdale, NJ: Erlbaum.

Svenson, O., Hedenborg, M., & Lingman, L. (1976). On children's heuristics for solving simple additions. *Scandinavian Journal of Educational Research*, *20*(1), 161–173.

Thevenot, C., & Barrouillet, P. (2006). Encoding numbers: Behavioral evidence for processing-specific representations. *Memory & Cognition*, *34*(4), 938–948.

Tschentscher, N., Hauk, O., Fischer, M. H., & Pulvermüller, F. (2012). You can count on the motor cortex: Finger counting habits modulate motor cortex activation evoked by numbers. *NeuroImage*, *59*(4), 3139–3148.

Van Harskamp, N. J., & Cipolotti, L. (2001). Selective impairments for addition, subtraction and multiplication. implications for the organisation of arithmetical facts. *Cortex*, *37*(3), 363–388.

Vingerhoets, G. (2014). Contribution of the posterior parietal cortex in reaching, grasping, and using objects and tools. *Frontiers in Psychology*, *5*, 151.

Von Aster, M. (2000). Developmental cognitive neuropsychology of number processing and calculation: Varieties of developmental dyscalculia. *European Child & Adolescent Psychiatry*, *9*, S41–S57.

Wasner, M., Moeller, K., Fischer, M. H., & Nuerk, H.-C. (2015). Related but not the same: Ordinality, cardinality and 1-to-1 correspondence in finger-based numerical representations. *Journal of Cognitive Psychology*, *27*(4), 426–441.

Wood, G., Willmes, K., Nuerk, H.-C., & Fischer, M. H. (2008). On the cognitive link between space and number: A meta-analysis of the SNARC effect. *Psychology Science Quarterly*, *50*(4), 489–525.

Woods, S. S., Resnick, L. B., & Groen, G. J. (1975). An experimental test of five process models for subtraction. *Journal of Educational Psychology*, *67*(1), 17–21.

Yu, X., Chen, C., Pu, S., Wu, C., Li, Y., Jiang, T., et al. (2011). Dissociation of subtraction and multiplication in the right parietal cortex: Evidence from intraoperative cortical electrostimulation. *Neuropsychologia*, *49*(10), 2889–2895.

Zhou, X., Chen, C., Zang, Y., Dong, Q., Chen, C., Qiao, S., et al. (2007). Dissociated brain organization for single-digit addition and multiplication. *NeuroImage*, *35*(2), 871–880.

Chapter 6

Neurocognitive Architectures and the Nonsymbolic Foundations of Fractions Understanding

Mark Rose Lewis, Percival G. Matthews, Edward M. Hubbard
Department of Educational Psychology, University of Wisconsin, Madison, Wisconsin, USA

INTRODUCTION

Understanding fractions is critically important for the development of mathematical competence (National Mathematics Advisory Panel, 2008; Siegler, Fazio, Bailey, & Zhou, 2013). Fractions are a key component of elementary and middle-school mathematics curricula and serve as a foundational concept for algebra and higher mathematics (Booth & Newton, 2012). For instance, 10-year-olds' fraction knowledge predicts later high school algebra skills and overall math achievement (Siegler et al., 2012). Unfortunately, both children and adults—including teachers and highly educated professionals—struggle to understand fractions (Lipkus, Samsa, & Rimer, 2001; National Mathematics Advisory Panel, 2008; Newton, 2008; Post, Harel, Behr, & Lesh, 1991; Reyna & Brainerd, 2008; Stigler, Givvin, & Thompson, 2010).

Research in mathematics education and cognitive psychology has identified several common difficulties faced by learners regarding fractions. One of the most important is a difficulty accessing the holistic magnitudes of symbolic fractions (Bonato, Fabbri, Umiltà, & Zorzi, 2007; Kallai & Tzelgov, 2009, 2012; Stafylidou & Vosniadou, 2004)—a difficulty that even expert mathematicians face (Obersteiner, Van Dooren, Van Hoof, & Verschaffel, 2013). Understanding fraction magnitudes is an important piece of conceptual knowledge that predicts accuracy on fraction computation, estimation, and word problems (Hecht, 1998; Hecht, Close, & Santisi, 2003; Hecht & Vagi, 2010; Siegler et al., 2013; Siegler & Pyke, 2013). Perhaps more importantly,

Development of Mathematical Cognition. http://dx.doi.org/10.1016/B978-0-12-801871-2.00006-X

understanding the magnitude of symbolic fractions predicts general mathematics achievement (Bailey, Hoard, Nugent, & Geary, 2012; Siegler & Pyke, 2013; Siegler, Thompson, & Schneider, 2011). Unfortunately, misunderstandings of fraction magnitude are all too common. For example, Mack (1990) found that typical 6th graders often claimed that 1/8 is greater than 1/6. Similarly, when a nationally representative sample of 8th graders was asked whether 12/13 + 7/8 was closest to 1, 2, 19, or 21, more chose 19 and 21 than 2 (Carpenter, Corbitt, Kepner, Lindquist, & Reys, 1981). Outside the classroom, the failure to understand fraction magnitude even affects fast-food choices, with as many as half the participants in a consumer focus group reporting that they thought the A&W 1/3 pound burger was smaller than McDonald's famous "quarter pounder." They therefore chose the smaller McDonald's burger, even though they preferred the taste of the A&W burger (Taubman, 2007, p. 62–63). This real-world example illustrates a clear failure to recognize that the magnitude of a fraction emerges from the relation between its two components (i.e., the numerator and denominator) rather than from the magnitudes of the individual components themselves.

A second common misunderstanding, known as "the whole-number bias," is closely linked to the first. This bias is so-named because students' struggles with fractions often reflect the misapplication of concepts and procedures that are appropriate for whole numbers but inappropriate for fractions (see Ni & Zhou, 2005 for a review). This bias is reflected not only in the kinds of computational errors people make, but also in their verbal reports. For example, one child in Mack's (1990) study described fraction addition as follows: "Well, you go across. You add the top numbers together and the bottom numbers together" (p. 23).

A third common misunderstanding is the failure to recognize that fractions have different properties than whole numbers (Siegler et al., 2013; Vamvakoussi & Vosniadou, 2010). To name a few of these properties: (a) fractions, unlike whole numbers, have no direct successor; (b) there is an infinite number of fractions between any two numbers; and (c) multiplying by a positive fraction sometimes produces a result that is smaller than the original operands, while dividing sometimes produces a result that is larger, depending on the magnitude of the multiplier or divisor.

Each of these misunderstandings can pose serious impediments for everything from acquiring a sense of fraction magnitude to acquiring a robust understanding of how arithmetic with fractions works. Designing effective instructional approaches to rectify such common misunderstandings—or to prevent them from arising—may be highly effective for promoting robust fraction knowledge. In this chapter, we suggest that research into the perceptually based capacities of the human cognitive architecture for nonsymbolic ratios, and teaching methods that build on these neurocognitive systems, may help provide a pathway for the amelioration and prevention of these common problems.

FUNDAMENTAL LIMITATIONS OF THE HUMAN COGNITIVE ARCHITECTURE

The pervasive problems with learning fractions have led some researchers to claim that fractions, unlike whole numbers, are largely incompatible with the human cognitive architecture (e.g., Bonato et al., 2007; Dehaene, 1997/2011; Feigenson, Dehaene, & Spelke, 2004; Gallistel & Gelman, 1992; Geary, 2007; Gelman & Williams, 1998). According to these "innate constraint" accounts (Ni & Zhou, 2005), whole-number symbols and concepts are relatively easy for most children to learn because they are supported by core neurocognitive representational systems with long phylogenetic histories (Dehaene & Cohen, 2007; Feigenson et al., 2004; Piazza, 2010). One proposed core system is the approximate number system (ANS). The ANS, situated in the intraparietal sulcus (IPS), endows humans—and indeed many nonhuman animals—with the ability to represent the approximate number of discrete elements in a set (Brannon & Roitman, 2003; Dehaene, Dehaene-Lambertz, & Cohen, 1998; Piazza, 2010; see Volume 1, Geary et al., 2015). It is widely hypothesized that early learning of whole-number concepts builds upon this system (e.g., Dehaene, 1997/2011; Feigenson et al., 2004; Nieder, 2005; Piazza, 2010; but see Ansari, Chapter 2). Innate constraints theorists go on to argue that the ANS—although naturally suited to whole numbers—is poorly suited for supporting the representation, conceptualization, and knowledge of fraction magnitudes.

In contrast to this pessimistic view, emerging evidence suggests that the human cognitive architecture may contain systems ideally suited for supporting fraction learning. In fact, there may exist neurocognitive startup tools (Piazza, 2010) that can be exploited to cultivate a robust understanding of key fraction concepts, like magnitude, and to help decrease the prevalence of common difficulties. On our view, many misunderstandings of fractions do not arise from a fundamental incompatibility between fractions concepts and the human cognitive architecture, but rather because current teaching methods do not effectively recruit and leverage our existing perceptual abilities. In particular, similar to Siegler et al. (2011, 2013), we suggest that teaching methods that fail to build an understanding of fractions as magnitudes lead to the conceptual misunderstandings noted above, and to rote (mis)application of procedures for manipulating fractions. Our account complements that of Siegler and colleagues by suggesting that nonsymbolic ratio processing might be a critical neurocognitive primitive for understanding fractions as magnitudes and by exploring how these primitives are linked to systems for symbolic fraction understanding.

In this chapter, we integrate the evidence currently scattered across neuroscience, psychology, and education research to lay out an agenda for developing an educational neuroscience framework for a *ratio processing system* (RPS, similar to the ANS) that may provide a nonsymbolic foundation for fractions understanding. This account may eventually deepen our understanding of this fundamental component of mathematical cognition and inform educational practice.

A COMPETING VIEW: THE RATIO PROCESSING SYSTEM

A growing body of evidence demonstrates that humans and nonhuman primates possess a ratio processing system (RPS), a set of neurocognitive architectures that support the representation and processing of nonsymbolic ratios. Indeed, sensitivity to the magnitudes of nonsymbolic ratios has been demonstrated in nonhuman primates (Vallentin & Nieder, 2008, 2010; Woodruff & Premack, 1981), preverbal infants (McCrink & Wynn, 2007), elementary school-age children (Boyer, Levine, & Huttenlocher, 2008; Duffy, Huttenlocher, & Levine, 2005; Meert, Grégoire, Seron, & Noël, 2013; Sophian, 2000; Spinillo & Bryant, 1999), typically developing adults (Hollands & Dyre, 2000; Meert, Grégoire, Seron, & Noël, 2011; Stevens & Galanter, 1957), and individuals with limited number vocabularies and formal arithmetic skills (McCrink, Spelke, Dehaene, & Pica, 2013). This widespread competence—even among innumerate subjects and nonhuman animals—demonstrates that these architectures are present even in the absence of formal education and likely predate the emergence of *Homo sapiens* (Jacob, Vallentin, & Nieder, 2012).

Although symbolic fractions are not usually introduced formally until grades 3 and 4 (National Governors Association Center for Best Practices, & Council of Chief State School Officers, 2010), children far younger than this demonstrate some understanding of nonsymbolic ratios (Jeong, Levine, & Huttenlocher, 2007; Sophian, 2000; Spinillo & Bryant, 1991, 1999). McCrink & Wynn (McCrink & Wynn, 2007) demonstrated that 6-month-old infants who were habituated to specific nonsymbolic ratios (e.g., a 2:1 ratio composed of yellow Pacmen intermixed with blue pellets) subsequently looked longer at novel ratio stimuli that differed by a factor of two (e.g., a 4:1 ratio). By 4 years of age, children can order pictures of part-whole ratios based on magnitude (Goswami, 1995) and perform above chance on tasks that require the addition and subtraction of nonsymbolic part-whole ratios (Mix, Levine, & Huttenlocher, 1999) even though much older children struggle to perform analogous tasks with symbolic fractions (Mack, 1990). Moreover, this competence with nonsymbolic ratios seems to be flexible and abstract, allowing children to draw proportional analogies even when materials are visually dissimilar (e.g., drawing analogies between pizzas and chocolates and glasses of lemonade) (Goswami, 1989; Singer-Freeman & Goswami, 2001).

Recent findings suggest this early sensitivity to nonsymbolic ratios may be supported by dedicated fronto-parietal circuitry. Neuroimaging studies of adult humans and single-unit physiology studies of nonhuman primates have revealed a fronto-parietal network that is sensitive to the holistic magnitudes of nonsymbolic ratios (see Figure 1). Evidence using fMRI-adaptation methods (Grill-Spector, Henson, & Martin, 2006; Grill-Spector & Malach, 2001; Krekelberg, Boynton, & van Wezel, 2006) with humans provides indirect evidence of neuronal tuning to specific ratios (Jacob & Nieder, 2009b). When humans are repeatedly presented with the same nonsymbolic ratio (composed of

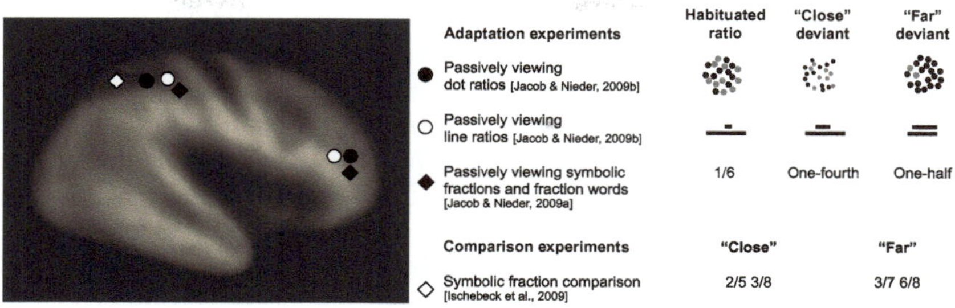

FIGURE 1 Approximate locations of brain areas showing distance-dependent activation to the holistic magnitudes of nonsymbolic ratios and symbolic fractions. In adaptation experiments, the repeated presentation of the same ratio (e.g., 1:2 instantiated by varying sets of black and gray dots) leads to neural adaptation reflected by reduced BOLD signals in these regions. The subsequent presentation of a new ratio leads to signal recovery, which is dependent on the distance between the holistic magnitudes of the adapted and deviant ratio—with stronger BOLD recovery for ratios that are more distant from the adapted ratio. In the one study of symbolic fraction comparison, right IPS showed increased activation when participants were comparing fraction pairs that were closer together. Brain image from CARET (Van Essen, 2005; Van Essen et al., 2001). Example of nonsymbolic ratios adapted from (Jacob & Nieder, 2009a, 2009b).

dots or lines), populations of neurons in the prefrontal cortex and parietal cortex near the IPS adapt to that ratio, resulting in lower blood oxygen level-dependent (BOLD) signals. Presentation of a novel ratio activates a different population of neurons, leading to recovery from adaptation, reflected in stronger BOLD signals. The strength of this recovery is dependent on the holistic distance between the novel ratio and the adapted ratio (see values shown in Figure 1). Similarly, direct recordings from rhesus monkeys (*Macaca mulatta*) trained to perform delayed match-to-sample tasks with nonsymbolic line ratios have shown that some neurons in monkey prefrontal and parietal regions are selectively tuned to specific visuospatial ratios (Vallentin & Nieder, 2010). Within this network, subpopulations of neurons show classical distance-dependent responses to particular ratios. For example, a population of neurons tuned to 3/5 will activate strongly in response to 3/5, less strongly for close but nonidentical ratios (e.g., 4/7 or 5/8), and weakly for more distant ratios (e.g., 1/4 or 8/9). Notably, this tuning is not sensitive to the absolute sizes of components used to instantiate a ratio but instead tracks the holistic magnitude corresponding to the relationship between two given components.

HOW THE RPS MAY INFLUENCE FRACTION LEARNING

The findings reviewed earlier suggest that the RPS may exist as a phylogenetically ancient system that processes ratios of objects, complementary to the ANS with its role in supporting whole-number processing. Hence, neurocognitive architectures tuned to the magnitudes of nonsymbolic ratios—such as the relative lengths of two lines or the relative areas of two figures—are present even before learners receive fraction instruction. On this account, children are equipped with cognitive architectures that support ratio magnitude concepts prior to formal education in the same way that their innate ability to estimate approximate numerosities (sets of countable objects) equips them to learn about whole numbers (e.g., Nieder & Dehaene, 2009). We further propose that leveraging perceptual sensitivity to nonsymbolic ratio magnitudes can play a major role in fostering the developing understanding of symbolic fractions as magnitudes, which in turn will improve both conceptual and procedural fraction processing. We argue that making explicit connections between symbolic fractions and RPS representations of nonsymbolic ratio magnitude may prove to be a key step for promoting successful fraction learning.

Figure 2 presents our hypothesized model of how the RPS may influence learners' understanding of fractions. Within this model, we propose that guided experience with symbolic fractions in concert with nonsymbolic instantiations—both in and out of school—can help generate links between analog, RPS-based representations of magnitudes and fraction symbols, whether they be verbal labels such as "one-third" and "one-half" or orthographic symbols such as "2/3" and "3/4." In this way, the RPS can help provide an intuitive grounding for understanding fraction symbols as representing magnitudes. We hypothesize that the precision of these links is moderated by a learner's RPS acuity: higher acuity provides a more precise representation of magnitude, furnishing a more reliable grounding relative to the representations provided by lower acuity.

Note that according to this hypothetical model, these mechanisms should operate even when instruction does not explicitly attempt to capitalize on RPS abilities, albeit suboptimally. Visuospatial representations such as pies, number lines, and counters are used extensively to teach about fractions, typically in the context of a pedagogy that emphasizes both counting and partitioning. Even

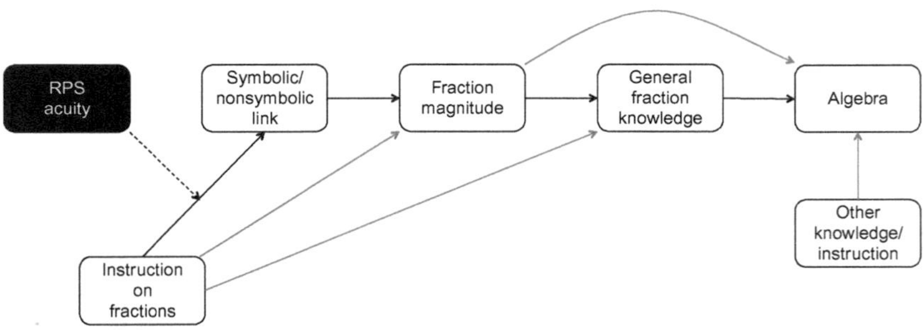

FIGURE 2 A conceptual model of how the RPS might influence fraction learning and subsequent algebra achievement. Instruction on fractions generates links between symbols and representations of nonsymbolic ratio magnitude. The precision of these links is moderated by a learner's nonsymbolic ratio acuity. These links form an important basis of a learner's understanding of the holistic magnitude of fractions, a key understanding that promotes the acquisition of procedural and conceptual fraction knowledge. In turn, general fraction knowledge supports algebra learning and achievement.

though counting and partitioning elicit whole-number concepts and procedures, RPS networks—which antedate formal educational experiences—still encode the holistic ratio magnitudes of these visuospatial representations. In this way, educational experiences continue to build symbol-to-percept links in an implicit fashion even when the RPS is not an explicit focus. Over time, these links strengthen, providing intuitive knowledge of fraction magnitudes, precisely the sort of magnitude knowledge that Siegler et al. (2011) have shown to be predictive of fractions arithmetic proficiency and overall scores on mathematics achievement tests. However, when such instruction focuses too early on strategies that involve partitioning and counting, learners may activate incompatible ANS representations and whole-number schemas, leading to less effective learning. We argue that instruction that seeks to optimally leverage the continuous representations of nonsymbolic ratio magnitude conferred by the RPS can make the symbol-percept links stronger and at earlier points in development.

Note that our model does not assume that this sort of perceptual training alone will result in developing fractions mastery. It is important to recall that it takes years for young children to gain explicit knowledge of the cardinal values of natural numbers even after memorizing the count sequence (e.g., Briars & Siegler, 1984; Le Corre & Carey, 2007, Piazza, 2010). Given the inherent complexity of fractions concepts, our hypothesis should not be taken to support massive and discontinuous reorganization of knowledge structures; instead, we hypothesize that symbol-percept links can form an important basis of a learner's understanding of the holistic magnitude of fractions. Grounding the meaning of fractions symbols in intuitively accessible representations of magnitude can help provide a key understanding that promotes the acquisition of procedural and conceptual fraction knowledge. Because fraction knowledge has been found to be a critical gatekeeper of subsequent algebra achievement (Booth & Newton, 2012; Siegler et al., 2012), this model also suggests that the quality of these links and the fraction understanding they support could have effects on students' algebra learning downstream.

Three key predictions flow from the RPS model. First, the model suggests that individual differences in the functioning of nonsymbolic ratio circuits should predict individual differences in symbolic fraction competence, similar to studies demonstrating that individual differences in the acuity of the ANS predict mathematics achievement (Halberda, Mazzocco, & Feigenson, 2008; Libertus, Feigenson, & Halberda, 2011; Mazzocco, Feigenson, & Halberda, 2011b; but see De Smedt, Noël, Gilmore, & Ansari, 2013) and dyscalculia (Mazzocco, Feigenson, & Halberda, 2011a; Mussolin, Mejias, & Noël, 2010; Piazza et al., 2010; Price, Holloway, Räsänen, Vesterinen, & Ansari, 2007). However, even when ANS acuity correlates with mathematical skills, it does so less strongly than do symbolic number skills (e.g., Chu, vanMarle, & Geary, 2013; Lyons, Price, Vaessen, Blomert, & Ansari, 2014), suggesting that symbolic representations and the strength of the links between the ANS and symbolic representations are important for predicting mathematical ability (see De Smedt et al., 2013 for a review). Similarly, our model focuses not only on RPS acuity, but also on the strength of the links between the RPS and symbolic fractions. To date no studies have explored whether individual differences in ratio processing acuity can predict either typical mathematical achievement or a learning disability.

Second, the model suggests that the strength and accuracy of the symbolic-nonsymbolic links that learners generate are influenced by education/experience and that the quality of these links may be amenable to targeted interventions. One reason that students may not be generating strong links between symbolic and intuitive nonsymbolic representations of fraction magnitude is that most fraction curricula place an emphasis on activities such as partitioning that may activate incompatible ANS representations and whole-number schemas. One way to rectify this situation and to improve learners' links may be to exploit perceptual learning paradigms (see Figure 5) to help learners associate fraction symbols with relational magnitudes rather than with discrete numerosities.

Finally, at the neural level, this model suggests that RPS circuits may be involved in the processing of both nonsymbolic ratios and symbolic fractions for adults. A handful of neuroimaging studies suggest that the same fronto-parietal cortical networks implicated in the representation of nonsymbolic ratios are also involved in representing and processing symbolic fractions. Fronto-parietal activation patterns are observed during symbolic fraction addition and subtraction (Schmithorst & Brown, 2004). Even simple comparison of symbolic fractions activates a network of fronto-parietal areas, with distance-dependent activation in a right IPS area (Ischebeck, Schocke, & Delazer, 2009) similar to that activated during nonsymbolic ratio tasks. Finally, repeated presentation of symbolic fractions (Arabic numerals and fraction words) leads to adaptation and distance-dependent recovery in frontal and parietal regions (Jacob & Nieder, 2009a), showing that these networks are recruited even in the absence of specific task demands. Despite this tantalizing evidence, no studies have directly investigated whether the representation and processing of symbolic fractions directly engage nonsymbolic RPS circuitry. For example, when adults represent the symbolic fraction 1/4, does this engage populations of neurons that are tuned to a nonsymbolic ratio of equivalent value? Of course, this also means that no studies have looked at this from a developmental perspective. That is, even if 1/4 engages RPS circuitry, how does this engagement change with age and experience?

EMERGING BEHAVIORAL AND NEUROIMAGING EVIDENCE FOR RPS MODEL PREDICTIONS

We are currently investigating several key predictions of the RPS model using a combination of behavioral methods, targeted training, and fMRI. In one strand of research, we are investigating whether individual differences in nonsymbolic ratio acuity predict individual differences in symbolic fraction knowledge and algebra achievement. In a second strand we are investigating whether the quality of learners' symbolic-to-nonsymbolic links is amenable to intervention, and whether improving these links will improve performance on a measure of fraction knowledge. In a third strand, we are using an fMRI adaptation paradigm (see Figure 1 above and Study 3 to follow) to test whether adults' representations of symbolic fractions recruit the same circuitry as their representations of nonsymbolic ratio magnitudes. We summarize the results of three studies representing each of these strands below.

Study 1—Correlations between RPS acuity and fraction knowledge. To investigate whether individual differences in RPS acuity predict fraction knowledge and subsequent algebra achievement, we have constructed four novel measures of RPS acuity (see Figure 3) similar to those used to measure ANS acuity. In each of these tasks, the participant's task was simply to choose the

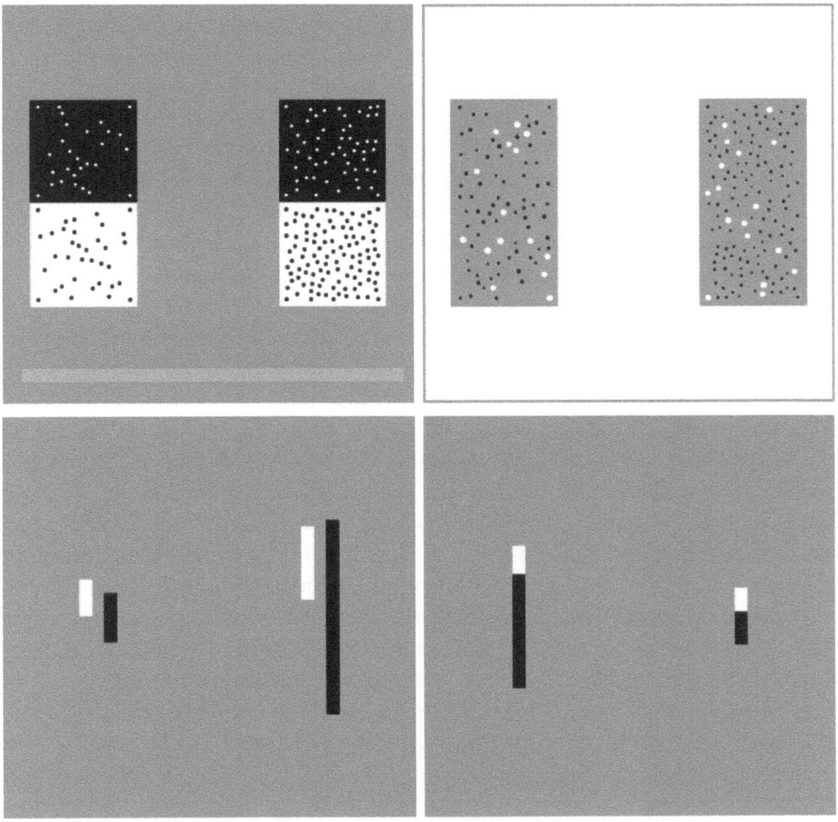

FIGURE 3 Sample stimuli from the RPS comparison battery. Participants were instructed to choose the larger of two ratios formed by sets of dots or pairs of lines. After receiving 10 practice trials, participants quickly acclimated to the novel instructions.

larger of two nonsymbolic ratios. In addition to this battery of RPS tasks, participants completed an array of cognitive tasks, including: (1) three measures of symbolic fraction knowledge (symbolic fraction comparison, number-line estimation with fraction stimuli, and a paper-and-pencil fraction knowledge assessment); (2) two tasks to control for differences in participants' acuity for processing nonsymbolic ratio components (a standard ANS acuity task and a line-length discrimination task); and (3) a flanker task[1] to control for differences in inhibitory control that may affect participants' ability to suppress componential processing when evaluating the magnitudes of ratio stimuli. Finally, we requested access to participants' algebra subtest scores from their university entrance examinations.

The key research question for Study 1 was whether RPS acuity predicts symbolic fraction knowledge and algebra achievement above and beyond ANS acuity and other control measures. To date, 178 undergraduate students have completed the study, and the results have been consistent with model predictions. For example, the ability to successfully compare the magnitudes of nonsymbolic line ratios predicts symbolic fraction comparison ability (see Figure 4), even after accounting for performance on control tasks. We have found that RPS acuity similarly predicts number-line estimation ability and performance on the fraction knowledge assessment. We are currently developing match-to-sample protocols that can be used to test for the same correlations among school-age children. These protocols are less dependent on language, which will better permit us to investigate the development of nonsymbolic ratio discrimination over time and its connection to fraction learning. This developmental question will ultimately prove to be a critical test of our theory.

Study 2—Training symbolic-nonsymbolic links. Whereas Study 1 provides evidence that the acuity of a learner's RPS is significantly related to fraction learning, the correlational nature of the evidence does not allow us to address questions of causality. Additionally, Study 1 does not directly assess the quality of the symbolic-to-nonsymbolic links that are a key feature of our hypothesis regarding how RPS acuity could affect symbolic fraction knowledge. In Study 2, we are exploring these questions by testing a perceptually based training paradigm in which participants improve the quality of their symbolic-nonsymbolic links by learning to associate the magnitude of nonsymbolic ratios with corresponding symbolic fractions of the same value. Our initial pilot studies used three 20-minute computerized training sessions spread across three days in which participants learned these associations by completing either cross-format match-to-sample tasks with feedback or by completing cross-format comparison tasks with feedback (see Figure 5 for an illustration). These tasks were

1. The flanker task is a classic test of inhibitory control, in which participants are asked to indicate the direction in which a central arrow points while ignoring an array of "flanking" arrows. The difference between reaction times for congruent trials (e.g., picking "right" for >>>>>>) and incongruent trials (e.g., "right" for << > <<) is the key measure.

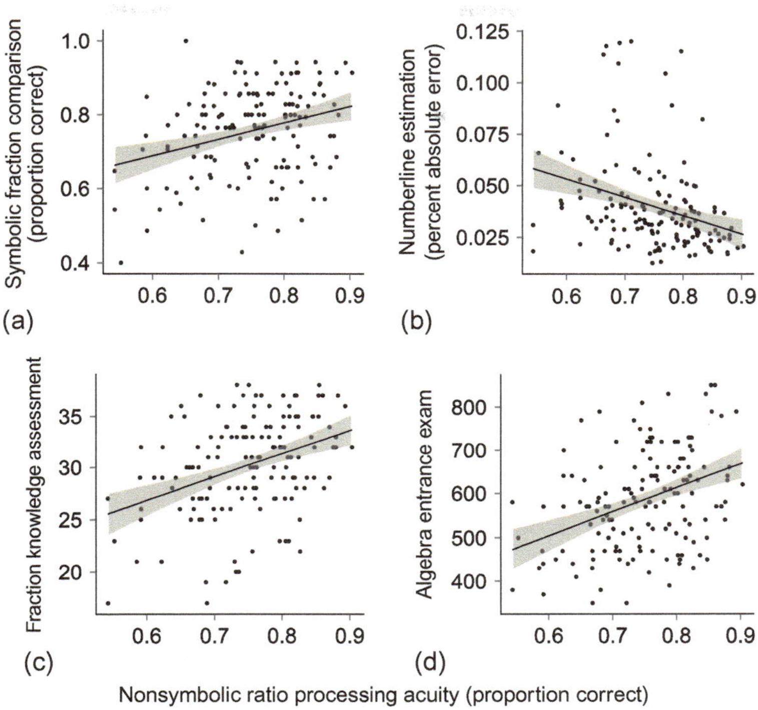

FIGURE 4 Correlations between RPS acuity and fraction knowledge as well as general mathematical knowledge. Participants' performance on the RPS comparison tasks significantly predicts fraction knowledge and algebra achievement. Participants with higher acuity are better at symbolic fraction comparison (a, higher proportion correct), better at placing a fraction on a number line (b, lower percent absolute error), score higher on a paper-and-pencil assessment of fraction knowledge (c), and have higher university algebra entrance exam scores (d).

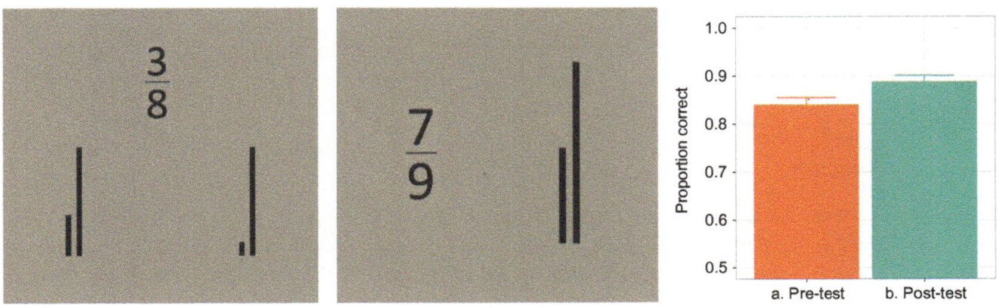

FIGURE 5 Training paradigms and pilot results. Left: In match-to-sample training, participants learn to associate symbolic fraction with their corresponding nonsymbolic ratio magnitudes by selecting the matching line ratio. Middle: In comparison training, participants learn the relative magnitude of symbolic fractions by deciding whether a symbolic fraction or a line ratio is larger. Right: These training paradigms produced significant pretraining to posttraining gains in performance on symbolic fraction comparison, a measure of fraction knowledge.

adaptive: as participants improved in accuracy, the program reduced the distance between target and distractor ratios, with the goal of increasing the precision of participants' links between symbolic fractions and their corresponding nonsymbolic ratio magnitudes.

The pilot data shown in Figure 5 were collected from eight graduate students, and the results have been encouraging despite the small size of our initial sample. These highly educated students showed statistically significant gains in performance on symbolic fraction comparison tasks as a result of the training, suggesting that improving the links between nonsymbolic and symbolic representations can indeed improve processing of symbolic fractions. We are currently running a full version of this experiment with undergraduate students and plan to use these results to inform similar training interventions with school-age children.

Study 3—fMRI adaptation. In Study 3, we are investigating the neural signature of these symbolic-to-nonsymbolic links. Specifically, we are investigating whether links between representations of nonsymbolic ratio magnitudes and symbolic fraction representations result in the same sort of distance-dependent recovery from adaptation (Figure 6) that has been observed between nonsymbolic numerosity (dot arrays) and symbolic whole numbers (Piazza, Pinel, Le Bihan, & Dehaene, 2007). To date, six adults and one 10-year-old child have completed this study. During the study, participants were repeatedly presented with a specific nonsymbolic ratio magnitude. In one run, participants viewed a series of line ratios in which the component line lengths varied but the shorter line was always 3/10 as long as the longer line, while in the other, the adapting ratios were always 7/10. During each run, subjects were also presented with 15 deviant magnitudes, presented in each of two notations (symbolic, nonsymbolic) and distances (near, far). For example, after presenting a stream of 7/10

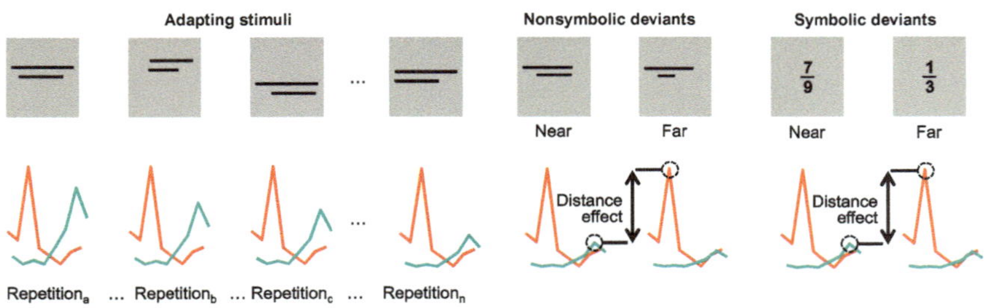

FIGURE 6 Predicted neural adaptation. The repeated presentation of a stimulus (in this case, a nonsymbolic ratio with magnitude 0.70 (top row, left) adapts neural circuits that respond to that magnitude, reducing their subsequent response and associated BOLD signal in IPS. Broadly tuned neural circuits that represent the approximate nonsymbolic ratio magnitude 7/9 and the symbolic fraction "7/9" are significantly habituated by the repeated presentation of the nonsymbolic ratio .7, but neural circuits that represent "1/3" are not. As a result, when the near deviant nonsymbolic ratio magnitude 7/9 or the symbolic fraction "7/9" are presented, adapted neural circuits do not respond strongly, resulting in little signal recovery. However, when the far deviant nonsymbolic ratio magnitude 1/3 or the symbolic fraction "1/3" is presented, nonadapted circuits tuned to the magnitude 1/3 respond strongly, resulting in strong BOLD signal recovery and a robust distance effect.

stimuli, a near symbolic deviant might be the fraction 7/9 and a far symbolic deviant might be the fraction 1/3. Likewise, after presenting a stream of 3/10 stimuli, a near nonsymbolic deviant might be a line ratio in which the shorter line is 1/4 as long as the longer line, and a far nonsymbolic deviant might be a line ratio in which the shorter line is 5/7 as long as the longer line (see Figure 6).

Providing initial confirmation of the model predictions, we find the expected distance effect, with larger responses to far deviants than to close deviants, in the right mid-IPS (Figure 7). This is true both when we analyze the data collapsed across line ratios and symbolic fractions (Figure 7A) and when we examine the critical cross-notation rebound effect by restricting our analyses to digits (Figure 7B). The fact that adaptation transfers from nonsymbolic line ratios to symbolic fractions in a distance-dependent manner suggests that adults have made the links between symbolic fractions and the more basic RPS system that represents fraction magnitude, as predicted by our model. These links are apparent even in the absence of explicit task demands to map between different dimensions (indeed, participants were performing a completely irrelevant task in which they merely detected the brief dimming of the fixation cross), which is considered to be a key method for isolating representations (see Cohen Kadosh & Walsh, 2009 and commentaries contained therein). Building on this basic foundation, we will begin to explore individual differences in RPS sensitivity and the strength of the links between the RPS system and fractions, in adults and children. In particular, it will be critical to examine developmental changes in these links as a result of formal instruction in fractions, by comparing, for example, 2nd, 5th, and 8th graders, who are at different stages of fraction instruction.

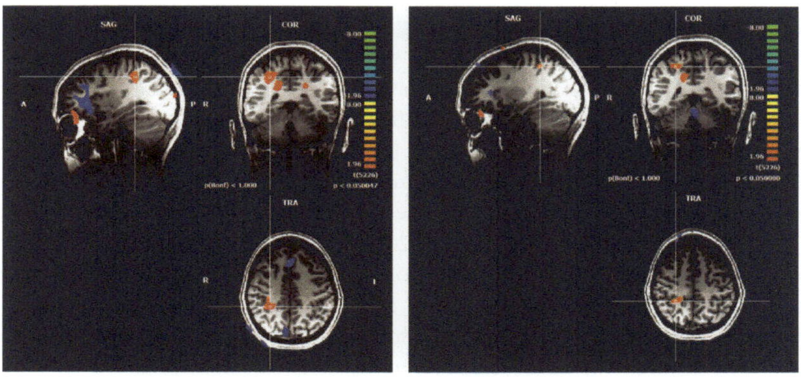

Lines and Digits (29, −37, 39) Digits Only (26, −38, 44)

FIGURE 7 Preliminary fMRI adaptation results showing regions that demonstrated a distance effect (increased responses for far deviants > close deviants) collapsed across both the line ratios and symbolic fractions (A) and examining cross-notation dishabituation from line ratios to symbolic fractions only (B). Maps show a fixed effects analysis ($n=6$) thresholded at $p<0.05$, uncorrected, with a minimum cluster of 30 voxels. The coordinates below the plots indicate peak activation in the two analyses, and are highly similar, differing only by a few millimeters.

OPEN QUESTIONS

Although the work we reviewed earlier has begun to delimit the nature of the RPS and its relationship with symbolic fractions processing, much more remains to be done. We believe that the RPS may represent a core numerical competence that has yet to be optimally leveraged to support fraction learning. However, a number of key questions need to be answered to build a scientific understanding of this ability that both deepens our understanding of mathematical cognition and produces knowledge that can inform fraction instruction. Our first set of questions involves basic neuroscientific inquiry into the phylogenesis and ontogenesis of the brain's ratio processing system and its contribution and connectivity to other brain networks. A second set of questions involves the potential role of the RPS in both individual differences in mathematical abilities and dyscalculia. A third set of questions involves developing and testing a host of ways to improve fraction learning by more directly leveraging the RPS, thereby imbuing symbolic fractions with meaning. Finally, addressing these first three questions allows us to ask a fourth, more nuanced, set of questions: When in symbolic fractions learning is the RPS most important? Are its effects primarily limited to helping young children acquire basic understandings of fraction magnitude, or can leveraging the RPS help even more advanced students improve their proficiency with mathematical operations on fractions?

CHARTING THE DEVELOPMENT AND ARCHITECTURE OF THE RPS

Although there is substantial neuroimaging data on the development of parietal cortex generally (for a recent review, see Giedd et al., 2015) and the development of the ANS (e.g., Cantlon, Brannon, Carter, & Pelphrey, 2006; Izard, Dehaene-Lambertz, & Dehaene, 2008) there is no developmental neuroimaging evidence regarding the ontogenesis of the RPS. The only neuroscientific evidence regarding the brain's representation of nonsymbolic ratios comes from fMRI studies of adult humans and single-unit studies of trained monkeys (Jacob & Nieder, 2009b; Vallentin & Nieder, 2008, 2010). Charting the brain's ability to represent nonsymbolic ratios prior to, during, and after formal education on fractions is a critical step in not only understanding how these architectures prepare children to learn about symbolic fractions but also in illuminating how these architectures may themselves be reshaped by formal education.

Another key question regards how RPS architectures relate to systems that are tuned to the magnitudes of the components that create a nonsymbolic ratio (e.g., the length of the two lines in a line ratio or the numerosity of the two sets of dots in a dot ratio). Our current studies statistically control for the effects of these component processing systems, such as the ANS, but do not directly investigate pathways relating the two. Distance-dependent activation for component dimensions such as physical size and numerosity are observed in IPS regions (Cohen Kadosh et al., 2005; Pinel, Piazza, Le Bihan, & Dehaene, 2004) near areas associated with the processing of nonsymbolic ratios, raising the possibility that the RPS is a higher order extension of one or more of these representational systems. Determining how these systems interact may help to illuminate early differences in children's understanding of continuous and discrete ratios (Boyer et al., 2008; Jeong et al., 2007; Spinillo & Bryant, 1999) and may also help to inform the long-standing educational debate about the relative efficacy of different pedagogical representations of fractions (Cramer & Wyberg, 2009). Arguments for using different types of representations such as number lines, area models, or counters are usually justified by appealing to higher level ratio number schemas such as measurement, partitioning, or equal sharing, but the choice of fraction representations may also be informed by lower level architectural considerations. For example, if the RPS is robustly connected to systems tuned to continuous magnitudes but not to systems tuned to discrete magnitudes, it may be most effectively leveraged by using continuous as opposed to discrete stimuli.

LEVERAGING THE RPS TO SUPPORT FRACTION LEARNING

There is good reason to believe that conventional fraction instruction fails to effectively leverage the brain's ability to represent nonsymbolic fractional magnitudes. The first stages of conventional fraction instruction usually rely on partitioning or equal-sharing (Ni & Zhou, 2005; Pitkethly & Hunting, 1996; Siegler et al., 2010). Such approaches involve dividing a figure into equal parts or partitioning a set of items (e.g., partitioning 12 candies into four equally numerous groups) and often fundamentally rely on counting. To identify the fraction illustrated by a partially shaded figure, a child counts the number of shaded parts, assigns this to the numerator and counts the total number of parts and assigns this to the denominator (Davydov & Tsvetkovich, 1991). As a result of this reliance on counting, the early stages of conventional fraction training may encourage the reappropriation of count-based, whole-number schemas rather than harnessing the capabilities of the RPS. This reappropriation can have serious consequences (Mack, 1990; Ni & Zhou, 2005; Siegler et al., 2013). Indeed, Mack (1995) found that partitioning approaches often led 3rd- and 4th-grade students to overgeneralize their whole-number knowledge to fractions. This overgeneralization prevents children from grasping fraction concepts and can lead to common procedural errors such as saying that 12/13 + 7/8 is closer to 19 and 21 than to 2 (Carpenter et al., 1981).

We propose that fraction education may be improved by designing instruction that more directly leverages the RPS while reducing the misapplication of whole-number schemas. Following Feigenson et al. (2004), we argue that acquiring number concepts is easy when they are supported by core systems of representation and hard when this acquisition goes beyond the limits of a core system. However, we disagree with their conclusion that core number systems are incompatible with rational number concepts. Instead, we argue that ratio brain architectures might naturally support fraction concepts that the ANS cannot. Explicitly leveraging the RPS may help discourage the misapplication of whole-number concepts and build a more generative foundation for future learning than for conventional instruction.

We are not necessarily proposing a complete reformulation of fraction instruction, but rather offering a new theoretical basis for (1) imagining how fraction education can be better grounded in children's preexisting abilities, and (2) systematically developing and testing modifications to conventional fractions instruction. Educators have long used nonsymbolic referents to teach fractions, justifying their use as attempts to ground understanding in children's informal knowledge (e.g., their knowledge of sharing) (Mack, 1990; Siegler et al., 2010) or to better illustrate the formal logic of rational number mathematics (Davydov & Tsvetkovich, 1991; Moss & Case, 1999; Wu, 2008). Here, we argue that an alternative way of conceptualizing early fraction education is as a process of building upon children's preexisting abilities to perceive and represent magnitudes corresponding to nonsymbolic ratios. From this perspective, fraction learning does not need to start from scratch or require onerous abstraction; instead, fraction learning can build upon the solid foundation provided by nonsymbolic RPS architectures.

In practice, changes emerging from this perspective may appear small, but their impact may be profound. For example, this perspective suggests that it may be fruitful to replace or supplement conventional nonsymbolic referents composed of discrete, countable elements (e.g., pies and wedges or small sets) with uncountable nonsymbolic ratios such as pairs of lines or uncountable sets of dots. Because these types of ratios are inherently uncountable, their use should help prevent the inappropriate application of whole-number knowledge (see also Boyer et al., 2008; Jeong et al., 2007). Other changes might include the adoption of targeted interventions such as the training paradigm we are employing in Experiment 2. The take-home message is clear: if the brain of the elementary school child—like that of the human adult or the nonhuman primate—is able to represent the holistic magnitudes of these nonsymbolic ratios, pedagogies based on this capacity may also help children build an intuitive understanding of fraction magnitude that serves as a generative foundation for future learning.

We refer to a "generative foundation" because the conceptual foundation built by leveraging the RPS may have effects far beyond cultivating an intuitive understanding of fraction magnitude. Building a stronger, more grounded understanding of fractions in the early stages of learning can support future learning of fractions and related concepts (e.g., decimals, percent, and measure) throughout elementary and middle school and can help prepare students for algebra (Booth & Newton, 2012; Siegler et al., 2012). Building a stronger foundation can also facilitate the profound reorganization of numerical reasoning that Siegler et al. (2011, 2013) have attributed to the acquisition of fraction concepts. In comparison to count-based methods, which may bind thought in terms of whole numbers, proper engagement of the RPS can potentially help students develop a clearer understanding of the relational properties of numbers, enabling them to better grasp key mathematical and scientific concepts such as ratio, rate, and probability. Of course, at this early point, these intriguing possibilities remain speculative and stand in need of experimental verification before they can guide classroom practices. It is our hope that the current chapter will spur such experimental work.

RPS AND DYSCALCULIA?

If the RPS does indeed serve as a core system upon which mathematical concepts and learning are built, deficits in nonsymbolic ratio acuity at the behavioral and neural levels might provide deeper explanations of atypical math achievement including developmental dyscalculia. Developmental dyscalculia is a severe difficulty in learning mathematics that affects 5-7% of the population (Butterworth, 2005; Butterworth, Varma, & Laurillard, 2011). One potential source of dyscalculia is a deficit in the ANS characterized by reduced nonsymbolic acuity. Some researchers have found that children with dyscalculia are often less accurate than IQ-matched peers in performing simple nonsymbolic comparison tasks such as choosing the more numerous set of dots (Mazzocco et al., 2011a; Piazza et al., 2010, for contrasting accounts see Gilmore et al., 2013; Szűcs, Devine, Soltesz, Nobes, & Gabriel, 2013). These behaviorally measured deficits are often observed in parallel at the neural level, with dyscalculic children showing reduced IPS activation during comparison tasks (Chapter 7; Kucian et al., 2011; Price et al., 2007). If RPS deficits are also associated with dyscalculia, deficits in nonsymbolic ratio acuity at the behavioral and neural levels may also differentiate children with dyscalculia from their typically developing peers.

RPS deficits may be comorbid with ANS deficits. Alternatively, the two deficits may dissociate, resulting in distinct subtypes of dyscalculia. Under this scenario, children with both deficits might experience the most profound difficulties, children with ANS deficits alone might exhibit specific difficulty learning whole-number facts and procedures,[2] and children with RPS deficits alone might succeed in early whole-number learning but struggle when they

2. It is important to note that whole-number deficits are related at least in part to retrieving math facts from long-term memory (Geary, 1993) and this in turn may be related to frontal-hippocampal problems (e.g., Qin et al., 2014). Hence, mathematical learning difficulties, at both a behavioral and anatomical level are likely to be more complex than just ANS difficulties (Fias et al., 2013; Szűcs & Goswami, 2013). Similar complexity is also likely to be present for fractions learning.

encounter fractions and higher levels of mathematics that depend on rational number concepts. This pattern would suggest that the group with ANS deficits alone might still be able to learn about fractions, but not through traditional approaches that stress whole-number concepts and use discrete referents that rely on the ANS. For the other two groups, it would suggest that they might need even more remediation when learning fraction concepts. Finally, if the two systems are dissociated, it raises the possibility that deficits in one system might be remediated in part by relying more heavily on the remaining functional system. For instance, children with impairments specific to the ANS might be more efficiently taught about number magnitudes using ratio representations of whole-number concepts as opposed to using discrete whole-number representations.

SUMMARY AND CONCLUSIONS

Recent research has revealed a set of neurocognitive architectures—the ratio processing system or RPS—that are tuned to the holistic magnitudes of nonsymbolic ratios. We have proposed that the ability to represent ratio/fraction magnitudes conferred by the RPS may support a better understanding of fractions as relative magnitudes, and may even help children to achieve the conceptual insight that fractions represent magnitudes. Specifically, we hypothesize that learning can be improved by increasing the degree to which instruction leverages this brain network, helping ground children's understanding of symbolic fractions in their preexisting abilities to represent holistic ratio magnitudes intuitively. This improved understanding of fractions as magnitudes, in turn, will support better conceptual understanding of fractions and more appropriate application of arithmetic procedures with fractions. A number of key questions need to be answered in order to build an educationally relevant model of the RPS that is sufficiently detailed to deepen our understanding of mathematical cognition and to usefully inform real-world educational interventions. Despite these open questions, compelling evidence is accumulating that the RPS does indeed exist, and is another example of what Piazza (2010) calls a neurocognitive startup tool for supporting mathematical cognition. Early research from our lab suggests that the RPS may be critically important for learning about fractions. Future work delimiting its function and its engagement during symbolic fraction learning holds significant promise for enriching our understanding of human numerical cognition.

ACKNOWLEDGMENTS

This work was supported by NSF REAL 1420211 (Matthews and Hubbard, co-PIs) and NIH 1R03HD081087-01 (Matthews). The authors also gratefully acknowledge the support of the Wisconsin Alumni Research Foundation (WARF).

REFERENCES

Bailey, D. H., Hoard, M. K., Nugent, L., & Geary, D. C. (2012). Competence with fractions predicts gains in mathematics achievement. *Journal of Experimental Child Psychology, 113*(3), 447–455. http://dx.doi.org/10.1016/j.jecp.2012.06.004.

Bonato, M., Fabbri, S., Umiltà, C., & Zorzi, M. (2007). The mental representation of numerical fractions: Real or integer? *Journal of Experimental Psychology. Human Perception and Performance, 33*(6), 1410–1419. http://dx.doi.org/10.1037/0096-1523.33.6.1410.

Booth, J. L., & Newton, K. J. (2012). Fractions: Could they really be the gatekeeper's doorman? *Contemporary Educational Psychology, 37*(4), 247–253. http://dx.doi.org/10.1016/j.cedpsych.2012.07.001.

Boyer, T. W., Levine, S. C., & Huttenlocher, J. (2008). Development of proportional reasoning: Where young children go wrong. *Developmental Psychology, 44*(5), 1478–1490. http://dx.doi.org/10.1037/a0013110.

Brannon, E. M., & Roitman, J. D. (2003). Nonverbal representations of time and number in animals and human infants. In W. H. Meck (Ed.), *Functional and neural mechanisms of interval timing* (pp. 143–182). Boca Raton, FL: CRC Press.

Briars, D., & Siegler, R. S. (1984). A featural analysis of preschoolers' counting knowledge. *Developmental Psychology, 20*(4), 607–618. http://dx.doi.org/10.1037/0012-1649.20.4.607.

Butterworth, B. (2005). Developmental dyscalculia. In J. Campbell (Ed.), Handbook of mathematical cognition (pp. 455–467). New York: Psychology Press.

Butterworth, B., Varma, S., & Laurillard, D. (2011). Dyscalculia: From brain to education. *Science, 332*(6033), 1049–1053. http://dx.doi.org/10.1126/science.1201536.

Cantlon, J. F., Brannon, E. M., Carter, E. J., & Pelphrey, K. A. (2006). Functional imaging of numerical processing in adults and 4-y-old children. *Public Library of Science, Biology, 4*(5), e125. http://dx.doi.org/10.1371/journal.pbio.0040125.

Carpenter, T. P., Corbitt, M. K., Kepner, H., Lindquist, M. M., & Reys, R. (1981). *Results from the Second Mathematics Assessment of the National Assessment of Educational Progress.* Reston, VA: National Council of Teachers of Mathematics.

Chu, F. W., vanMarle, K., & Geary, D. C. (2013). Quantitative deficits of preschool children at risk for mathematical learning disability. *Frontiers in Psychology, 4*, http://dx.doi.org/10.3389/fpsyg.2013.00195, Article 195.

Cohen Kadosh, R., Henik, A., Rubinsten, O., Mohr, H., Dori, H., van de Ven, V., et al. (2005). Are numbers special?: The comparison systems of the human brain investigated by fMRI. *Neuropsychologia, 43*(9), 1238–1248. http://dx.doi.org/10.1016/j.neuropsychologia.2004.12.017.

Cohen Kadosh, R., & Walsh, V. (2009). Numerical representation in the parietal lobes: Abstract or not abstract? *Behavioral and Brain Sciences, 32*(3-4), 313–328. http://dx.doi.org/10.1017/S0140525X09990938.

Cramer, K., & Wyberg, T. (2009). Efficacy of different concrete models for teaching the part-whole construct for fractions. *Mathematical Thinking and Learning, 11*(4), 226–257. http://dx.doi.org/10.1080/10986060903246479.

Davydov, V. V., & Tsvetkovich, Z. (1991). On the objective origin of the concept of fractions. *Focus on Learning Problems in Mathematics, 12*(1), 13–64.

De Smedt, B., Noël, M. P., Gilmore, C., & Ansari, D. (2013). How do symbolic and non-symbolic numerical magnitude processing relate to individual differences in children's mathematical skills? A review of evidence from brain and behavior. *Trends in Neuroscience and Education, 2*(2), 48–55. http://dx.doi.org/10.1016/j.tine.2013.06.001.

Dehaene, S. (1997/2011). The number sense: How the mind creates mathematics. New York: Oxford University Press, ISBN: 978-0199753871.

Dehaene, S., & Cohen, L. (2007). Cultural recycling of cortical maps. *Neuron*, *56*(2), 384–398. http://dx.doi.org/10.1016/j.neuron.2007.10.004.

Dehaene, S., Dehaene-Lambertz, G., & Cohen, L. (1998). Abstract representations of numbers in the animal and human brain. *Trends in Neurosciences*, *21*(8), 355–361. http://dx.doi.org/10.1016/S0166-2236(98)01263-6.

Duffy, S., Huttenlocher, J., & Levine, S. (2005). It is all relative: How young children encode extent. *Journal of Cognition and Development*, *6*(1), 51–63. http://dx.doi.org/10.1207/s15327647jcd0601_4.

Feigenson, L., Dehaene, S., & Spelke, E. (2004). Core systems of number. *Trends in Cognitive Sciences*, *8*(7), 307–314. http://dx.doi.org/10.1016/j.tics.2004.05.002.

Fias, W., Menon, V., & Szucs, D. (2013). Multiple components of developmental dyscalculia. *Trends in Neuroscience and Education*, *2*(2), 43–47. http://dx.doi.org/10.1016/j.tine.2013.06.006.

Gallistel, C. R., & Gelman, R. (1992). Preverbal and verbal counting and computation. *Numerical Cognition*, *44*(1–2), 43–74. http://dx.doi.org/10.1016/0010-0277(92)90050-R.

Geary, D. C. (1993). Mathematical disabilities: Cognitive, neuropsychological, and genetic components. *Psychological Bulletin*, *114*(2), 345–362. http://dx.doi.org/10.1037/0033-2909.114.2.345.

Geary, D. C. (2007). Development of mathematical understanding. In D. Kuhn, R. S. Siegler, W. Damon, & R. M. Lerner (Eds.), *Handbook of child psychology: Vol. 2* (6th ed., pp. 777–810). Hoboken, NJ: John Wiley & Sons, Inc., Retrieved from http://dx.doi.org/10.1002/9780470147658.chpsy0218.

Geary, D. C., Berch, D. B., & Mann Koepke, K. (Eds.), (2015). *Evolutionary origins and early development of number processing*. San Diego, CA: Academic Press.

Gelman, R., & Williams, E. M. (1998). Enabling constraints for cognitive development and learning: Domain specificity and epigenesis. In D. Kuhn, & R. Siegler (Eds.), *Cognition, perception, and language: Vol. 2* (pp. 575–630). New York: John Wiley & Sons Inc.

Giedd, J. N., Raznahan, A., Alexander-Bloch, A., Schmitt, E., Gogtay, N., & Rapoport, J. L. (2015). Child psychiatry branch of the National Institute of Mental Health longitudinal structural magnetic resonance imaging study of human brain development. *Neuropsychopharmacology*, *40*(1), 43–49. http://dx.doi.org/10.1038/npp.2014.236.

Gilmore, C., Attridge, N., Clayton, S., Cragg, L., Johnson, S., Marlow, N., et al. (2013). Individual differences in inhibitory control, not non-verbal number acuity, correlate with mathematics achievement. *PloS One*, *8*(6). http://dx.doi.org/10.1371/journal.pone.0067374, e67374.

Goswami, U. (1989). Relational complexity and the development of analogical reasoning. *Cognitive Development*, *4*(3), 251–268. http://dx.doi.org/10.1016/0885-2014(89)90008-7.

Goswami, U. (1995). Transitive relational mappings in three- and four-year-olds: The analogy of Goldilocks and the three bears. *Child Development*, *66*(3), 877–892. http://dx.doi.org/10.1111/j.1467-8624.1995.tb00911.x.

Grill-Spector, K., Henson, R., & Martin, A. (2006). Repetition and the brain: Neural models of stimulus-specific effects. *Trends in Cognitive Sciences*, *10*(1), 14–23. http://dx.doi.org/10.1016/j.tics.2005.11.006.

Grill-Spector, K., & Malach, R. (2001). fMR-adaptation: A tool for studying the functional properties of human cortical neurons. *Acta Psychologica*, *107*(1–3), 293–321. http://dx.doi.org/10.1016/S0001-6918(01)00019-1.

Halberda, J., Mazzocco, M. M. M., & Feigenson, L. (2008). Individual differences in non-verbal number acuity correlate with maths achievement. *Nature*, *455*(7213), 665–668. http://dx.doi.org/10.1038/nature07246.

Hecht, S. A. (1998). Toward an information-processing account of individual differences in fraction skills. *Journal of Educational Psychology*, *90*(3), 545–559. http://dx.doi.org/10.1037/0022-0663.90.3.545.

Hecht, S. A., Close, L., & Santisi, M. (2003). Sources of individual differences in fraction skills. *Journal of Experimental Child Psychology*, *86*(4), 277–302. http://dx.doi.org/10.1016/j.jecp.2003.08.003.

Hecht, S. A., & Vagi, K. J. (2010). Sources of group and individual differences in emerging fraction skills. *Journal of Educational Psychology*, *102*(4), 843–859. http://dx.doi.org/10.1037/a0019824.

Hollands, J. G., & Dyre, B. P. (2000). Bias in proportion judgments: The cyclical power model. *Psychological Review*, *107*(3), 500–524. http://dx.doi.org/10.1037/0033-295X.107.3.500.

Ischebeck, A., Schocke, M., & Delazer, M. (2009). The processing and representation of fractions within the brain: An fMRI investigation. *NeuroImage*, *47*(1), 403–413. http://dx.doi.org/10.1016/j.neuroimage.2009.03.041.

Izard, V., Dehaene-Lambertz, G., & Dehaene, S. (2008). Distinct cerebral pathways for object identity and number in human infants. *Public Library of Science, Biology*, *6*(2), e11. http://dx.doi.org/10.1371/journal.pbio.0060011.

Jacob, S. N., & Nieder, A. (2009a). Notation-independent representation of fractions in the human parietal cortex. *The Journal of Neuroscience*, *29*(14), 4652–4657. http://dx.doi.org/10.1523/JNEUROSCI.0651-09.2009.

Jacob, S. N., & Nieder, A. (2009b). Tuning to non-symbolic proportions in the human frontoparietal cortex. *European Journal of Neuroscience*, *30*(7), 1432–1442. http://dx.doi.org/10.1111/j.1460-9568.2009.06932.x.

Jacob, S. N., Vallentin, D., & Nieder, A. (2012). Relating magnitudes: The brain's code for proportions. *Trends in Cognitive Sciences*, *16*(3), 157–166. http://dx.doi.org/10.1016/j.tics.2012.02.002.

Jeong, Y., Levine, S. C., & Huttenlocher, J. (2007). The development of proportional reasoning: Effect of continuous versus discrete quantities. *Journal of Cognition and Development*, *8*(2), 237–256. http://dx.doi.org/10.1080/15248370701202471.

Kallai, A. Y., & Tzelgov, J. (2009). A generalized fraction: An entity smaller than one on the mental number line. *Journal of Experimental Psychology. Human Perception and Performance*, *35*(6), 1845–1864. http://dx.doi.org/10.1037/a0016892.

Kallai, A. Y., & Tzelgov, J. (2012). When meaningful components interrupt the processing of the whole: The case of fractions. *Acta Psychologica*, *139*(2), 358–369. http://dx.doi.org/10.1016/j.actpsy.2011.11.009.

Krekelberg, B., Boynton, G. M., & van Wezel, R. J. A. (2006). Adaptation: From single cells to BOLD signals. *Trends in Neurosciences*, *29*(5), 250–256. http://dx.doi.org/10.1016/j.tins.2006.02.008.

Kucian, K., Grond, U., Rotzer, S., Henzi, B., Schönmann, C., Plangger, F., et al. (2011). Mental number line training in children with developmental dyscalculia. *NeuroImage*, *57*(3), 782–795. http://dx.doi.org/10.1016/j.neuroimage.2011.01.070.

Le Corre, M., & Carey, S. (2007). One, two, three, four, nothing more: An investigation of the conceptual sources of the verbal counting principles. *Cognition*, *105*(2), 395–438. http://dx.doi.org/10.1016/j.cognition.2006.10.005.

Libertus, M. E., Feigenson, L., & Halberda, J. (2011). Preschool acuity of the approximate number system correlates with school math ability. *Developmental Science*, *14*(6), 1292–1300. http://dx.doi.org/10.1111/j.1467-7687.2011.01080.x.

Lipkus, I. M., Samsa, G., & Rimer, B. K. (2001). General Performance on a Numeracy Scale among Highly Educated Samples. *Medical Decision Making*, *21*(1), 37–44. http://dx.doi.org/10.1177/0272989X0102100105.

Lyons, I. M., Price, G. R., Vaessen, A., Blomert, L., & Ansari, D. (2014). Numerical predictors of arithmetic success in grades 1-6. *Developmental Science*, *17*(5), 714–726. http://dx.doi.org/10.1111/desc.12152.

Mack, N. K. (1990). Learning fractions with understanding: Building on informal knowledge. *Journal for Research in Mathematics Education*, 16–32.

Mack, N. K. (1995). Confounding whole-number and fraction concepts when building on informal knowledge. *Journal for Research in Mathematics Education*, 422–441.

Mazzocco, M. M. M., Feigenson, L., & Halberda, J. (2011a). Impaired acuity of the approximate number system underlies mathematical learning disability (dyscalculia). *Child Development*, *82*(4), 1224–1237. http://dx.doi.org/10.1111/j.1467-8624.2011.01608.x.

Mazzocco, M. M. M., Feigenson, L., & Halberda, J. (2011b). Preschoolers' precision of the approximate number system predicts later school mathematics performance. *PloS One*, *6*(9), e23749. http://dx.doi.org/10.1371/journal.pone.0023749.

McCrink, K., Spelke, E. S., Dehaene, S., & Pica, P. (2013). Non-symbolic halving in an Amazonian indigene group. *Developmental Science*, *16*(3), 451–462. http://dx.doi.org/10.1111/desc.12037.

McCrink, K., & Wynn, K. (2007). Ratio abstraction by 6-month-old infants. *Psychological Science*, *18*(8), 740–745. http://dx.doi.org/10.1111/j.1467-9280.2007.01969.x.

Meert, G., Grégoire, J., Seron, X., & Noël, M.-P. (2011). The mental representation of the magnitude of symbolic and nonsymbolic ratios in adults. *The Quarterly Journal of Experimental Psychology*, *65*(4), 702–724. http://dx.doi.org/10.1080/17470218.2011.632485.

Meert, G., Grégoire, J., Seron, X., & Noël, M.-P. (2013). The processing of symbolic and nonsymbolic ratios in school-age children. *PloS One*, *8*(11). http://dx.doi.org/10.1371/journal.pone.0082002, e82002.

Mix, K. S., Levine, S. C., & Huttenlocher, J. (1999). Early fraction calculation ability. *Developmental Psychology*, *35*(1), 164–174. http://dx.doi.org/10.1037/0012-1649.35.1.164.

Moss, J., & Case, R. (1999). Developing children's understanding of the rational numbers: A new model and an experimental curriculum. *Journal for Research in Mathematics Education*, *30*(2), 122–147. http://dx.doi.org/10.2307/749607.

Mussolin, C., Mejias, S., & Noël, M.-P. (2010). Symbolic and nonsymbolic number comparison in children with and without dyscalculia. *Cognition*, *115*(1), 10–25. http://dx.doi.org/10.1016/j.cognition.2009.10.006.

National Governors Association Center for Best Practices, & Council of Chief State School Officers. (2010). *Common core state standards*. Washington, DC: National Governors Association Center for Best Practices, Council of Chief State School Officers.

National Mathematics Advisory Panel. (2008). *Foundations for success: The final report of the National Mathematics Advisory Panel*. Washington, DC: U.S. Department of Education.

Newton, K. J. (2008). An extensive analysis of preservice elementary teachers' knowledge of fractions. *American Educational Research Journal*, *45*(4), 1080–1110. http://dx.doi.org/10.3102/0002831208320851.

Ni, Y., & Zhou, Y.-D. (2005). Teaching and learning fraction and rational numbers: The origins and implications of whole number bias. *Educational Psychologist*, *40*(1), 27–52. http://dx.doi.org/10.1207/s15326985ep4001_3.

Nieder, A. (2005). Counting on neurons: The neurobiology of numerical competence. *Nature Reviews Neuroscience*, *6*(3), 177–190.

Nieder, A., & Dehaene, S. (2009). Representation of number in the brain. *Annual Review of Neuroscience*, *32*, 185–208.

Obersteiner, A., Van Dooren, W., Van Hoof, J., & Verschaffel, L. (2013). The natural number bias and magnitude representation in fraction comparison by expert mathematicians. *Learning and Instruction*, *28*, 64–72. http://dx.doi.org/10.1016/j.learninstruc.2013.05.003.

Piazza, M. (2010). Neurocognitive start-up tools for symbolic number representations. *Trends in Cognitive Sciences*, *14*(12), 542–551. http://dx.doi.org/10.1016/j.tics.2010.09.008.

Piazza, M., Facoetti, A., Trussardi, A. N., Berteletti, I., Conte, S., Lucangeli, D., et al. (2010). Developmental trajectory of number acuity reveals a severe impairment in developmental dyscalculia. *Cognition*, *116*(1), 33–41. http://dx.doi.org/10.1016/j.cognition.2010.03.012.

Piazza, M., Pinel, P., Le Bihan, D., & Dehaene, S. (2007). A magnitude code common to numerosities and number symbols in human intraparietal cortex. *Neuron*, *53*(2), 293–305. http://dx.doi.org/10.1016/j.neuron.2006.11.022.

Pinel, P., Piazza, M., Le Bihan, D., & Dehaene, S. (2004). Distributed and overlapping cerebral representations of number, size, and luminance during comparative judgments. *Neuron*, *41*(6), 983–993. http://dx.doi.org/10.1016/S0896-6273(04)00107-2.

Pitkethly, A., & Hunting, R. (1996). A review of recent research in the area of initial fraction concepts. *Educational Studies in Mathematics*, *30*(1), 5–38.

Post, T. R., Harel, G., Behr, M., & Lesh, R. (1991). Intermediate teachers' knowledge of rational number concepts. In E. Fennema, T. Carpenter, & S. Lamon (Eds.), Integrating research on teaching and learning mathematics (pp. 177–198). New York: State University of NY Press.

Price, G. R., Holloway, I., Räsänen, P., Vesterinen, M., & Ansari, D. (2007). Impaired parietal magnitude processing in developmental dyscalculia. *Current Biology*, *17*(24), R1042–R1043. http://dx.doi.org/10.1016/j.cub.2007.10.013.

Qin, S., Cho, S., Chen, T., Rosenberg-Lee, T., Geary, D. C., & Menon, V. (2014). Hippocampal-neocortical functional reorganization underlies children's cognitive development. *Nature Neuroscience*, *17*(9), 1263–1269. http://dx.doi.org/10.1038/nn.3788.

Reyna, V. F., & Brainerd, C. J. (2008). Numeracy, ratio bias, and denominator neglect in judgments of risk and probability. *Learning and Individual Differences*, *18*(1), 89–107. http://dx.doi.org/10.1016/j.lindif.2007.03.011.

Schmithorst, V. J., & Brown, R. D. (2004). Empirical validation of the triple-code model of numerical processing for complex math operations using functional MRI and group Independent Component Analysis of the mental addition and subtraction of fractions. *NeuroImage*, *22*(3), 1414–1420. http://dx.doi.org/10.1016/j.neuroimage.2004.03.021.

Siegler, R. S., Carpenter, T., Fennell, F., Geary, D., Lewis, J., Okamoto, Y., et al. (2010). *Developing effective fractions instruction for kindergarten through 8th grade: A practice guide (NCEE #2010-4039)*. Washington, DC: National Center for Education Evaluation and Regional Assistance, Institute of Education Sciences, U.S. Department of Education.

Siegler, R. S., Duncan, G. J., Davis-Kean, P. E., Duckworth, K., Claessens, A., Engel, M., et al. (2012). Early predictors of high school mathematics achievement. *Psychological Science*, *23*(7), 691–697. http://dx.doi.org/10.1177/0956797612440101.

Siegler, R. S., Fazio, L. K., Bailey, D. H., & Zhou, X. (2013). Fractions: The new frontier for theories of numerical development. *Trends in Cognitive Sciences*, *17*(1), 13–19. http://dx.doi.org/10.1016/j.tics.2012.11.004.

Siegler, R. S., & Pyke, A. A. (2013). Developmental and individual differences in understanding of fractions. *Developmental Psychology*, *49*(10), 1994–2004. http://dx.doi.org/10.1037/a0031200.

Siegler, R. S., Thompson, C. A., & Schneider, M. (2011). An integrated theory of whole number and fractions development. *Cognitive Psychology*, *62*(4), 273–296. http://dx.doi.org/10.1016/j.cogpsych.2011.03.001.

Singer-Freeman, K. E., & Goswami, U. (2001). Does half a pizza equal half a box of chocolates?: Proportional matching in an analogy task. *Cognitive Development*, *16*(3), 811–829. http://dx.doi.org/10.1016/S0885-2014(01)00066-1.

Sophian, C. (2000). Perceptions of proportionality in young children: Matching spatial ratios. *Cognition*, *75*(2). http://dx.doi.org/10.1016/S00100277(00)00062-7.

Spinillo, A. G., & Bryant, P. (1991). Children's proportional judgments: The importance of "half.". *Child Development*, *62*(3), 427–440. http://dx.doi.org/10.1111/j.1467-8624.1991.tb01542.x.

Spinillo, A. G., & Bryant, P. E. (1999). Proportional Reasoning in Young Children: Part-Part Comparisons about Continuous and Discontinuous Quantity. *Mathematical Cognition*, *5*(2), 181–197. http://dx.doi.org/10.1080/135467999387298.

Stafylidou, S., & Vosniadou, S. (2004). The development of students' understanding of the numerical value of fractions. *Learning and Instruction*, *14*(5), 503–518. http://dx.doi.org/10.1016/j.learninstruc.2004.06.015.

Stevens, S. S., & Galanter, E. H. (1957). Ratio scales and category scales for a dozen perceptual continua. *Journal of Experimental Psychology*, *54*(6), 377–411. http://dx.doi.org/10.1037/h0043680.

Stigler, J. W., Givvin, K. B., & Thompson, B. J. (2010). What community college developmental mathematics students understand about mathematics. *MathAMATYC Educator*, *1*(3).

Szűcs, D., Devine, A., Soltesz, F., Nobes, A., & Gabriel, F. (2013). Developmental dyscalculia is related to visuo-spatial memory and inhibition impairment. *Cortex*, *49*(10), 2674–2688. http://dx.doi.org/10.1016/j.cortex.2013.06.007.

Szűcs, D., & Goswami, U. (2013). Developmental dyscalculia: Fresh perspectives. *Trends in Neuroscience and Education*, *2*(2), 33–37. http://dx.doi.org/10.1016/j.tine.2013.06.004.

Taubman, A. A. (2007). *Threshold Resistance: The Extraordinary Career of a Luxury Retailing Pioneer*. New York: Harper Business, ISBN: 978-0061235375.

Vallentin, D., & Nieder, A. (2008). Behavioral and prefrontal representation of spatial proportions in the monkey. *Current Biology*, *18*(18), 1420–1425. http://dx.doi.org/10.1016/j.cub.2008.08.042.

Vallentin, D., & Nieder, A. (2010). Representations of visual proportions in the primate posterior parietal and prefrontal cortices. *European Journal of Neuroscience*, *32*(8), 1380–1387. http://dx.doi.org/10.1111/j.1460-9568.2010.07427.x.

Vamvakoussi, X., & Vosniadou, S. (2010). How many decimals are there between two fractions? Aspects of secondary school students' understanding of rational numbers and their notation. *Cognition and Instruction*, *28*(2), 181–209. http://dx.doi.org/10.1080/07370001003676603.

Van Essen, D. C. (2005). A Population-Average, Landmark- and Surface-based (PALS) atlas of human cerebral cortex. *NeuroImage*, *28*(3), 635–662. http://dx.doi.org/10.1016/j.neuroimage.2005.06.058.

Van Essen, D. C., Drury, H. A., Dickson, J., Harwell, J., Hanlon, D., & Anderson, C. H. (2001). An integrated software suite for surface-based analyses of cerebral cortex. *Journal of the American Medical Informatics Association*, *8*(5), 443–459. http://dx.doi.org/10.1136/jamia.2001.0080443.

Woodruff, G., & Premack, D. (1981). Primative mathematical concepts in the chimpanzee: Proportionality and numerosity. *Nature*, *293*(5833), 568–570. http://dx.doi.org/10.1038/293568a0.

Wu, H. (2008). *Fractions, decimals, and rational numbers*. Retrieved from www.math.berkeley.edu/~wu/NMPfractions4.pdf.

Chapter 7

Developmental Dyscalculia and the Brain*

Karin Kucian
Center for MR-Research and Children's Research Center, University Children's Hospital Zurich, Zurich, Switzerland

INTRODUCTION

Numerical abilities are essential for many aspects of day-to-day living, and they are becoming even more crucial with the increasing role of technology in contemporary society. Low numeracy skills reduce employment prospects and mental and physical health of individuals, and widespread innumeracy can compromise the economic status of countries (Gross, Hudson, & Price, 2009; Parsons & Bynner, 2005). Importantly, profound difficulties with numeracy are very common with a prevalence rate between 3% and 6% (Kaufmann & von Aster, 2012; Shalev & von Aster, 2008).

Such severe difficulties in numeracy can occur as: (a) an acquired disorder due to loss of an established ability, or (b) developmentally, as an inability to acquire numerical concepts or procedures. The former is easily explained by changes in numerical abilities before and after a damaging event (e.g., a stroke). In contrast, the latter, referred to as developmental dyscalculia (DD), can only be explained by difficulties acquiring numerical competencies that are far from fully understood. Usually, only children who perform in the lowest 10% fall into this category of DD, often called specific mathematical learning disorder (MLD), although as many as 25% of children have persistent difficulties in learning mathematics (Geary, Hoard, Byrd-Craven, Nugent, & Numtee, 2007; Murphy, Mazzocco, Hanich, & Early, 2007). These children show profound difficulties in numerical understanding and deserve special help. In contrast to the labels of DD or mathematics disorder, some authors refer to all children below the 25th percentile as having a specific math learning disability. In addition to DD, MLD, or math learning disability, other terms for these difficulties include

* Portions of this chapter were adapted from Kucian, Kaufmann, and von Asterr (2014) and Kucian and von Asterr (2015).

Development of Mathematical Cognition. http://dx.doi.org/10.1016/B978-0-12-801871-2.00007-1

acalculia and anarithmetria, and there is no consensus on their use in the literature. In the current chapter I will use DD or MLD to describe the clinically diagnosed cases and mathematical learning disability to the subclinical group of children with severe problems learning mathematics.

Despite the relatively high occurrence of specific numerical learning disorders or disabilities, only a few research projects focus on this clearly high priority area. It is important to gain a clear understanding about typical as well as atypical development of numerical competencies on behavioral and neuronal levels to foster the acceptance as a disorder and raise public awareness for the need to provide targeted educational and therapeutic support tailored to affected children.

In the following sections, I am going to outline both general and specific points that have to be considered in the definition and diagnosis of DD and describe the atypical developmental trajectories of cognitive number representations associated with DD. Because recent findings suggest that DD is related to alterations in the brain (e.g., see meta-analysis by Kaufmann, Wood, Rubinsten, & Henik, 2011), neuronal underpinnings of DD are reviewed. Finally, based on current behavioral and neuronal knowledge about DD, general conclusions and future perspectives are presented.

DEVELOPMENTAL DYSCALCULIA

DD is a specific learning disability affecting the development of arithmetical skills with estimated prevalence rates between 3% and 6% (Kaufmann & von Aster, 2012; Shalev & von Aster, 2008). Recent studies claim that more girls than boys are affected (Fischbach et al., 2013), but the literature is still controversial, and opposite findings have also been reported (Desoete, Roeyers, & De Clercq, 2004; Reigosa-Crespo et al., 2012). Independent of gender, DD seems to be an enduring learning disorder persisting into adulthood (Grond, von Aster, O'Gorman, Martin, & Kucian, 2014; Kohn, Wyschkon, Ballaschk, Ihle, & Esser, 2013; Shalev, Manor, & Gross-Tsur, 2005), attributable both to genetic and environmental factors (Alarcon, DeFries, Light, & Pennington, 1997; Baron-Cohen et al., 2014; Shalev et al., 2001). The first large-scale genetically sensitive investigation of number sense has recently revealed that basic numerical understanding is only modestly heritable (32%), with individual differences being primary explained by nonshared environmental influences (68%) (Tosto et al., 2014). However, this heritability estimate may not be applicable to all forms of DD (see Chapter 11, for a thorough review of genetic contributions to mathematical learning and learning disorders).

DD is a heterogeneous disorder resulting from individual deficits in basic numerical or arithmetic functioning at behavioral, cognitive/neuropsychological, and neuronal levels (Kaufmann et al., 2013; Rubinsten & Henik, 2009). However, it has to be considered that difficulties in numerical understanding can reflect individual differences in both numerical and nonnumerical functions. Children with DD can have problems in one or several of the quantitative abilities in Table 1 (Geary, 1993, 2004; Heine, Engl, Thaler, Fussenegger, & Jacobs, 2012).

Regarding domain-general abilities, impairments in visuospatial working memory have often been reported in children with DD (Ashkenazi, Rosenberg-Lee, Metcalfe, Swigart, & Menon, 2013; Rotzer et al., 2009; Szucs, Devine, Soltesz, Nobes, & Gabriel, 2013). However, this is not actually the case for many children with a math learning disability (Geary, Hoard, Nugent, & Bailey, 2012; Geary et al., 2007). Hence, difficulties in visuospatial working memory likely holds for only some subsets of DD children.

Generally, comorbidities are much more common than pure DD, which applies to only a minority of cases. Rubinsten and Henik (2009) propose different frameworks for the origin of DD or mild learning disabilities and their cognitive deficits. First, a unique core deficit in processing numerosities (i.e., apprehending the approximate quantity of collections of items) is assumed because of distinctive abnormalities in areas of the parietal cortex that support the processing of numerical magnitudes. Butterworth, Varma, and Laurillard (2011) similarly emphasize a core deficit in understanding sets and their numerosities, which the

TABLE 1 Examples of Difficulties in Numerical Skills Frequently Shown by Children with DD

Magnitude processing	Difficulties of nonsymbolic magnitude estimation or comparisons, problems in subitizing—the immediate enumeration of small quantities without counting
Counting skills	Lack of one-to-one correspondence of numbers and objects, no understanding of irrelevance of counting order
Transcoding	Between number words, digits, and quantities
Spatial number representation	Problems in number-line tasks and immature characteristics of spatial number representation (logarithmic rather than linear)
Arithmetic	Immature and inefficient calculation procedures, such as counting or calculation with fingers
Arithmetical fact knowledge	Deficits in the establishment and retrieval of mathematical facts
Procedural knowledge	Mistakes while executing written calculations, deficits in the use of arithmetic symbols
Domain general, nonnumerical skills important for number processing	Working memory, attentional processes, or executive functions (planning, control, inhibition, monitoring, shifting)

authors claim to be fundamental to all aspects of elementary school mathematics. However, other research has challenged that a unique cognitive deficit is a possible cause of DD. For instance, Rousselle and Noël (2007) found that children with mathematical disabilities were only impaired when comparing Arabic digits, but not when comparing nonsymbolic magnitudes, suggesting a deficit in mapping numerals to the magnitudes they represent and not a deficit in the ability to represent these magnitudes.

Finally, a single restricted biological deficit as the origin of DD is not in line with current neurobiological findings which indicate functional and structural abnormalities in a widespread network across cortical and subcortical brain regions (Fias, Menon, & Szucs, 2013; Kucian et al., 2014; see also Chapter 8). Taken together, the heterogeneous behavioral deficits, the common occurrence of comorbidities, and the neuronal underpinnings of DD make alternative frameworks rather plausible. For example, Rubinsten and Henik (2009) suggest that multiple cognitive deficits in DD might be explained by multiple brain dysfunctions or that dysfunction in central brain areas, such as the parietal lobes, could result in multiple cognitive or behavioral manifestations.

As described earlier, arithmetical difficulties are frequently associated with other learning problems, most often in reading or writing (dyslexia "or dysgraphia") (Landerl & Moll, 2010) or attentional deficits (ADHD) (Rubinsten, 2009; von Aster, Schweiter, & Weinhold Zulauf, 2007). Probably as a consequence of these multiple learning difficulties found in DD, many children develop additional psychiatric disorders, such as anxieties (see Kohn, Richtmann et al., 2013 for a new math anxiety questionnaire for German-speaking children), depression, or aggressive behavior. As a consequence, such secondary symptoms of DD can lead to refusal to attend school and impairments in other school topics.

In sum, quantitative abilities comprise a multitude of different numerical and nonnumerical competencies, and children with DD differ in their individual profiles of strength and weaknesses in these competencies. Therefore, a diagnosis of DD should be based on multidimensional assessments tracking different numerical and arithmetical processes and relevant domain-general abilities, as well as neurological and socioemotional functioning.

Diagnosis of Developmental Dyscalculia

According to Kaufmann and von Aster (2012) a detailed diagnostic evaluation is needed when dyscalculia is suspected in order to take proper account of the complexity of this learning disorder and to produce an accurate picture of the affected child's particular strengths and weaknesses. In view of the multiplicity of the functional components comprising this learning disorder and the wide range of potential comorbidities, it is clear that the diagnostic evaluation must go beyond the strictly mathematical components to include a thorough personal, familial, and scholastic developmental history. The evaluation of the child's general cognitive development must consider nonnumerical cognitive domains, social factors, emotional well-being, and, where appropriate, findings from neurology and neuroimaging. Although most children who struggle with numbers and calculation are diagnosed at school age, children who are potentially at risk could be identified much earlier.

In children of kindergarten age, specific precursor skills can be identified that have been found to be reliable predictors of later calculation abilities (Aunio & Niemivirta, 2010; Stock, Desoete, & Roeyers, 2010). For instance, the understanding of quantity, subitizing, mastery of counting skills, or the identification of Arabic numerals can already be impaired in preschool children. However, not every preschooler who exhibits impairments in one or more of these domains will necessarily develop DD. At this early age it has to be taken into account that there are substantial individual differences in rates of development, and therefore children differ also in the mastery of numerical skills. In other words, some children might simply be slow learners but not necessarily dyscalculic.

In the International Classification of Diseases (ICD, 10th revision) of the World Health Organization (World Health Organization, 2010) as well as in the former version of the *Diagnostic and Statistical Manual of Mental Disorders* (DSM, 4th revision) of the American Psychiatric Association (APA, 2013), DD has been defined as a domain-specific learning disorder that emerges at an early stage of development and cannot be explained by inappropriate schooling or deficient learning opportunities. Domain specificity commonly has been described by a discrepancy between general intelligence and specific ability. This concept implies a rather static and endogenous origin of the disorder, which is challenged by recent research and the power of environmental factors, ranging from cultural circumstances (e.g., nature and extent of schooling, nature of the counting system) to the effects of pre-/postnatal illness or socioemotional adversity (Kaufmann et al., 2013; Kucian et al., 2014). In addition, there is growing empirical evidence that children with numerical difficulties and average as well as below-average intellectual levels show identical patterns of difficulties in math learning (Ehlert, Schroeders, & Fritz-Stratmann, 2012; Moser Opitz, 2007).

It should also be kept in mind that numerical understanding, mathematical capabilities, and general cognitive skills measured by intelligence tests are partly overlapping constructs (for more detailed discussion see Ehlert et al., 2012). Many intelligence test batteries include subtests requiring numerical abilities. Therefore, it is very difficult, if not impossible, to disentangle cognitive functions related to number processing and/or psychometric intelligence. Accordingly, in the recently published 5th revision of the DSM (APA, 2013) the specific developmental disorders of scholastic skills have been grouped together as dimensions within a single category, in each of which there can be a greater or lesser degree of impairment; the psychometric demonstration of a discrepancy to otherwise normal intelligence is no longer required. This change has been introduced to take better account of the heterogeneity of these disorders with respect to performance profiles and comorbidities, and thus to improve the clinical utility of the diagnosis.

A recent approach focused on diagnosing DD on the basis of individual differences in brain activation (Dinkel, Willmes, Krinzinger, Konrad, & Koten, 2013). Despite the promising results obtained in this study, the authors emphasized that brain-based diagnostics are not as good as standard diagnostic tests that are based on behavioral measures only, but as the following sections demonstrate, might be a valuable extension in the near future.

What Neuroimaging Is Telling Us about Developmental Dyscalculia

Converging evidence is growing that DD is associated with different alterations in brain function and brain structure. Recent work in the field of DD has examined the neural aspects of this learning disorder by means of contemporary brain imaging techniques such as electrophysiology and magnetic resonance imaging (MRI). Using these methods (see below), we are able to generate high-resolution anatomical images of our brains, examine fiber tracts, gain metabolic insights, observe brain activation, or measure temporal processes while participants are performing a numerical task.

Magnetic Resonance Imaging (MRI)

MRI produces brain images non-invasively by a powerful magnet and radiofrequency. Different MRI acquisition methods provide information about various aspects of our brains. The recording of high resolution anatomical brain images allows differentiation between gray and white brain matter and investigation of focal differences in brain structures. Alternatively, diffusion tensor imaging (DTI) enables the measurement of the integrity of fiber connections between different brain regions, and magnetic resonance spectroscopy (MRS) measures brain chemistry to study changes of various brain metabolites. The latter can provide information on alterations of neurotransmitter concentrations, energy balance of brain cells, the integrity of cell membranes, or neuronal loss. DTI provides information about white matter microstructure noninvasively. Diffusion of water molecules is restricted by axons in white matter. This directional dependence of diffusion can be quantified as fractional anisotropy (FA), which reflects both the size and number of myelinated axons, and the coherence of axonal orientation. Reduced FA is a marker of white matter integrity, and can point to a disruption or disorganization of white matter tracts, which has been demonstrated in a variety of diseases (details about DTI can be found in a review article by Le Bihan et al., 2001). Finally, functional magnetic resonance imaging (fMRI) uses the change in oxygen levels of the blood in active brain areas to create images of brain regions that are active during a specific task (see also Table 2).

TABLE 2 Brain Imaging Methods

Method	Procedure	Images	Advantages/Disadvantages
Magnetic resonance imaging (MRI)	Powerful magnet and radio-frequency	• Brain activity measurement by oxygenation changes of blood • Anatomical brain structure • Fiber tracts by diffusion tensor images (DTI) • Metabolites by magnetic resonance spectroscopy (MRS)	+ Noninvasive + Whole brain measurement + High spatial resolution – Low temporal resolution – Expensive – Sensitive to head motion
Positron emission tomography (PET)	Injection of radioactive tracer into the blood system	Brain activity measurement by oxygenation changes of blood	– Invasive + Whole brain measurement + Very high spatial resolution – Low temporal resolution – Expensive
Electroencephalography (EEG)	Electric fields are measured with electrodes placed on the scalp	Brain activity measurement of depolarization of synchronously active neurons	+ Noninvasive + Whole brain measurement + High-temporal resolution – Low-spatial resolution + Portable

(Continued)

TABLE 2 Brain Imaging Methods—Cont'd

Method	Procedure	Images	Advantages/Disadvantages
Magnetoencephalography (MEG)	Magnetic fields are measured with super-conductive coils placed on the scalp	Brain activity measurement of depolarization of synchronously active neurons	+ Noninvasive + Whole brain measurement + High-temporal resolution – Low-spatial resolution
Near infrared spectroscopy (NIRS)	Near infrared light is shined through the skull	Brain activity measurement by oxygenation changes of blood	+ Noninvasive – Only part of the brain can be measured + High-temporal resolution – Low-spatial resolution – Only brain surface can be measured + Cheap + Portable

Positron Emission Tomography

Positron emission tomography (PET) imaging enables the visualization of biochemical and physiological functions of the brain. A radioactive tracer is injected into the blood system. Areas of high radioactivity indicate high amounts of radioactive-labeled oxygen, and therefore are associated with brain activity; similar to the principle of fMRI, it is assumed that active regions are flooded with oxygenated blood (see also Table 2).

Electroencephalography/Magnetoencephalography (MEG)

Electrophysiological (EEG) methods register the electrical activity of neurons noninvasively. Depolarization of synchronously active neurons create electrical and magnetic fields that can be recorded at the scalp. While EEG measures the changes of the electric field with electrodes placed on the scalp, MEG records magnetic field changes by an arrangement of super-conductive coils. In contrast to fMRI and PET, which provide high spatial resolution but lower temporal resolution, electrophysiological methods measure across larger regions of the brain but can detect changes of brain activation in the millisecond range. See also Table 2.

Near Infrared Spectroscopy

In near infrared spectroscopy (NIRS), near infrared light is shined through the skull; the light travels through the outer layers of the brain and is measured by a nearby receiver as it leaves the brain. By measuring the quantity of returning photons, one can infer the spectral absorption of the underlying tissue and make some conclusions about its average oxygenation and deoxygenation. Therefore, NIRS can be used for noninvasive assessment of brain function by detecting changes in oxygen concentrations in the blood, which are associated with neural activity. However, it should be noted that a disadvantage of NIRS is that due to the shallow depth of penetration, only changes in the brain activity of surface areas can be measured (see also Table 2).

Neuronal Correlates of Developmental Dyscalculia

Numbers in the Adult Brain

In the last few years, we have gained a clearer picture of functional processes in the typical adult brain during number processing and calculation, with brain imaging revealing the need for an entire neuronal network to carry out such operations. Within this distributed network the intraparietal sulcus (IPS) has been confirmed as the core locus of numerical processing. The IPS is activated whenever numerical magnitude is implicated, even in the absence of numerical task demands where, for instance, subjects may only have to look at changing amounts of dots (for review see Butterworth & Walsh, 2011; Kaufmann et al., 2005; Notebaert, Nelis, & Reynvoet, 2011).

Numerical and arithmetical processing also engage domain-general capacities such as memory, attentional, perceptual, motor, and spatial functions (Cho et al., 2012; Rivera, Reiss, Eckert, & Menon, 2005). Particularly, the prefrontal cortex has been associated with general cognitive processes with emphasis on its role in monitoring, strategy choice, and planning or manipulating information, as required in calculation tasks (Christoff & Gabrieli, 2000).

Taken together, number processing and calculation is a demanding cognitive ability which is processed by a complex neuronal network. In addition to the key areas for numerical cognition located in the parietal lobes, the prefrontal cortices and regions associated with the dorsal and ventral visual pathways, as well as subcortical areas and the cerebellum all play a significant role in numerical tasks (for reviews, see Arsalidou & Taylor, 2011; Menon, 2014). Figure 1

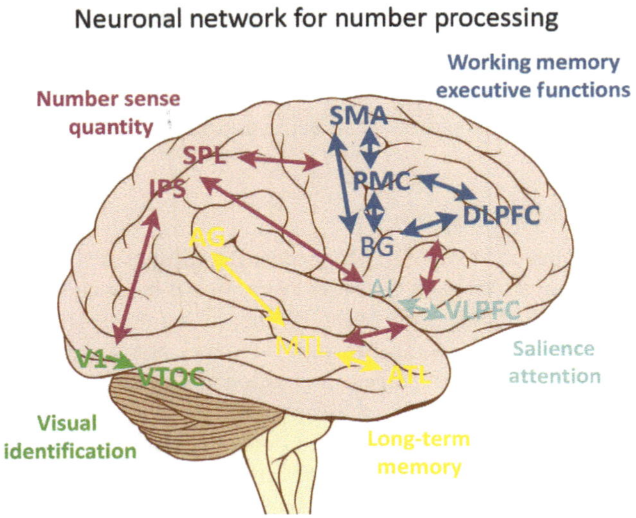

FIGURE 1 Illustrative model of the neuronal network for number processing and calculation. Regions in the occipital cortex (shown in green, primary visual cortex (V1), ventral temporal-occipital cortex (VTOC)) decode the visual input (magnitudes, Arabic digits, etc.). Together with the intraparietal sulcus (IPS) and superior parietal lobe (SPL) a visuospatial representation of quantity is formed. Regions in the parietal lobe (shown in pink) represent the core areas of numerical understanding, also referred to as the site of our number sense in the brain. Working memory and executive function circuits (shown in blue) involving parietal areas (IPS, supramarginal gyrus) and prefrontal cortex (premotor cortex (PMC), supplementary motor area (SMA), dorsolateral prefrontal cortex (DLPFC)) together with subcortical areas (basal ganglia (BG)) create a hierarchy of short-term representations that allow manipulation of numerical input over several seconds. Executive functions, such as planning, control, problem solving, or reasoning also play a relevant role during numerical problem solving. Long-term memory formation (shown in yellow) of numerical facts is anchored in the angular gyrus (AG) in the parietal cortex as well as in temporal regions including the anterior temporal (ATL) and medial temporal lobe (MTL), which consists of important subcortical areas such as the hippocampus and the amygdala. Finally, prefrontal processes related to saliency and attentional functions including the anterior insula (AI) and ventrolateral prefrontal cortex (VLPFC) guide and maintain behavior in goal-directed problem solving. Subcortical structures (BG, MTL) are indicated by finer fonts. *Figure is adapted from Menon (2014).*

illustrates a model of the neuronal network for number processing and calculation adapted from Menon (2014).

Typical Development of Number Representations in the Brain

During development, neurocognitive systems are still immature, and functional differentiation of specific brain regions has either not yet taken place or remains incomplete in children (Berl, Vaidya, & Gaillard, 2006; Karmiloff-Smith, 1998; Kaufmann, Kucian, & von Aster, 2014; Kucian & Kaufmann, 2009). With respect to the development of numerical cognition, it has been argued that with increasing age and experience, number representations and their interconnections undergo qualitative changes (Kucian & Kaufmann, 2009).

Converging evidence suggests an anterior-posterior shift of brain activity associated with number processing and arithmetic during development (Ansari, Garcia, Lucas, Hamon, & Dhital, 2005; Kucian, von Aster, Loenneker, Dietrich, & Martin, 2008; Rivera et al., 2005). These changes are thought to reflect decreasing reliance on domain-general, supporting (frontal) processing mechanisms perhaps related to the use of counting for enumerating and simple arithmetic, and increasing functional specialization of number-relevant (fronto-parietal) brain regions (Kaufmann et al., 2014). Moreover, the effectiveness of calculation potentially depends on the development of more sophisticated strategies requiring automated retrieval of arithmetical facts and processing of quantities as well as Arabic digits.

These maturational changes are reflected in performance differences and accompanied brain functions. On the one hand, an increased and more focal recruitment of brain areas, such as the intraparietal sulcus, seems to be associated with number processing as it becomes more automatized. On the other hand, the stronger reliance on numerical fact retrieval is assumed to be reflected by an increased activation of left hemispheric perisylvian brain areas and the contribution of hippocampal-prefrontal circuits (Cho, Ryali, Geary, & Menon, 2011; Cho et al., 2012; Qin et al., 2014). For instance, Qin et al. (2014) showed that children's increased use of memory-based strategies was characterized by increased hippocampal engagement, decreased prefrontal-parietal engagement, and increased functional connectivity between the hippocampus and multiple cortical regions (see also Figure 1).

It is not simply age and level of numerical competence that influences patterns of brain activation; it is the specific numerical skill needed to solve the problem, which constitutes a special challenge in comparing different imaging studies. Hence, findings derived from mature brain systems are not readily applicable to developing brain systems (Karmiloff-Smith, 1998; Kaufmann et al., 2011). Immature brain systems show higher inter-individual variance, which is even more pronounced in children with a specific developmental disorder, such as dyscalculia.

Despite the relatively high prevalence of DD, only a few imaging studies have addressed the question of neuronal correlates of this specific learning deficit. Reasons for this might be that the examination of children is demanding, especially brain imaging, which requires special care, patience, and expertise when measuring young subjects. For instance, data quality of many imaging methods is dependent on complete motionlessness during recording, which is very difficult for many children. Another challenge lies in the heterogeneous nature and high comorbidity rates of DD. Despite these obstacles, few studies have met the challenge and provided initially important insights into the neural correlates of DD. In the following sections, these studies of brain function and the effects of development or intervention on neuronal systems are summarized, along with findings from brain structure and metabolism.

Deficient Functional Networks

Functional MRI or electrophysiological studies that include children with DD are comparatively scarce, as noted. In addition, the studies that have been conducted have used different paradigms and tasks, including symbolic (e.g., Arabic digits) or nonsymbolic (e.g., dot patterns) numerical stimuli, active tasks where children were asked to give an answer, or passive adaptation paradigms where no answer was required, and either basic number processing or arithmetic calculations tested at different difficulty levels. All these variables have a significant impact on corresponding brain functions. Furthermore, the use of different definitional criteria and diagnostic tools for classifying children with DD across studies yielded different neuropsychological profiles, thus further limiting comparisons of obtained results. As a consequence of all these diverse factors, it remains difficult to draw a clear picture of the functional brain correlates of DD. Nevertheless, current findings indicate consistent dysfunctions of the parietal lobe during number processing and calculation (Ashkenazi, Rosenberg-Lee, Tenison, & Menon, 2012; Berteletti, Prado, & Booth, 2014; Kaufmann, Vogel, Starke, Kremser, & Schocke, 2009; Kaufmann, Vogel, Starke, Kremser, Schocke, et al., 2009; Kucian, Loenneker, Martin, & von Aster, 2011; Kucian et al., 2006; Kucian, Grond, et al., 2011; Mussolin et al., 2010; Price, Holloway, Räsänen, Vesterinen, & Ansari, 2007; Rotzer et al., 2009; Soltesz, Szucs, Dékány, Márkus, & Csépe, 2007); for a review, see the meta-analysis of Kaufmann et al. (2011).

Aberrant Brain Activation in Number-Related Areas in DD

The majority of studies have shown a reduced level of brain activation in parietal areas in DD (Berteletti et al., 2014; Kucian, Grond, et al., 2011; Kucian et al., 2006; Mussolin et al., 2010; Price et al., 2007; Rotzer et al., 2009). Although these findings are based on different numerical tasks ranging from rather basic number sense (nonsymbolic distance effect Price et al., 2007) to the understanding of ordinality (Kucian, Grond, et al., 2011), from symbolic number comparison (Mussolin et al., 2010) to arithmetic (multiplication or approximate addition, Berteletti et al., 2014; Kucian et al., 2006), and even on nonnumerical skills that are processed by overlapping networks (spatial working memory, Rotzer et al., 2009), coincident reduction of brain activation in the parietal lobe has been reported. These findings suggest deficits in children with DD in the core region for number processing (see also Figure 1) and might be seen as a direct neuronal correlate of MLDs.

Brain activation in the parietal lobe is related to working memory load. Numerical tasks with varying difficulty levels usually induce increased brain activation for more complex problems (e.g., Price et al., 2007). However, brain activation in people with DD does not seem to be modulated in the same way as found in typically developing people. Some studies have demonstrated a lack of modulation of parietal brain activation related to task complexity with greater activation for more difficult problems in DD (Ashkenazi et al., 2012, 2013; De Smedt, Holloway, & Ansari, 2011; Mussolin et al., 2010; Price et al., 2007; Soltesz et al., 2007).

An electrophysiological study by Soltesz et al. (2007) revealed no differences between adolescents with or without DD. However, late EEG signals appeared over several parietal electrodes in typically achieving adolescents that did not emerge for adolescents with DD. The authors suggested that early, more automatic processing of digits is similar in both groups, and that differences arose later during more complex controlled processing.

In contrast to reported reduction of brain activation in the parietal lobe, two studies have reported enhanced activity in parietal regions (Kaufmann, Vogel, Starke, Kremser, Schocke, et al., 2009; Rosenberg-Lee et al., 2014). In addition, Rosenberg-Lee et al. (2014) demonstrated hyperconnectivity, rather than reduced connectivity, between the intraparietal sulcus and the prefrontal and parietal cortices, which adds further support to the claims that the intraparietal sulcus is a major locus of dysfunction in DD. However, the authors argue that the associated dysfunctions may range from increased parietal activation reflecting compensatory mechanisms (Kaufmann, Vogel, Starke, Kremser, Schocke, et al., 2009) to inappropriate connectivity, rather than gross underactivation (Rosenberg-Lee et al., 2014).

In sum, the most prominent differences between dyscalculics and controls are evident in or in close vicinity to the intraparietal sulcus, indicating a deficient neuronal representation of numerosity (see also Figure 1). However, hypo- or hyperactivation, exact locations within the parietal lobes, and their laterality vary among studies. Divergent results might derive from differences between studies regarding inclusion criteria of children with DD, conducted paradigms, and fMRI parameters. Corroborating evidence that the parietal lobe indeed plays a crucial role in dyscalculia comes from simulated interruption of right parietal activity by transcranial magnetic stimulation (TMS), which evokes dyscalculic-like behavioral characteristics in typically developing adult volunteers (Cohen Kadosh et al., 2007; also see Chapter 10).

Aberrant Brain Activation in Domain-General Areas in DD

It has to be kept in mind that even the easiest arithmetic problem poses a substantial cognitive challenge in terms of neuronal processing and networks from different brain areas required for a successful solution (see also Figure 1). Hence, it is not surprising that dysfunctions in areas other than the parietal lobes might contribute to the occurrence of DD. There is growing evidence that the development of specific math difficulties might also be due to deficits in other or additional brain regions. In particular, several imaging studies have reported impaired brain activation in distributed areas of the frontal lobe, including the superior, middle, and inferior frontal gyrus, medial prefrontal gyrus, the paracentral frontal lobe, the cingulate gyrus, or the insula (Kaufmann et al., 2011; Kucian, Grond, et al., 2011; Kucian, Loenneker, et al., 2011; Kucian et al., 2006; Mussolin et al., 2010; Price et al., 2007; Rosenberg-Lee et al., 2014; Rotzer et al., 2009). These frontal areas are thought to subserve domain-general executive functions. A variety of cognitive control, regulation, or managing processes are included in the term executive functions, such as planning, working memory, attention, problem solving, inhibition, mental flexibility, or initiation and monitoring of actions. Many are mandatory for different components of number processing and calculation, including an accurate and automated access to mental number representations.

In line with aberrant brain activation in domain-general areas, different studies have reported deficits in various facets of executive functions in DD, such as working memory (Ashkenazi et al., 2013; Geary et al., 2012; Kaufmann, Lochy, Drexler, & Semenza, 2004; Rotzer et al., 2009; Szucs et al., 2013), recruiting attention (Ashkenazi & Henik, 2010; Geary et al., 2012), or problem solving (Osmon, Smerz, Braun, & Plambeck, 2006; Soltesz et al., 2007). Moreover, there is growing knowledge of brain regions related to executive processes. For instance, the right inferior frontal gyrus, supplementary motor area, and basal ganglia are necessary for successful inhibition and, whereas medial frontal brain regions situated on the medial surface of the hemisphere are commonly associated with monitoring behavior (detection of errors, conflict processing), middle frontal regions are supposed to be related to adjusting behavior after conflicts or errors (Verbruggen & Logan, 2008). Future research will have to clarify the relationship between impairments of different areas within the frontal lobes and decelerated executive functioning in DD.

Similar to the lack of brain modulation in parietal areas, Berteletti et al. (2014) described an absence of brain modulation by problem size (i.e., easier vs. harder problems, such as $2+1$ vs. $8+9$) in verbal regions of the frontal lobe, suggesting that children with math difficulties do not consistently retrieve solutions verbally even for smaller problems. In terms of reported difficulties in spatial working memory in DD, Ashkenazi et al. (2013) showed that children with math difficulties likewise show no modulation of brain activity in the prefrontal cortex during arithmetic problem solving (e.g., $3+4=8$ correct or incorrect?) related to visuospatial working memory performance. The authors conclude that, unlike typically developing peers, children with math difficulties do not use visuospatial working memory resources appropriately during arithmetic problem solving.

In addition to these deficits in frontal regions, occipito-temporal areas and deep brain structures also seem to be affected in children with DD. For instance, the left fusiform and lingual gyrus were reported to be deficient in dyscalculic children, presumably associated with complications in Arabic digit processing (Kaufmann et al., 2011; Price et al., 2007; Rosenberg-Lee et al., 2014). Abnormal activation in subcortical areas, such as the thalamus, basal ganglia, hippocampus, or the parahippocampal gyrus is present in children with DD, which even broadens the impaired network to include a disruption of cortico-subcortical loops (Ashkenazi et al., 2013; Delazer et al., 2004; Kucian, Grond, et al., 2011; Kucian et al., 2006). On the basis of the fact retrieval deficits that are found in some children with DD (Geary, 1993) and the role of the hippocampus in the development of memories for arithmetic facts (Qin et al., 2014), the hippocampus might also be involved in some forms of DD. However, this remains to be assessed.

Compensatory Mechanisms in DD

As regards the wide range of reported abnormalities in neuronal networks for number processing and calculation, it might appear surprising that children with DD are still able to perform some numerical tasks quite well or even at the same level as their typically achieving peers. Different neuroimaging studies have reported equivalent behavioral performance, despite differences in brain activation patterns between children with and without DD (Kaufmann, Vogel, Starke, Kremser, & Schocke, 2009; Kovas et al., 2009; Kucian, Loenneker, et al., 2011; Kucian et al., 2006; Mussolin et al., 2010; Soltesz et al., 2007). Such results raise the question of possible compensatory mechanisms that children with DD might develop to cope with their learning disability. In fact, it has been reported that affected children show stronger activation in certain brain regions that might account for such compensatory mechanisms, including areas adjacent to regions in the frontal lobes (supplementary motor area, post-central gyrus, superior frontal lobe, paracentral frontal lobe, anterior cingulate cortex), the left lingual gyrus, the thalamus, and even inferior parietal areas (Kaufmann, Vogel, Starke, Kremser, & Schocke, 2009; Kaufmann et al., 2011; Kovas et al., 2009; Kucian, Loenneker, et al., 2011; Mussolin et al., 2010).

Kaufmann and colleagues (2011) argued that increased activation of more anterior parietal areas reaching into the post-central gyrus might be a result of the dependence on finger-based strategies (e.g., using fingers to keep track of counting during problem solving) when solving simple arithmetical tasks. Moreover, stronger activation in frontal areas suggests that dyscalculic children need to exert more effort than their typically achieving peers to solve the same numerical problems. To achieve the same level of performance, children with DD show greater engagement of visual working memory, attention, or response selection than their typically achieving peers (Kaufmann, Vogel, Starke, Kremser, & Schocke, 2009; Kaufmann et al., 2011; Kucian, Loenneker, et al., 2011; Mussolin et al., 2010).

To summarize, based on existing literature, it can be concluded that the activation pattern in children with DD is less precise, and key deficits are apparent in core regions for number processing, which predominantly include parietal regions. However, other cortical and subcortical regions that contribute to numerical cognition can also be affected. The stronger recruitment of supporting areas associated with working memory, attention, monitoring, updating, or finger representation may reflect compensatory mechanisms in dyscalculic children. The increased need for supportive functions might be explained by deficient development and automated access to adaptive number representations in children with DD.

Changes of Brain Function Due to Development and Intervention

As outlined in the previous section, there is clear evidence for differences in brain function between typically developing children and children with DD. However, as these findings represent only momentary snapshots in time, studies examining the developmental trajectories of children with DD would provide greater insights into the dynamic processes of regular development, learning, or remediation.

On a behavioral basis, several longitudinal studies indicate that children with DD do improve over time, but most of them still perform more poorly than their peers (Desoete, Ceulemans, De Weerdt, & Pieters, 2012; Geary, Hamson, & Hoard, 2000; Geary et al., 2012; Kohn et al., 2013; Landerl, 2013; Shalev et al., 2005). Hence, dyscalculia can be seen as a learning disorder that persists into adulthood. In line with the improvements observed on a behavioral level in children with DD over time, an initial study has recently examined developmental changes in functional brain networks from a longitudinal perspective (Grond et al., 2014). Findings indicated a continuation in neuronal development of number representation in children with DD over a four-year period. Brain imaging results suggested an increasing age-related activation in parietal regions, pointing to a delayed development of number processing areas. Second, increased activation in frontal areas was observed over time, indicating a possible use of compensatory mechanisms or different but less effective problem-solving strategies. On the one hand, these findings are promising as they clearly show that functional brain networks positively develop in children with DD. On the other hand, it is important to note that children with DD still showed less-focused activation patterns underscoring that the deficiencies do not vanish with time.

Our brain is a highly plastic organ and adapts constantly—not only to typical development, but also to special stimulation such as remediation. Up to now, only one study has tracked changes in brain function that followed specific intervention in children with DD (Kucian, Loenneker, et al., 2011). Children were trained with a computer-based intervention for five weeks. The aim of the training was to improve number representations and the close link of numbers to spatial properties on the internal mental number line. Specifically, either an Arabic digit, an addition problem, a subtraction problem, or a number of dots were shown on a spaceship, and the children were asked to land the spaceship at the corresponding location on a 0–100 number line. Results showed that children with and without DD enhanced their spatial number representations. In addition, a positive transfer effect was demonstrated by improved arithmetic abilities, although the children were never trained in exact arithmetic. In addition to behavioral improvements, the training was followed by a modulation of brain functions. Functional MRI depicted a decrease in supporting brain activation mainly in the frontal lobes after the training. Therefore, it seems that after completion of the training, the fMRI task puts less demand on quantity processing and working memory and requires less attentional effort. A follow-up examination five weeks after training revealed a significant increase of activity in parietal areas for the children with DD, including the intraparietal sulcus, bilaterally. Since the intraparietal sulcus is known to play a pivotal role in number representation, the authors speculate that time for consolidation after training was needed to establish neuronal representation.

There is an urgent need to transform basic knowledge about neurocognitive development into evidence-based, practical applications for special needs education and remediation. Hence, further studies examining the effects and neuronal correlates of different intervention strategies are warranted to develop adaptive and differential indications for the particular needs of individual children (see detailed discussion in Cohen Kadosh, Dowker, Heine, Kaufmann, & Kucian, 2013; Ise, Dolle, Pixner, & Schulte-Körne, 2012). This may also include techniques such as transcranial direct current stimulation (TDCS) that are supposed to facilitate learning of numerical abilities by specific stimulation of relevant brain regions during cognitive training (Cohen Kadosh, Soskic, Iuculano, Kanai, & Walsh, 2010, Chapter 10).

Abnormal Neuronal Macro- and Microstructures

Compared to brain imaging studies investigating differences in brain function in DD, structural or metabolic correlates of DD have received less attention from developmental neuroscientists. The acquisition of high-resolution anatomical brain scans by means of MRI allows the examination of structural brain differences between subjects with and without DD. In more detail, these images offer an opportunity to differentiate between gray and white brain matter, and to compare structural differences in these volumes.

MRI also enables the measurement of connections between different brain regions using DTI. Recent findings point to possible disconnections between brain areas in children with DD. Both findings of macrostructures (gray and white matter volume), as well as microstructures (integrity of fiber connections) are discussed in the following sections.

Brain Structure

Brain anatomy can be examined by voxel-based morphometry (VBM), which is a neuroimaging analysis technique that allows investigation of focal differences in brain structure. VBM registers every brain to a template, which gets rid of most of the large differences in brain anatomy among people. Then the brain images are segmented into gray and white matter and cerebrospinal fluid. Finally, the local concentration of gray or white matter is compared between groups of subjects at every voxel. So far, only two studies approached the systematic examination of white and gray matter differences between children with DD and age- and gender-matched control children (Rotzer et al., 2008; Rykhlevskaia, Uddin, Kondos, & Menon, 2009). Results point to deficits in gray matter in the intraparietal sulcus and adjacent regions, including the superior parietal lobe (see Figure 2). Similarly, reduced gray matter has been reported in the parietal lobe in adolescents with very low birth weight and difficulties learning mathematics (Isaacs, Edmonds, Lucas, & Gadian, 2001). Because the parietal lobe in general, and particularly the intraparietal sulcus, are presumed to represent core

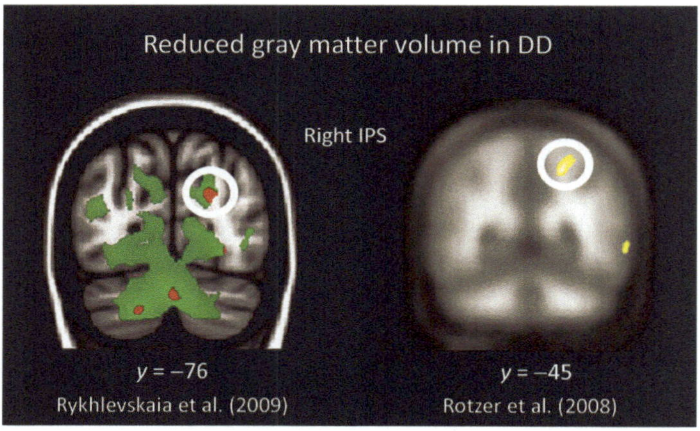

FIGURE 2 Reduced gray matter volume has been reported in the right IPS in children with DD by Rykhlevskaia et al. (2009) and Rotzer et al. (2008). On the left side reduced gray matter volume in 8.8-year-old children with DD is illustrated in red at $p<0.001$ (in green at $p<0.01$) by Rykhlevskaia et al. (2009). On the right side reduced gray matter volume in 9.3-year-old children with DD is illustrated in yellow at $p<0.001$ by Rotzer et al. (2008). Image is adapted with permission from Elsevier for figure from Rotzer et al. (2008) and with kind permission from Vinod Menon for figure from Rykhlevskaia et al. (2009).

regions implicated in the development of numerical skills, reduced gray matter volumes in these areas might represent the morphometric correlate of atypical number development. Therefore, gray matter volume in parietal areas seems to be a crucial indicator for the accurate development of numerical skills.

The intraparietal sulcus is thought to be the neural substrate most closely associated with DD. However, recent results also argue for a more distributed neuronal pattern. For example, reported gray matter deficits in children with DD suggest widespread but subtle malformations in the brain. Reduced gray matter volume has also been reported in the anterior cingulate cortex, the left inferior frontal gyrus, and the dorsolateral prefrontal cortex (Rotzer et al., 2008). Based on the fact that an important component in the development of arithmetic skills is the growth of memory for numerical information, the authors argue that deficits in these frontal areas may be of major importance in the development of dyscalculia. Similarly, VBM analysis has revealed that individual differences in the arithmetic scores of typically achieving children are significantly and positively correlated with the gray matter volume in the left intraparietal sulcus (Li, Hu, Wang, Weng, & Chen, 2013).

The ventral visual stream seems to be affected as well (Rykhlevskaia et al., 2009). In particular, children with DD showed deficits in the fusiform gyrus, parahippocampal gyrus, and right anterior temporal cortex which might hinder the development of semantic memory representations important for rapid numerical fact retrieval (Rykhlevskaia et al., 2009).

Regarding white matter integrity, the results of existing studies report reduced white matter volume in children with DD, but regions are not consistent (Rotzer et al., 2008; Rykhlevskaia et al., 2009). Whereas Rotzer et al. (2008) reported white matter deficits in the left frontal lobe and in the right parahippocampal gyrus in dyscalculic children, Rykhlevskaia et al. (2009) found reduced white matter volume in the right temporo-parietal region and the splenium of the corpus callosum. Deficits in these regions may have a negative influence on fact retrieval and spatial memory processing, as well as on visuospatial processes during the acquisition of some types of mathematical problem solving (Rotzer et al., 2008; Rykhlevskaia et al., 2009). Taken together, DD is characterized by reduced gray and white matter volume in areas that are thought to play a key role for the domain-specific development of numerical representations and abilities, as well as in regions important for the development of domain-general abilities.

Fiber Connections

Brain regions are linked and communicate by a complicated network formed by short fiber connections among different cortical and subcortical regions. Successful cognitive performance relies on the development and establishment of such well-organized networks in the brain. Fast and accurate connections between the different brain regions are crucial for efficient transfer and adjustment of information.

Thus far, only two DTI studies have examined microstructural differences in white matter between children with DD and typically achieving children (Kucian et al., 2013; Rykhlevskaia et al., 2009). The results emphasize the evidence of impaired fiber tracts connecting different brain areas of the fronto-parietal network known to be responsible for number processing and calculation. More specifically, Rykhlevskaia et al. (2009) reported long-range white matter projection fibers linking the right fusiform gyrus with temporo-parietal white matter as a specific source of vulnerability in children with DD. However, Rykhlevskaia et al. (2009) performed specific analyses of fiber tracts only in regions that showed reduced white matter volume in a prior analysis of brain structure. Taken together, the present results indicate the importance of deficient fiber connections as "a" neurological correlate of dyscalculia, along with additional fiber tracts that might be affected.

A more recent study (Kucian et al., 2013) examined the white matter integrity on a whole brain analysis of children with DD. Results highlighted a deficit of the superior longitudinal fasciculus, a fiber tract that connects parietal, frontal, and temporal areas. In particular, the superior longitudinal fasciculus seemed to be affected in parts that are adjacent to key areas for number processing, namely the intraparietal sulcus. The authors concluded that the development of axonal coherence and/or myelination projecting to/from the parietal lobe might be impaired or delayed in dyscalculia, leading to the possibility that DD might be seen as a disconnection syndrome (Figure 3).

Comparably, white matter integrity correlates significantly with mathematical abilities in typically developing children (Li et al., 2013; Tsang, Dougherty, Deutsch, Wandell, & Ben-Shachar, 2009; van Eimeren, Niogi, McCandliss, Holloway, & Ansari, 2008). Van Eimeren et al. (2008) reported that white matter microstructures in the left hemisphere seemed to be related to individual mathematical competencies in typically developing children. Particularly, the posterior part of the left inferior longitudinal fasciculus correlated specifically with achievement. White matter integrity in the inferior longitudinal fasciculus as part of the visual stream might be related to the efficiency with which children are able to process Arabic numerals presented in calculation tests (van Eimeren et al., 2008). Another study explicitly investigated the role of the anterior portion of the superior longitudinal fasciculus, a pathway connecting parietal and frontal areas (Tsang et al., 2009). The authors chose this tract based on consistent findings of brain activation in a fronto-parietal network during number processing and calculation. DTI analysis revealed a positive correlation of white matter integrity with arithmetic approximation skills in typically achieving children. Accordingly, lower approximation ability would be explained by deficiencies in one or several characteristics of efficient fiber tracts connecting parietal with frontal areas (Tsang et al., 2009). In line with this account, Li et al. (2013) reported that FA values in the tract pathways connecting left IPS were related to children's arithmetic achievement.

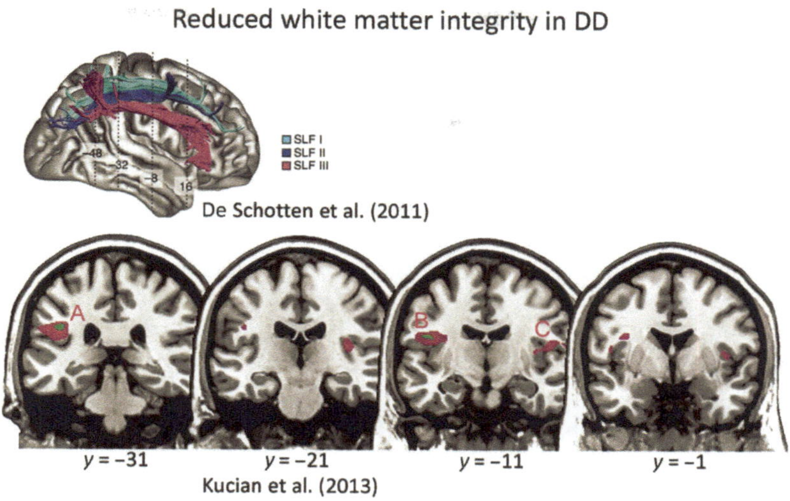

FIGURE 3 At the upper part, a three-dimensional reconstruction of the three branches of the superior longitudinal fasciculus (SLF) are illustrated (De Schotten et al., 2011). At the bottom, coronal slices show regions in purple where children with DD show reduced fractional anisotropy (FA) compared to control children at $p < 0.001$: *A* left SLF, close to parietal lobe; *B* left SLF, close to central operculum and secondary somatosensory cortex; *C* right SLF, close to insula. Within this pattern of reduced FA, increased radial diffusivity (RD) in dyscalculic children is marked in green. Figure is adapted with kind permission from Springer Science and Business Media (Kucian et al., 2013) and from Macmillan Publishers Ltd: Nature Neuroscience (De Schotten et al., 2011), copyright (2011).

Finally, a DTI study examining children with velocardiofacial syndrome, a condition associated with low arithmetic skills, reported reduced white matter integrity adjacent to left inferior parietal lobe and intraparietal sulcus (Barnea-Goraly, Eliez, Menon, Bammer, & Reiss, 2005). FA values in this area correlated significantly with arithmetic ability in this region, suggesting an important role of fiber projections from and to the left inferior parietal lobe.

In sum, little is currently known concerning specific impairments of brain connections that are probable neuronal correlates of DD. Moreover, results from typically achieving children or children with math difficulties due to other neurodevelopmental disorders are difficult to assign to DD. Nevertheless, the evidence to date provides additional support for the hypothesis that DD may be attributable to a disconnection or reduced connection between areas important for number processing.

Neurometabolites

As described in a previous section and Table 2, MRS provides a measure of brain chemistry that can be used as a noninvasive technique for studying changes in brain metabolites of different disorders. To our knowledge, the only report to date of this technique being used for investigating dyscalculia was a case study of an 18-year-old man with DD (Levy, Reis, & Grafman, 1999). MRS revealed specific and stable changes in a left temporo-parietal region near the angular gyrus, with a significant decrease in choline and creatinine levels, as well as a mild focal decrease in N-acetyl-aspartate. The authors concluded that changes in these brain metabolites are due to an alteration in cell density, impaired cellular energy metabolism in these neurons, and only a little general neuronal loss (Levy et al., 1999). It is difficult to draw strong conclusions based on only this one case study, but results point to significant abnormalities in brain chemistry near the angular gyrus, a region that has also been reported to play an important role in calculation and numerical fact retrieval (Grabner, Ansari, Koschutnig, Reishofer, & Ebner, 2011; Grabner, Ansari, et al., 2009; Grabner, Ischebeck, et al., 2009), and invites further research using MRS in children with DD.

CONCLUSIONS AND FUTURE DIRECTIONS

Although specific attributes of both brain morphology and functional activity have been shown to be correlated with certain patterns of mathematics ability, such neuroscientific evidence does not, of course, fully explain the nature and origins of DD. Brain imaging studies to date have provided some important albeit still rather vague answers to questions concerning whether, and if so how, the brains of children diagnosed with DD differ from those of their typically achieving peers. What we have learned so far is that DD is a complex and heterogeneous phenomenon that affects different components of mathematical development and requires interdisciplinary research and remediation. Moreover, comorbidities in children with DD is the rule rather than the exception. Keeping these thoughts in mind, it is not surprising that reported imaging findings are not limited to only a few brain areas and are sometimes even contradictory.

Because basic number processing and calculation depend on a complicated network that includes different and widespread brain areas, deficits in brain function, brain structure, interconnections, or metabolism have been reported across almost the entire network. Reported differences of affected brain regions are possibly explained by the heterogeneous character of dyscalculia. Regarding this wide range of abnormalities in neural networks for number processing and calculation in DD, it is difficult to draw a clear picture. However, based on the existing literature it can be concluded that the activation pattern of children with DD is less precise, and key functional and structural deficits are apparent in core regions for number processing, which mainly comprise parietal areas. However, other cortical and subcortical regions that contribute to numerical cognition can also be affected. The stronger recruitment of supporting areas associated with working memory, attention, monitoring, updating, or finger representation are supposed to reflect compensatory mechanisms in dyscalculic children, but could also reflect deficits in these domain-general skills that might contribute to the development of DD.

Our knowledge and understanding of the behavioral and neural characteristics of DD as well as the public's general awareness of this disorder has increased steadily. Although we are gaining a clearer picture of this learning disability, we should expand our perspectives on dyscalculia. First, current findings from neuroscience clearly illustrate that number processing and calculation require the integration of different co-activated brain networks. On the one hand, empirically grounded knowledge has highlighted the important role of the intraparietal sulcus. On the other hand, there is a wide range of reported, but up to now mostly neglected differences in brain function or structure in many other cortical and subcortical brain areas between typical and atypical developmental trajectories, such as frontal brain areas or the hippocampus. Therefore, future studies should focus on other brain regions in addition to the parietal lobe.

Second, restricting most research to the study of pure DD is missing the point that DD children with additional comorbidities are the rule rather than the exception. Although it is essential to keep one's studied populations as homogenous as possible to draw clear conclusions, future research should focus more on children who are representative of the most commonly occurring form of the disorder in the general population, namely DD children with comorbid disorders. In contrast, our understanding of DD might advance to the next level if we do not pool all children with DD together. Instead, dividing children into different subgroups with more homogenous difficulty profiles would probably permit a more precise identification of the underlying causes and behavioral consequences of DD. Also, following the behavioral studies, identifying the similarities and differences in the brain regions associated with severe and less severe forms of DD will be informative (e.g., Geary et al., 2012; Murphy et al., 2007).

Third, it is still unclear to what extent the link between brain regions activated in number processing and calculation might explain the neuronal deficits in DD. In other words, future studies should go beyond the mere localization of abnormalities and offer information on connectivity between brain networks and how such networks differ between groups of individuals. Fourth, present methods allow us to gain insights into DD from different perspectives, such as brain function, brain structure, brain metabolism, or the time course of neuronal processes by means of electrophysiology. The challenge now will be to integrate these findings by adopting multidimensional approaches to the study of DD. Although such investigations will be very demanding, they can provide an opportunity to define clearer models of the underlying impairments.

Fifth, secondary effects like specific math anxiety have been disregarded to a vast degree in behavioral studies of children with DD as well as in the investigation of neuronal underpinnings and in intervention studies. However, on the grounds that many children with DD suffer from accompanying anxieties, one should not neglect the significant negative effects of anxiety symptoms on learning. Moreover, it is unclear to what extent observed brain activation increases or decreases, or even brain structural alterations are due to elevated anxiety levels or corresponding emotional control mechanisms. Finally, our future efforts should pursue the goal of improving the appropriate support for children with DD. This would start with a uniform definition and diagnosis of DD, followed by a regulated compensatory approach to any disadvantages and remediation that takes into account the heterogeneous nature of each child's disorder.

REFERENCES

Alarcon, M., DeFries, J. C., Light, J. G., & Pennington, B. F. (1997). A twin study of mathematics disability. *Journal of Learning Disabilities*, *30*, 617–623.

Ansari, D., Garcia, N., Lucas, E., Hamon, K., & Dhital, B. (2005). Neural correlates of symbolic number processing in children and adults. *Neuroreport*, *16*, 1769–1773.

APA. (2013). *Diagnostic and statistical manual of mental disorders*. Washington, DC: American Psychiatric Association.

Arsalidou, M., & Taylor, M. J. (2011). Is 2+2=4? Meta-analyses of brain areas needed for numbers and calculations. *NeuroImage*, *54*, 2382–2393.

Ashkenazi, S., & Henik, A. (2010). Attentional networks in developmental dyscalculia. *Behavioral and Brain Functions*, *6*, 2.

Ashkenazi, S., Rosenberg-Lee, M., Metcalfe, A. W., Swigart, A. G., & Menon, V. (2013). Visuo-spatial working memory is an important source of domain-general vulnerability in the development of arithmetic cognition. *Neuropsychologia*, *51*, 2305–2317.

Ashkenazi, S., Rosenberg-Lee, M., Tenison, C., & Menon, V. (2012). Weak task-related modulation and stimulus representations during arithmetic problem solving in children with developmental dyscalculia. *Developmental Cognitive Neuroscience*, *2*(Suppl. 1), S152–S166.

Aunio, P., & Niemivirta, M. (2010). Predicting children's mathematical performance in grade one by early numeracy. *Learning and Individual Differences*, *20*, 427–435.

Barnea-Goraly, N., Eliez, S., Menon, V., Bammer, R., & Reiss, A. L. (2005). Arithmetic ability and parietal alterations: A diffusion tensor imaging study in velocardiofacial syndrome. *Cognitive Brain Research*, *25*, 735–740.

Baron-Cohen, S., Murphy, L., Chakrabarti, B., Craig, I., Mallya, U., Lakatosova, S., et al. (2014). A genome wide association study of mathematical ability reveals an association at chromosome 3q29, a locus associated with autism and learning difficulties: A preliminary study. *PLoS One*, *9*, e96374.

Berl, M. M., Vaidya, C. J., & Gaillard, W. D. (2006). Functional imaging of developmental and adaptive changes in neurocognition. *NeuroImage*, *30*, 679–691.

Berteletti, I., Prado, J., & Booth, J. R. (2014). Children with mathematical learning disability fail in recruiting verbal and numerical brain regions when solving simple multiplication problems. *Cortex*, *57*, 143–155.

Butterworth, B., Varma, S., & Laurillard, D. (2011). Dyscalculia: From brain to education. *Science*, *332*, 1049–1053.

Butterworth, B., & Walsh, V. (2011). Neural basis of mathematical cognition. *Current Biology*, *21*, R618–R621.

Cho, S., Metcalfe, A. W., Young, C. B., Ryali, S., Geary, D. C., & Menon, V. (2012). Hippocampal-prefrontal engagement and dynamic causal interactions in the maturation of children's fact retrieval. *Journal of Cognitive Neuroscience*, *24*, 1849–1866.

Cho, S., Ryali, S., Geary, D. C., & Menon, V. (2011). How does a child solve 7 + 8? Decoding brain activity patterns associated with counting and retrieval strategies. *Developmental Science*, *14*, 989–1001.

Christoff, K., & Gabrieli, J. D. E. (2000). The frontopolar cortex and human cognition: Evidence for a rostrocaudal hierarchical organization within the human cortex. *Psychobiology*, *28*, 168–186.

Cohen Kadosh, R., Cohen Kadosh, K., Schuhmann, T., Kaas, A., Goebel, R., Henik, A., et al. (2007). Virtual dyscalculia induced by parietal-lobe TMS impairs automatic magnitude processing. *Current Biology*, *17*, 689–693.

Cohen Kadosh, R., Dowker, A., Heine, A., Kaufmann, L., & Kucian, K. (2013). Interventions for improving numerical abilities: Present and future. *Trends in Neuroscience and Education*, *2*, 85–93.

Cohen Kadosh, R., Soskic, S., Iuculano, T., Kanai, R., & Walsh, V. (2010). Modulating neuronal activity produces specific and long-lasting changes in numerical competence. *Current Biology*, *20*, 2016–2020.

De Schotten, M. T., Dell'Acqua, F., Forkel, S. J., Simmons, A., Vergani, F., Murphy, D. G. M., et al. (2011). A lateralized brain network for visuospatial attention. *Nature Neuroscience*, *14*, 1245–1246.

De Smedt, B., Holloway, I. D., & Ansari, D. (2011). Effects of problem size and arithmetic operation on brain activation during calculation in children with varying levels of arithmetical fluency. *NeuroImage*, *57*, 771–781.

Delazer, M., Domahs, F., Lochy, A., Karner, E., Benke, T., & Poewe, W. (2004). Number processing and basal ganglia dysfunction: A single case study. *Neuropsychologia*, *42*, 1050–1062.

Desoete, A., Ceulemans, A., De Weerdt, F., & Pieters, S. (2012). Can we predict mathematical learning disabilities from symbolic and non-symbolic comparison tasks in kindergarten? Findings from a longitudinal study. *British Journal of Educational Psychology*, *82*, 64–81.

Desoete, A., Roeyers, H., & De Clercq, A. (2004). Children with mathematics learning disabilities in Belgium. *Journal of Learning Disabilities*, *37*, 50–61.

Dinkel, P. J., Willmes, K., Krinzinger, H., Konrad, K., & Koten, J. W., Jr. (2013). Diagnosing developmental dyscalculia on the basis of reliable single case fMRI methods: Promises and limitations. *PLoS One*, *8*, e83722.

Ehlert, A., Schroeders, U., & Fritz-Stratmann, A. (2012). Criticism of the discrepancy criterion in the diagnosis of dyslexia and dyscalculia. *Lernen und Lernstörungen*, *1*, 169–184.

Fias, W., Menon, V., & Szucs, D. (2013). Multiple components of developmental dyscalculia. *Trends in Neuroscience and Education*, *2*, 43–47.

Fischbach, A., Schuchardt, K., Brandenburg, J., Klesczewski, J., Balke-Melcher, C., Schmidt, C., et al. (2013). Prävalenz von Lernschwächen und Lernstörungen: Zur Bedeutung der Diagnosekriterien. *Lernen und Lernstörungen*, *2*, 65–76.

Geary, D. C. (1993). Mathematical disabilities: Cognitive, neuropsychological, and genetic components. *Psychological Bulletin*, *114*, 345–362.

Geary, D. C. (2004). Mathematics and learning disabilities. *Journal of Learning Disabilities*, *37*, 4–15.

Geary, D. C., Hamson, C., & Hoard, M. (2000). Numerical and arithmetical cognition: A longitudinal study of process and concept deficits in children with learning disability. *Journal of Experimental Child Psychology*, *77*, 236–263.

Geary, D. C., Hoard, M. K., Byrd-Craven, J., Nugent, L., & Numtee, C. (2007). Cognitive mechanisms underlying achievement deficits in children with mathematical learning disability. *Child Development*, *78*, 1343–1359.

Geary, D. C., Hoard, M. K., Nugent, L., & Bailey, D. (2012). Mathematical cognition deficits in children with learning disabilities and persistent low achievement: A five-year prospective study. *Journal of Educational Psychology*, *104*, 206–223.

Grabner, R. H., Ansari, D., Koschutnig, K., Reishofer, G., & Ebner, F. (2011). The function of the left angular gyrus in mental arithmetic: Evidence from the associative confusion effect. *Human Brain Mapping*, *34*, 1013–1024.

Grabner, R. H., Ansari, D., Koschutnig, K., Reishofer, G., Ebner, F., & Neuper, C. (2009). To retrieve or to calculate? Left angular gyrus mediates the retrieval of arithmetic facts during problem solving. *Neuropsychologia*, *47*, 604–608.

Grabner, R. H., Ischebeck, A., Reishofer, G., Koschutnig, K., Delazer, M., Ebner, F., et al. (2009). Fact learning in complex arithmetic and figural-spatial tasks: The role of the angular gyrus and its relation to mathematical competence. *Human Brain Mapping*, *30*, 2936–2952.

Grond, U., von Aster, M., O'Gorman, R., Martin, E., & Kucian, K. (2014). How does the mathematical brain develop? A longitudinal fMRI study in children with and without dyscalculia. In *Poster presented at the EARLI, Meeting of the SIG 22 Neuroscience and Education, Göttingen*.

Gross, J., Hudson, C., & Price, D. (2009). *The long term costs of numeracy difficulties*. London: Every Child a Chance Trust and KPMG.

Heine, A., Engl, V., Thaler, V. M., Fussenegger, B., & Jacobs, A. M. (2012). *Neuropsychologie von entwicklungsstörungen schulischer fertigkeiten*. Göttingen: Hogrefe.

Isaacs, E. B., Edmonds, C. J., Lucas, A., & Gadian, D. G. (2001). Calculation difficulties in children of very low birthweight: A neural correlate. *Brain*, *124*, 1701–1707.

Ise, E., Dolle, K., Pixner, S., & Schulte-Körne, G. (2012). Effektive Förderung rechenschwacher Kinder: Eine Metaanalyse. *Kindheit und Entwicklung*, *21*, 181–192.

Karmiloff-Smith, A. (1998). Development itself is the key to understanding developmental disorders. *Trends in Cognitive Sciences*, *2*, 389–398.

Kaufmann, L., Koppelstaetter, F., Delazer, M., Siedentopf, C., Rhomberg, P., Golaszewski, S., et al. (2005). Neural correlates of distance and congruity effects in a numerical Stroop task: An event-related fMRI study. *NeuroImage*, *25*, 888–898.

Kaufmann, L., Kucian, K., & von Aster, M. (2014). Development of the numerical brain. In A. Dowker & R. Cohen Kadosh (Eds.), *Oxford handbook of numerical cognition*. Oxford: Oxford University Press.

Kaufmann, L., Lochy, A., Drexler, A., & Semenza, C. (2004). Deficient arithmetic fact retrieval-storage or access problem? A case study. *Neuropsychologia*, *42*, 482–496.

Kaufmann, L., Mazzocco, M., Dowker, A., von Aster, M., Göbel, S. M., Grabner, R. H., et al. (2013). Dyscalculia from a developmental and differential perspective. *Frontiers in Developmental Psychology*, *4*, 516.

Kaufmann, L., Vogel, S., Starke, M., Kremser, C., & Schocke, M. (2009). Numerical and nonnumerical ordinality processing in children with and without developmental dyscalculia: Evidence from fMRI. *Cognitive Development*, *24*, 486–494.

Kaufmann, L., Vogel, S., Starke, M., Kremser, C., Schocke, M., & Wood, G. (2009). Developmental dyscalculia: Compensatory mechanisms in left intraparietal regions in response to nonsymbolic magnitudes. *Behavioral and Brain Functions*, *5*, 35.

Kaufmann, L., & von Aster, M. (2012). The diagnosis and management of dyscalculia. *Deutsches Ärzteblatt International*, *109*, 767–778.

Kaufmann, L., Wood, G., Rubinsten, O., & Henik, A. (2011). Meta-analyses of developmental fMRI studies investigating typical and atypical trajectories of number processing and calculation. *Developmental Neuropsychology*, *36*, 763–787.

Kohn, J., Wyschkon, A., Ballaschk, K., Ihle, W., & Esser, G. (2013). Verlauf von Umschriebenen Entwicklungsstörungen: Eine 30-Monats-Follow-up-Studie. *Lernen und Lernstörungen*, *2*, 77–89.

Kohn, J., Richtmann, V., Rauscher, L., Kucian, K., Käser, T., Grond, U., et al. (2013). Math anxiety interview (MAI): First psychometric properties. *Lernen und Lernstörungen*, *2*, (3)177–189.

Kovas, Y., Giampietro, V., Viding, E., Ng, V., Brammer, M., Barker, G. J., et al. (2009). Brain correlates of non-symbolic numerosity estimation in low and high mathematical ability children. *PLoS One*, *4*, e4587.

Kucian, K., Ashkenazi, S. S., Hänggi, J., Rotzer, S., Jäncke, L., Martin, E., et al. (2013). Developmental dyscalculia: A dysconnection syndrome? *Brain Structure and Function*, *219*, 1721–1733.

Kucian, K., Grond, U., Rotzer, S., Henzi, B., Schonmann, C., Plangger, F., et al. (2011). Mental number line training in children with developmental dyscalculia. *NeuroImage*, *57*, 782–795.

Kucian, K., & Kaufmann, L. (2009). A developmental model of number representation. *Behavioral and Brain Sciences*, *32*, 340–341.

Kucian, K., Kaufmann, L., & von Aster, M. (2014). Brain correlates of numerical disabilities. In M. S. Cohen, & A. Dowker (Eds.), *Oxford handbook of numerical cognition*. Oxford: Oxford University Press.

Kucian, K., Loenneker, T., Dietrich, T., Dosch, M., Martin, E., & von Aster, M. (2006). Impaired neural networks for approximate calculation in dyscalculic children: A functional MRI study. *Behavioral and Brain Functions*, *2*, 31.

Kucian, K., Loenneker, T., Martin, E., & von Aster, M. (2011). Non-symbolic numerical distance effect in children with and without developmental dyscalculia: A parametric fMRI study. *Developmental Neuropsychology*, *36*, 741–762.

Kucian, K., & von Aster, M. (2015). Developmental dyscalculia. *European Journal of Pediatrics*, *174*, 1–13.

Kucian, K., von Aster, M., Loenneker, T., Dietrich, T., & Martin, E. (2008). Development of neural networks for exact and approximate calculation: A fMRI study. *Developmental Neuropsychology*, *33*, 447–473.

Landerl, K. (2013). Development of numerical processing in children with typical and dyscalculic arithmetic skills—A longitudinal study. *Frontiers in Developmental Psychology*, *4*, 459.

Landerl, K., & Moll, K. (2010). Comorbidity of learning disorders: Prevalence and familial transmission. *Journal of Child Psychology and Psychiatry, and Allied Disciplines*, *51*, 287–294.

Le Bihan, D., Mangin, J. F., Poupon, C., Clark, C. A., Pappata, S., Molko, N., et al. (2001). Diffusion tensor imaging: Concepts and applications. *Journal of Magnetic Resonance Imaging*, *13*, 534–546.

Levy, L. M., Reis, I. L., & Grafman, J. (1999). Metabolic abnormalities detected by ^{1}H-MRS in dyscalculia and dysgraphia. *Neurology*, *53*, 639–641.

Li, Y., Hu, Y., Wang, Y., Weng, J., & Chen, F. (2013). Individual structural differences in left inferior parietal area are associated with schoolchildrens' arithmetic scores. *Frontiers in Human Neuroscience*, *7*, e844.

Menon, V. (2014). Arithmetic in the child and adult brain. In R. Cohen Kadosh, & A. Dowker (Eds.), *The oxford handbook of numerical cognition*. Oxford: Oxford University Press.

Moser Opitz, E. (2007). *Rechenschwäche/Dyskalkulie. Theoretische klärungen und empirische studien an betroffenen schülerinnen und schülern*. Bern: Haupt.

Murphy, M. M., Mazzocco, M. M., Hanich, L. B., & Early, M. C. (2007). Cognitive characteristics of children with mathematics learning disability (MLD) vary as a function of the cutoff criterion used to define MLD. *Journal of Learning Disabilities*, *40*, 458–478.

Mussolin, C., De Volder, A., Grandin, C., Schlogel, X., Nassogne, M. C., & Noel, M. P. (2010). Neural correlates of symbolic number comparison in developmental dyscalculia. *Journal of Cognitive Neuroscience*, *22*, 860–874.

Notebaert, K., Nelis, S., & Reynvoet, B. (2011). The magnitude representation of small and large symbolic numbers in the left and right hemisphere: An event-related fMRI study. *Journal of Cognitive Neuroscience*, *23*, 622–630.

Osmon, D. C., Smerz, J. M., Braun, M. M., & Plambeck, E. (2006). Processing abilities associated with math skills in adult learning disability. *Journal of Clinical and Experimental Neuropsychology*, *28*, 84–95.

Parsons, S., & Bynner, J. (2005). *Does numeracy matter more?* London: Institute of Education.

Price, G. R., Holloway, I., Räsänen, P., Vesterinen, M., & Ansari, D. (2007). Impaired parietal magnitude processing in developmental dyscalculia. *Current Biology*, *17*, R1042–R1043.

Qin, S., Cho, S., Chen, T., Rosenberg-Lee, M., Geary, D. C., & Menon, V. (2014). Hippocampal-neocortical functional reorganization underlies children's cognitive development. *Nature Neuroscience*, *17*, 1263–1269.

Reigosa-Crespo, V., Valdes-Sosa, M., Butterworth, B., Estevez, N., Rodriguez, M., Santos, E., et al. (2012). Basic numerical capacities and prevalence of developmental dyscalculia: The Havana survey. *Developmental Psychology*, *48*, 123–135.

Rivera, S. M., Reiss, A. L., Eckert, M. A., & Menon, V. (2005). Developmental changes in mental arithmetic: Evidence for increased functional specialization in the left inferior parietal cortex. *Cerebral Cortex*, *15*, 1779–1790.

Rosenberg-Lee, M., Ashkenazi, S., Chen, T., Young, C. B., Geary, D. C., & Menon, V. (2014). Brain hyperconnectivity and operation-specific deficits during arithmetic problem solving in children with developmental dyscalculia. *Developmental Science*, *18*, 351–372.

Rotzer, S., Kucian, K., Martin, E., von Aster, M., Klaver, P., & Loenneker, T. (2008). Optimized voxel-based morphometry in children with developmental dyscalculia. *NeuroImage*, *39*, 417–422.

Rotzer, S., Loenneker, T., Kucian, K., Martin, E., Klaver, P., & von Aster, M. (2009). Dysfunctional neural network of spatial working memory contributes to developmental dyscalculia. *Neuropsychologia*, *47*, 2859–2865.

Rousselle, L., & Noël, M. P. (2007). Basic numerical skills in children with mathematics learning disabilities: A comparison of symbolic vs non-symbolic number magnitude processing. *Cognition*, *102*, 361–395.

Rubinsten, O. (2009). Co-occurrence of developmental disorders: The case of developmental dyscalculia. *Cognitive Development*, *24*, 362–370.

Rubinsten, O., & Henik, A. (2009). Developmental dyscalculia: Heterogeneity might not mean different mechanisms. *Trends in Cognitive Sciences*, *13*, 92–99.

Rykhlevskaia, E., Uddin, L. Q., Kondos, L., & Menon, V. (2009). Neuroanatomical correlates of developmental dyscalculia: Combined evidence from morphometry and tractography. *Frontiers in Human Neuroscience*, *3*, 51.

Shalev, R. S., Manor, O., & Gross-Tsur, V. (2005). Developmental dyscalculia: A prospective six-year follow-up. *Developmental Medicine and Child Neurology*, *47*, 121–125.

Shalev, R. S., Manor, O., Kerem, B., Ayali, M., Badichi, N., Friedlander, Y., et al. (2001). Developmental dyscalculia is a familial learning disability. *Journal of Learning Disabilities*, *34*, 59–65.

Shalev, R. S., & von Aster, M. (2008). *Identification, classification, and prevalence of developmental dyscalculia*. In *Encyclopedia of language and literacy development* (pp. 1–9). London, ON: Canadian Language and Literacy Research Network.

Soltesz, F., Szucs, D., Dékány, J., Márkus, A., & Csépe, V. (2007). A combined event-related potential and neuropsychological investigation of developmental dyscalculia. *Neuroscience Letters*, *417*, 181–186.

Stock, P., Desoete, A., & Roeyers, H. (2010). Detecting children with arithmetic disabilities from kindergarten: Evidence from a 3-year longitudinal study on the role of preparatory arithmetic abilities. *Journal of Learning Disabilities*, *43*, 250–268.

Szucs, D., Devine, A., Soltesz, F., Nobes, A., & Gabriel, F. (2013). Developmental dyscalculia is related to visuo-spatial memory and inhibition impairment. *Cortex*, *49*, 2674–2688.

Tosto, M., Petrill, S., Halberda, J., Trzaskowski, M., Tikhomirova, T., Bogdanova, O., et al. (2014). Why do we differ in number sense? Evidence from a genetically sensitive investigation. *Intelligence*, *43*, 35–46.

Tsang, J. M., Dougherty, R. F., Deutsch, G. K., Wandell, B. A., & Ben-Shachar, M. (2009). Frontoparietal white matter diffusion properties predict mental arithmetic skills in children. *Proceedings of the National Academy of Sciences of the United States of America*, *106*, 22546–22551.

van Eimeren, L., Niogi, S. N., McCandliss, B. D., Holloway, I. D., & Ansari, D. (2008). White matter microstructures underlying mathematical abilities in children. *Neuroreport*, *19*, 1117–1121.

Verbruggen, F., & Logan, G. D. (2008). Response inhibition in the stop-signal paradigm. *Trends in Cognitive Sciences*, *12*, 418–424.

von Aster, M., Schweiter, M., & Weinhold Zulauf, M. (2007). Rechenstörungen bei Kindern: Vorläufer, Prävalenz und psychische Symptome. *Zeitschrift für Entwicklungspsychologie und Pädagogische Psychologie*, *39*, 85–96.

World Health Organization. (2010). ICD-10 international statistical classification of diseases and related health problems: 10th revision. *Chapter V: Mental and behavioral disorders (F81.2)*: *(Vol. 2)*. Geneva: World Health Organization.

Chapter 8

Neurocognitive Components of Mathematical Skills and Dyscalculia

Wim Fias
Department of Experimental Psychology, Ghent University, Ghent, Belgium

INTRODUCTION

The search for the cognitive systems that support mathematical skills has a long history. It has been thought for some time that mathematical proficiency is determined by multiple cognitive components (e.g., Geary, 1993) comprising both cognitive functions (such as working memory) and mental representations (such as multiplication facts stored in memory). Some influential accounts emphasized cognitive functions, whereas others highlighted the representations. Geary (1993), for instance, proposed working memory to be a crucial factor explaining individual differences in math achievement. Dehaene (1992) specified a cognitive model that was based on three crucial types of representation: verbal, Arabic, and analog, each being the core code for specific numerical abilities: the verbal code for arithmetic tables, the Arabic code for multidigit operations, and the magnitude code for approximate calculation.

With the advent of neuroimaging techniques, attention has shifted away from this multiple component approach toward the search for quantity representation as a central determinant of math skill and dyscalculia. With brain imaging studies revealing consistent activation associated with experimental manipulation of quantity, the intraparietal sulcus (IPS), and in particular its more anterior regions forming the horizontal segment, have been taken as the brain region at the core of mathematical processing, the efficiency of which determines math skill (Butterworth, Varma, & Laurillard, 2011; Dehaene, Molko, Cohen, & Wilson, 2004). This view has been strengthened by demonstrations that this area is processing numerical magnitude in young children, as shown with functional magnetic resonance imaging (fMRI) in 4-year-old children (Cantlon, Brannon, Carter, & Pelphrey, 2006) and with EEG (Izard, Dehaene-Lambertz, & Dehaene, 2008) and functional near infrared spectroscopy (fNIRS) in children

Development of Mathematical Cognition. http://dx.doi.org/10.1016/B978-0-12-801871-2.00008-3

a few months old (Hyde, Boas, Blair, & Carey, 2010). Moreover the fact that many animal species show the same behavioral characteristics as humans when processing quantity, and that neurons in the homologous intraparietal areas of the macaque code numerical magnitude in a way that is compatible with the behavioral characteristics (Nieder & Miller, 2004; for reviews see Geary, Berch, & Mann Koepke, 2015), has been taken as strong evidence that the quantity-representing neural system situated in the intraparietal sulci constitutes the phylogenetic precursors of formal mathematical competencies.

Of course, in animals and in preverbal infants these representations are approximate rather than exact. That is why this system is often referred to as the approximate number system (ANS). The proposal that the ANS is the biologically based center of our intuitive understanding of number is strengthened by genetic studies showing that the contribution of parietal areas to mathematical tasks is heritable (Pinel & Dehaene, 2013) and that genetic disorders that are associated with math problems are accompanied by structural and functional impairments of the parietal areas (Molko et al., 2003; see also Chapter 12). The ANS has also been taken as an ontogenetic starting point for the development of more complex and schooling-induced formal arithmetic. It has been proposed that an impaired or inaccurate quantity representation system is taken to be the most important determinant of math skill and the core deficit of developmental dyscalculia. For instance, Halberda, Mazzocco, and Feigenson (2008) found that the precision with which the quantity of two sets of dots can be compared correlates retrospectively with scores on standardized math tests, but the strength of this relation is currently debated (Chen & Li, 2014; De Smedt, Noël, Gilmore, & Ansari, 2013; Fazio, Bailey, Thompson, & Siegler, 2014).

An important question that can be raised is how a single factor such as the ANS can be reconciled with the multifaceted nature of mathematical cognition and with the heterogeneity of developmental dyscalculia. Indeed, the major picture that emerges from the literature is that developmental dyscalculia is associated with many more cognitive deficits, is relatively heterogeneous, and often co-occurs with other developmental disorders. Although the exact numbers vary largely from study to study, it is clear that comorbidity of dyscalculia with dyslexia, ADHD, and developmental dyspraxia among other common childhood disorders is not exceptional (Shalev, Auerbach, Manor, & Gross-Tsur, 2000). Critically, brain imaging studies also suggest that multiple brain areas are recruited during the solving of mathematics problems. In addition to the ANS in the IPS, multiple areas are engaged even during the solving of numerical problems that are thought to tap the ANS directly, such as number comparison (Arsalidou & Taylor, 2011; Fias, Menon, & Szucs, 2013). Although the activations obtained with brain imaging are merely correlational with cognitive activity, and although clear relations with math skill have not always been reported, the involvement of a broad set of brain regions suggests a more complex picture than that implied by the ANS theory.

Demonstrating that number processing is subserved by multiple brain regions is an important first step, but in itself it leaves the functional specification of cognitive components unexplained. What is needed are neurocognitive theories that describe in detail the representations and processes that subserve arithmetic tasks and how they are implemented in neural systems. Such knowledge is needed to come to a deeper understanding of what determines math skill and difficulty. Below, I review some of the neurocognitive components that have been documented to be strongly related to number processing and are therefore plausible sources of variability in mathematical achievement.

ACCESSING QUANTITY REPRESENTATIONS

Single cell recordings in the IPS of macaque monkeys (*Macaca mulatta*) while they view dot displays of varying numerosity have established the precise characteristics of how quantity is neurally encoded. The most important property of these neurons is that they are tuned to a specific number (Nieder & Miller, 2004). This property is called number selectivity and means that the neurons have a preferred number and respond maximally when their favorite numerosity is displayed (e.g., 3 dots). The neural response decreases as a function of the numerical distance between the number displayed (hence a weaker response for 2 and 4 compared to 3 and an even weaker response to 1 and 5). Using the technique of fMRI adaptation, Piazza, Izard, Pinel, Le Bihan, and Dehaene (2004) were able to demonstrate that the same property characterizes the neural coding of number in the IPS of human adults, with a more precise tuning for symbolic numerals (e.g., "5", "8") than for numbers presented nonsymbolically as collections of dots (Piazza, Pinel, Le Bihan, & Dehaene, 2007).

The question that directly follows is how sensory input that conveys numerical information, either in nonsymbolic format or as number symbols (such as digits or number words) is transmitted to the number-selective coding in IPS. In other words, what type of preprocessing is needed to transform sensory input to access the parietal quantity representations. We initially addressed this question with computational modeling (Verguts & Fias, 2004).

Our starting point was that the transformation of neural activity in retinotopic visual areas induced by a collection of dots into a number-selective neural code is nonlinear. More concretely, when two dots are presented, then the neuron with the preference for 2 should respond, but when another dot is added, then the number 2 neuron should not respond, but rather the number 3 neuron should respond. Consequently, a direct coupling between sensory input and number-selective neurons is not a sufficient explanation. A computationally plausible way to accomplish such nonlinear transformation is to have an intermediate layer. We built an artificial network that comprised an additional layer of neurons between an input layer and an output layer. We used the back propagation learning algorithm (Rumelhart, McClelland, & the PDP research group, 1986), and we trained the network to give appropriate number-selective responses. After learning, the intermediate layer had learned to summate or accumulate the number of objects that were presented at input. More concretely, with more objects presented, more neurons became active. Based on this work, we hypothesized that for nonsymbolic quantities, the number-selective neurons in IPS can only be accessed via a preprocessing stage that converts visual input into an accumulation code, which is sensitive to number, but is not number-selective.

The biological reality of this hypothesis was later confirmed by the observation that neurons in the lateral intraparietal area of the macaque monkey followed the principles of accumulation coding (Roitman, Brannon, & Platt, 2007). This observation was further supported with a series of human fMRI studies demonstrating a pattern of brain activity that is consistent with the hypothesis of accumulation coding (Santens, Roggeman, Fias, & Verguts, 2010). A region of the parietal cortex that is functionally equivalent to the lateral intraparietal area of the monkey, but in humans located more medially, showed increasing activation with visual displays containing more dots. It was also shown that this region was more important during the processing of nonsymbolic numerosities compared to the processing of Arabic numerals, which is consistent with the model, as mapping symbols to number-selective neurons can be done directly and does not require preprocessing. We also demonstrated that, although accumulative activity in this region has been shown to be limited to four elements in visual short-term memory tasks (Todd & Marois, 2004), accumulation does not level off at four elements in the context of a task requiring enumeration (Knops, Piazza, Sengupta, Eger, & Melcher, 2014; Roggeman, Fias, & Verguts, 2010). To situate accumulation coding and number-selective coding with respect to each other, my colleagues and I performed an fMRI adaptation study, in which we could show that the processing stream from visual input, via number-sensitive to number-selective processing follows a posterior to anterior gradient along the dorsal visual processing stream, which runs from occipital to parietal cortex (see Figure 1; Roggeman et al., 2011).

It is clear that considerable processes precede the seemingly simple translation of dot displays into number-related neural codes. Although there is no direct evidence that explicitly supports this, it can reasonably be predicted that the integrity and quality of the accumulation system and by extension of the whole dorsal stream (i.e., visual information that is processed from occipital to parietal cortex) is a prerequisite for an efficiently functioning number-selective processing system. As the number-selective processing system forms the neural implementation of the ANS, this would imply that not only the number representation itself but also the mechanisms that provide access to it can potentially contribute to mathematics learning. It is not difficult to imagine that impoverished preprocessing may, through development, lead to an ANS low in precision.

Despite the fact that converting symbolic numbers (i.e., in Arabic or verbal format) into a number-selective neural code is less complicated, in the sense that it is a linearly separable problem, the pathways that are followed also need

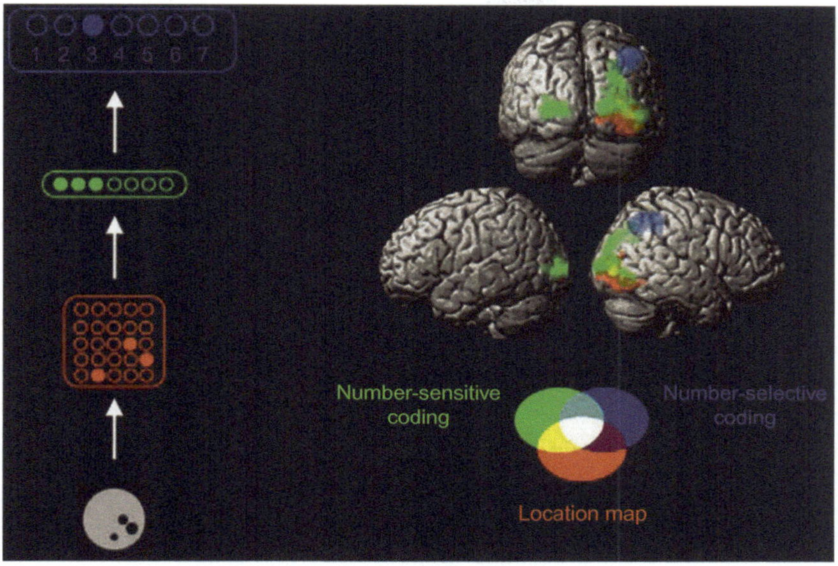

FIGURE 1 Three consecutive stages of processing visual numerosities are displayed. After visual input (consisting of a collection of dots), elements are first located in space (red). Next, a number-sensitive coding stage accumulates the information available in the location map (green). In a third stage, number-selective neurons identify the accumulated number of elements. *Adapted from Roggeman, Santens, Fias, and Verguts (2011).*

to be intact for normal development. The precise mechanisms of the numeral to number-selective coding pathway are not fully understood, but evidence is accumulating that an inferior temporal region in the ventral stream of the right hemisphere is involved (Shum et al., 2013). The precise properties of this visual number form area are not yet described. Crucially, the Arabic number system is a base-ten positional system in which numeral position determines the associated quantity (e.g., the 4 in 43 represents 40, but 4 in 34). Hence, determining position-specific digit identity is essential, thus structurally resembling—at least partially—the coding of position of letters in word reading. It might therefore be very revealing to explore to what extent the visual number form area (in the right hemisphere) and the visual word form area (in the left hemisphere) have common processing characteristics and to what extent they differ from each other. This information might open a window to understand how and why developmental dyscalculia regularly co-occurs with developmental dyslexia (Shalev et al., 2000). It also might constitute a more detailed functional elaboration of the idea that developmental dyscalculia, as expressed in poor ANS representations, might be attributed to initial deficiencies in mapping numerals to number magnitude (Noël & Rousselle, 2011; Rousselle & Noël, 2007; vanMarle, Chu, Li, & Geary, 2014).

WORKING MEMORY: THE ROLE OF SERIAL ORDER

Of course, number processing as it occurs in arithmetic or math consists of more than coding the numerical magnitude value of a numeral. It requires relating different numerals to each other, possibly producing new ones, as when computing the sum of two numbers. This requires a working memory that permits relevant numbers to be kept in mind, relating them to each other, and performing operations on them (Bull & Lee, 2014). Confirmation has been provided by studies showing that occupying working memory resources in dual-task situations may impair number processing tasks. This is true for verbal as well as for visuospatial working memory, depending on the task (see Raghubar, Barnes, & Hecht, 2010). Also, correlational approaches reveal a link between working memory and numerical task performance as well as general math skill (for a review, see Geary, 1993). Furthermore, brain imaging evidence shows that brain activity in the IPS during a working memory task is predictive of arithmetical performance two years later (Dumontheil & Klingberg, 2012).

Working memory is a cognitive function that consists of several subcomponents. Of course, an important overall distinction is between verbal working memory operating via a phonological loop and visuospatial working memory operating as a visuospatial sketchpad (Baddeley, 1986). The differential contributions of these two working memory modalities to mathematical processing have been described to some extent, with verbal working memory contributing more to rote verbal memory retrieval as in table-based multiplication and visuospatial working memory contributing more to strategy-based calculation as in column-wise addition (Imbo, Vandierendonck, & De Rammelaere, 2007). Yet, other basic subcomponents need to be distinguished. A first distinction that can be made is between item memory and order memory. Both aspects of information are important for efficient working memory and are based on distinct neural circuits (Majerus et al., 2010). Second, storage and operation are seldom distinguished in their relation to mathematical achievement. Overall, the specific relationships of these subcomponents of working memory with math skill have not been explored. Typically, investigations of working memory involvement in math skill are confined to relating capacity as a relatively global measure of working memory function to performance. Yet, there is reason to believe that the link between working memory and number processing can be described at a more refined functional level. I will focus on order working memory and attentional mechanisms in working memory respectively to illustrate this point.

Number processing shares neural resources with working memory. At a behavioral level, there is a close resemblance between processing the ordinal aspects of number (which is the larger number?) and the ordinal aspects of information that is maintained in working memory (which of the two items was further in the working memory sequence?). An important factor driving performance in both number comparison and working memory order judgments is the distance effect. It is more difficult to determine which of two numbers is the larger the closer the numbers in terms of their magnitude. Similarly, the closer two items are in working memory, the more difficult it is to determine which of the two items is positioned before (or after) the other (Marshuetz, 2005). In addition, Marshuetz, Smith, Jonides, DeGutis, and Chenevert's (2000) early brain imaging study of adults showed that judgments about serial order in working memory (Remember z-t-g-p-l. Are t-g in the correct order?), but not identity judgments (Did t belong in the working memory sequence?) recruit areas in the bilateral IPS similar to those activated in number processing.

In a study that used the same type of order and item judgment tasks, Majerus et al. (2010) went one step further. They presented either sequences of faces or of nonwords and engaged participants in an order judgment task or an item judgment task. It was discovered that the IPS of the right hemisphere was involved in order judgments with both types of stimuli, suggesting that the neural coding of serial position in working memory occurs at least partially in a modality-independent way. Modality-specific effects were observed for face stimuli in the right fusiform area and for letter strings in the left hemisphere language processing areas. These results suggest that working memory emerges from the use of modality-independent ordering processes geared toward sensory networks that underlie the processing and storage of modality-specific item information. The fact that the intraparietal sulci are involved in processing order in a stimulus-independent fashion is further corroborated by a study that investigated to extent to which the processing of numerical and of alphabetical order recruit the same brain areas. This study revealed that the brain activations induced by a number comparison task (which of two numbers is the largest?) and by an alphabet position comparison task (which of two letters is the furthest in the alphabet?) overlapped in a number of brain areas comprising the intraparietal sulci and premotor cortices (Fias, Lammertyn, Caessens, & Orban, 2007), which are highly similar to those brain areas that are active in the serial order working memory tasks as previously described.

Although suggestive of overlapping neural circuitry for the processing of order as expressed by numbers, letters, or position in working memory, these studies by themselves are not conclusive. Indeed, because of the fact that there are thousands of voxels in the brain and for methodological and statistical reasons, it is impossible to determine an overlap purely *de visu*. What is needed is a demonstration of the very same voxels being active in the different task contexts, in the very same subject and in the very same scanning session. That is exactly what Attout, Fias, Salmon, and Majerus (2014) aimed to do. They searched for the neural correlates of the distance effect as it was observed in three tasks addressing order: a number comparison task (is this number larger than 65?), an alphabet position task (are the two letters in alphabetic order?), and a working memory order task (do the two letters occur in the same order as you have just learned them?). It was found that the activity in the horizontal segment of the left IPS reflected the distance effect in all three tasks. The corresponding area in the right hemisphere was jointly activated by the number and working memory task, although it was not activated in the alphabetic task. Also in some superior frontal areas that are typically observed in working memory tasks, overlapping activations between the three tasks were observed.

Hence, generally speaking, brain imaging studies strongly suggest that the neural circuitry that is used to process numerals is strongly related to the processing of serial order in working memory. It is not clear at this point, however, which common operational mechanism is expressed in this overlapping neural circuitry. There are a several possibilities that one can consider. First, it is possible that numbers are somehow used to provide the serial position code of working memory. This can be done in a more or less explicit way as when using counting to tag items in working memory (Noël, 2009). However, it is also possible that the link is not made via procedural aspects of counting. Instead, it may be the representational characteristics of the number-selective neural system itself that are important. Botvinick and Watanabe (2007) showed that a computational model of working memory in which ordinal rank was coded with the same properties as number-selective neurons is able to predict behavioral serial position effects in working memory in great detail. Second, it is also meaningful to consider the possibility that a number is not processed in isolation but always in the context of other numbers, whereby working memory provides this context. By placing numbers in working memory together with other numbers and by doing this in a systematic way, numbers may acquire their functional meaning. A straightforward prediction of this idea is that number processing is to a large extent context dependent. One of the cases where a strong context dependency is observed is the relationship between number magnitude and space. That is what we will discuss in the next section.

Serial position in working memory links numbers to space. The representation and processing of number is tightly connected to the representation and processing of space. This is typically interpreted as an expression of the fact that numbers are represented as positions on a spatial mental number line typically running from left to right in left-to-right reading cultures and vice versa in right-to-left reading cultures (Dehaene, Bossini, & Giraux, 1993).

One of the most important pieces of evidence in support of this link between number and space comes from the SNARC effect, which is an acronym for spatial numerical association of response codes (Dehaene et al., 1993). It reflects faster left-hand than right-hand responses for small numbers and faster right-hand than left-hand responses for large numbers, expressing a clear association between small numbers and left and large numbers and right. In right-to-left reading cultures, these associations are reversed (Shaki & Fischer, 2008; Zebian, 2005). The SNARC effect isn't confined to one particular task but has been observed in parity (odd-even) judgment, magnitude comparison, and even in tasks that don't require processing of the number semantics at all, such as indicating whether a number is presented in italics or not, or even when a number is presented merely as a background against which shapes are presented and to which the shapes have to be responded (Fias, Lauwereyns, & Lammertyn, 2001). This suggests a high degree of automatic processing of number magnitude and its relation to space.

The SNARC effect is commonly interpreted as an expression of the fact that a number's magnitude is represented as a position on an oriented number line in memory (see Hubbard, Piazza, Pinel, & Dehaene, 2005). Yet, a number of observations suggest that the associations between number and space are more flexible than would be expected from a long-term memory representation. First, the SNARC effect is range dependent. Dehaene et al. (1993) and Fias, Brysbaert, Geypens, and Ydewalle (1996) showed that numbers 4 and 5 were preferentially responded to with the right hand when they were represented in the context of small numbers (i.e., the range from 0 to 5). However, when the same numbers were presented in the context of larger numbers (i.e., the range from 4 to 9), the numbers 4 and 5 showed faster left- than right-hand responses. So apparently, it is not long-term absolute numerical magnitude that drives the SNARC effect but rather the number's magnitude relative to the context in which it occurs. Second, it has been shown that the SNARC effect depends on visual imagery (Bächtold, Baumüller, & Brugger, 1998). When participants are asked to imagine number stimuli as being placed on a ruler, a typical SNARC effect is observed. Yet, when participants are instructed to imagine numbers as being positioned on a clockface, the SNARC effect reverses, with the small numbers associated with right and the large numbers associated with left, which is consistent with how numbers appear on a clock. Third, the SNARC effect has been shown to be flexibly dependent on reading habits (Shaki & Fischer, 2008). When Russian-Hebrew bilinguals had read a Russian text just before the SNARC effect was measured, a normal SNARC effect was observed, but after having read a Hebrew text, which is read from right to left, the SNARC effect was reversed. All these observations indicate that spatial coding is not inherently associated with numbers but that it is constructed during task execution, suggesting a crucial role of working memory.

A straightforward way to test the involvement of working memory is to use a dual-task design in which one task occupies working memory resources and then measure the SNARC effect while executing the main number processing task. Herrera, Macizo, and Semenza (2008) used this approach with a visuospatial working memory load and found that the SNARC effect in a number comparison task was abolished. Van Dijck, Gevers, and Fias (2009) extended this approach by adding a condition in which verbal working memory was loaded by asking participants to keep a string of consonants in memory and by including a parity judgment task. It was observed that working memory load indeed abolished the SNARC effect in the two tasks but that it occurred in a modality-dependent fashion. The parity judgment SNARC effect was abolished by a verbal working memory load but not by a visuospatial working memory load, the opposite being the case for the number comparison task.

The necessity of having working memory resources available for the SNARC effect to occur raises an important question: Which aspect of working memory processing determines the SNARC effect? As was highlighted earlier, working memory order tasks and number processing tasks share neural resources; hence a plausible candidate is the processing and representation of serial position (Fias, van Dijck, & Gevers, 2011). One possibility is that it is the serial position in working memory rather than number magnitude itself that is related to space. This can explain the SNARC effect if one makes the additional assumption that while executing a number task working memory is invoked to temporarily store number stimuli and their responses as part of the task set to facilitate and to optimize task performance (Monsell, 2003). Storing numbers in working memory as a function of magnitude is a very efficient way to fully exploit the limited working memory resources as it allows chunking (rather than having one number in each available slot, one can organize numbers in chunks [e.g., 1-2, 3-4, 5-6, etc.] and then allocate a chunk to a slot) and as it facilitates efficient access to specific numbers (e.g., if you have located 4 then you know that 5 is next to it).

To test the hypothesis that serial order in working memory is spatially coded and to determine whether this drives the SNARC effect, van Dijck and Fias (2011) designed the following experimental procedure. Participants were asked to encode and remember a sequence of numbers that was ordered arbitrarily (e.g., 6-3-1-9-8-4). During the retention interval, numbers were presented and participants had to perform a parity judgment task, but only if the presented number was part of the memory list. This latter restriction was imposed to make sure that participants accessed working memory to solve the task. Thereafter a recognition task was administered to verify whether the participants had correctly memorized the sequence. This cycle (encoding, parity judgment during retention interval and then recognition task) was then repeated each time with a different list of to-be-remembered numbers. With this design it is possible to separate numerical magnitude from serial order in working memory and thus to determine whether the left- or right-hand response times were modulated by serial position in working memory and/or numerical magnitude. It was clearly found that serial position in working memory was associated to space with initial items receiving faster left-hand response than right-hand responses, the opposite being true for items toward the end of the memory sequence. Number magnitude by itself was not systematically associated with the side of response.

To take this idea one step further, it was reasoned that the generator of the SNARC effect is not specifically number related and therefore does not have anything to do with numbers as such and that therefore the same effect can be obtained with whatever information is stored in working memory. This was tested in a follow-up experiment in which it was not sequences of numbers that were stored in working memory but rather fruit and vegetable names. And a fruit/vegetable categorization task was administered instead of a parity judgment task. Again an association was found between position in the sequence and space, with initial items associated with left and end items associated with right. Interestingly, the same participants were subsequently administered as a classic parity judgment task to measure their conventional SNARC effect. It was found that the size of the position-space association correlated significantly

with the size of the SNARC effect, corroborating the hypothesis that it is the serial position in working memory that drives the SNARC effect.

In sum, the temporary position-space associations are what drive the SNARC effect, rather than the long-term semantic representations of number to which the SNARC effect is traditionally ascribed. This provides a natural explanation for the fact that the SNARC effect has been shown to be flexibly dependent on the range of numbers used, on reading habits, and on visual imagery, as well as for the fact that the SNARC effect has also been demonstrated with nonnumerical ordinal information such as letters, days of the week, or months of the year (Gevers, Reynvoet, & Fias, 2003, 2004). It is suggested that efficient number processing does not merely entail retrieving the correct semantic information and corresponding responses, but that it mobilizes working memory resources to create a mental workspace in which information is systematically ordered to facilitate performance.

Workspace engages spatial attention. Number processing has been shown to engage mechanisms of spatial attention, the original idea being that spatial attention is used to move back and forth along the mental number line. The main sources of evidence are twofold. First, Fischer, Castel, Dodd, and Pratt (2003) used a variant of the Posner cueing paradigm. In this paradigm, a spatial cue typically is presented that directs attention to the left or the right. If a target is presented at the cued location, then target detection is facilitated. In Fischer et al.'s variant, a numeral was presented centrally as the cue. It was found that a small number cue facilitated detection of a left target and that a large number cue facilitated right target detection. Second, Zorzi, Priftis, and Umiltà (2002) examined left hemineglect patients' performance on a number bisection task. After brain damage to the right hemisphere, left hemineglect patients ignore the left side of visual space. This is truly an attentional deficit and is not caused by problems at the level of sensory input. As a consequence of their attentional bias, patients exhibit such behaviors as eating only the right side of their plate or shaving the right but not the left side of their face. In more formal tests, neglect patients show a rightward bias in a line bisection task: when asked to mark the midpoint of a line, patients deviate toward the right as they ignore the left side of the line. Analogous to this, Zorzi et al. (2002) asked left neglect patients which number is the midpoint between two other numbers. The patients produced answers that systematically deviated toward larger numbers. So, for instance, when asked which number is the midpoint between 1 and 9, neglect patients say 6 or 7 rather than the correct number 5. Importantly, this occurs in the absence of problems in mental arithmetic. Hence, patients have no difficulty computing the correct result of problems such as $(1+9)/2$, excluding the possibility that their number bisection error is due to an impaired capacity for mental arithmetic.

Recall the theoretical proposal that the link between numbers and space is not so much attributable to the properties of the long-term representation of number magnitude taking the form of a spatially oriented mental number line, but rather to the involvement of working memory to serially order numbers as a function of task requirements. Therefore one can wonder to what extent the number cueing paradigm and the number bisection task in neglect derive from spatial attention mechanisms that operate in the workspace in which serial position is spatially coded.

To test whether spatial attention operates in working memory, van Dijck, Abrahamse, Majerus, and Fias (2013) embedded the cueing paradigm of Fischer et al. (2003) in a working memory context. Participants first had to memorize a list of numbers in arbitrary order and during the retention interval they had to detect a left or right appearing target that was preceded by a number cue, only having to respond when the number belonged to the list they maintained in working memory (see Figure 2). The results show a cueing effect, not on the basis of the numerical magnitude of the number but of the position of the number in the memorized list: with a cue from the beginning of the list, left targets were detected faster, and with a cue from the end of the list, right targets were detected faster. This was the case, regardless of whether participants indicated target detection by a manual or a vocal response and was also replicated in some modified versions of this paradigm (for an overview see Abrahamse, van Dijck, Majerus, & Fias, 2014). The effect is strong and robust, in contrast to the original Fischer et al. study, which has been proven to be difficult to replicate (Zanolie & Pecher, 2014). It is highly likely that the reason that the effect is weak in the original paradigm is due to the fact that encoding in working memory was not required for task performance and that therefore subjects may

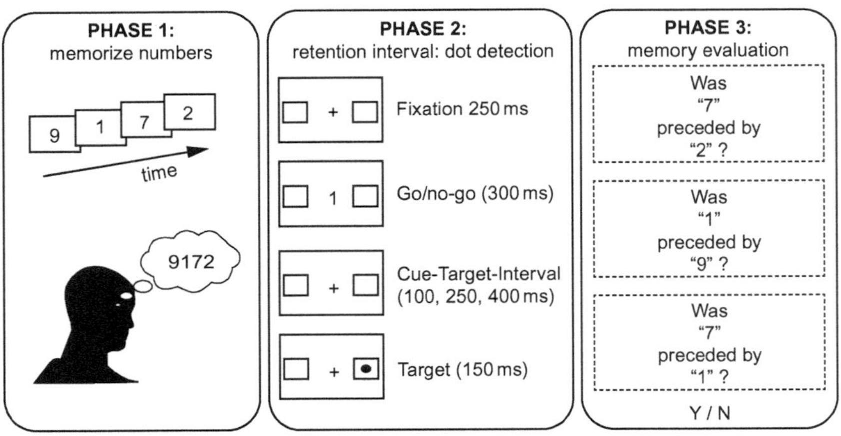

FIGURE 2 Participants first have to encode a series of numbers into working memory. During the retention interval they perform a dot detection task. After the fixation cross, a number cue is flashed briefly. If the number belongs to the memorized set, participants have to detect the appearance of a dot, in which case they have to press a button. After the retention interval a number of questions are asked to evaluate if the participants had correctly memorized the series of numbers. *Adapted from van Dijck et al. (2013).*

not have consistently stored numbers in working memory spontaneously (van Dijck, Abrahamse, Acar, Ketels, & Fias, 2014).

As far as the number bisection bias in neglect patients is concerned, there are also indications that working memory is involved. First, it has been shown that the bisection bias with lines does not correlate with the bisection bias observed with numbers (Rossetti et al., 2011). Even more importantly, the number bisection bias was only observed in patients whose lesion extended more anteriorly in frontal areas and whose working memory capacity was affected (Doricchi, Guariglia, Gasparini, & Tomaiuolo, 2005). Second, van Dijck, Gevers, Lafosse, Doricchi, and Fias (2011) reported on a single case who showed right spatial neglect after left hemisphere lesion as evidenced by a bias toward the left in a line bisection task. Surprisingly, however, this patient did not present a bias toward smaller numbers in the number bisection task as expected, but a bias toward larger numbers. Her SNARC effect was normal with small numbers associated with left and large numbers with right, indicating that her number-space associations were not reversed as they should have been if numbers are spatially arrayed on a mental number line. Further investigations of her working memory showed that she had pronounced problems with initial items of verbal sequences encoded in working memory, which is consistent with the direction of her number bisection bias. Of course, this is only a single case and it is difficult to judge at this point to what extent the involvement of a specific working memory deficit accounts for the pattern of number bisection bias in neglect patients. A study involving multiple cases shows that a working memory deficit is not the determining factor in all cases (Storer & Demeyere, 2014), but that study does not exclude the possibility that number bisection bias can be determined by multiple factors of which working memory is one (e.g., van Dijck, Gevers, Lafosse, & Fias, 2012).

In sum, it is clear that number processing, spatial attention, and working memory are tightly interwoven. Moreover, they recruit common brain areas, in particular, the intraparietal sulci. A crucial question, then, is to what extent this configuration of rather basic processes is a determining factor of mathematical skill. In a study that investigated number processing in a group of children with visuospatial disabilities, Bachot, Gevers, Fias, and Roeyers (2005) found that children (aged 7 to 12) with a small visuospatial working memory capacity and poor performance on number concept (e.g., 12 is 9 apples more than?) and complex addition tasks (26 + 63=) also exhibited a reduced SNARC effect in a number comparison task, compared to a control group that was matched on age and verbal intelligence. In contrast, evidence suggests that in adults there is an opposite relation between the SNARC effect obtained in a parity judgment task and mathematical proficiency (Hoffmann, Mussolin, Martin, & Schiltz, 2014), with smaller SNARC effects in math-proficient participants and no mediating effect of visuospatial working memory capacity. The authors attribute this to the fact that those proficient in math are more efficient in inhibiting the magnitude information, which is irrelevant in a parity judgment task (see also Hoffmann,

Pigat, & Schiltz, 2014). Yet, the Bachot and the Hoffmann studies remain essentially correlational in nature and don't directly address the link between space and mental arithmetic.

A recent study used fMRI to directly assess the involvement of the neural circuits of spatial processing in complex mental arithmetic (2-digit addition and subtraction). Knops, Thirion, Hubbard, Michel, and Dehaene (2009) used multivariate decoding to distinguish the patterns of activation in parietal regions between leftward versus rightward eye movements. Interestingly these same patterns also distinguished between subtraction (associated with leftward) and addition (associated with rightward), suggesting the involvement of mechanisms of spatial attention in mental arithmetic. The extent to which this reflects the movement of spatial attention over working memory positions or over long-term number representations needs to be determined. Recent theories of working memory even suggest that the distinction between working and long-term memory may not be that strict after all. Theories like the one of Cowan (1999) or Oberauer (2009) suggest that working memory builds on activations in long-term memory with mechanisms of attention operating on these preactivated segments of long-term memory (see also D'Esposito & Postle, 2015).

Clearly, much work needs to be done to establish the details of how number, space, and working memory interact and to test how it relates to mathematical proficiency. Whether or not the mechanism of orienting attention in a spatially oriented working memory sequence will prove to contribute to mathematical skills, it is an illustrative case of how constructing and testing functionally explicit models will permit progress to be made in advancing our understanding of *how* it is that working memory and mathematical skill might be related. It is also worth considering the extent to which ordinal working memory processing also contributes to other achievement domains such as language development (Majerus, Poncelet, Greffe, & Van der Linden, 2006), and eventually may provide a basis for explaining comorbidities between developmental dyscalculia and dyslexia.

EXECUTIVE FUNCTIONS

For efficient mathematical learning and performance, it is necessary to be endowed with good sensory processing systems (to estimate numerosity or to encode numerical symbols) and efficient systems to store, maintain, and access information in memory (e.g., to keep track of intermediate results of calculations). But of equal importance are the executive functions that control the way these information processing and storage systems are orchestrated to be optimally configured for dealing with requirements dictated by task and context (e.g., for removing the results of intermediate calculations from working memory when they are no longer necessary) (Hitch, 1978).

Numerous studies have confirmed the contribution of executive functions to efficient arithmetic processing. Correlational studies have found mathematical and arithmetical skill to be associated with individual differences in executive function. An excellent example is the finding by Bull, Johnston, and Roy (1999) that arithmetic performance is correlated with scores on the Wisconsin card sorting task, suggesting the involvement of inhibitory process in mental arithmetic. Other studies have used dual-task paradigms to show that arithmetic performance deteriorates when participants simultaneously perform a task that consumes executive resources. This has been shown to be the case for multidigit addition (Imbo, Vandierendonck, & De Rammelaere, 2007) and multiplication (Imbo, Vandierendonck, & Vergauwe, 2007), but, intriguingly, also for single-digit arithmetic (e.g., De Rammelaere, Stuyven, & Vandierendonck, 2001). The latter finding suggests that the impact of executive functions also comprises relatively basic levels of processing and is not restricted to complex capacities such as the storage of intermediate outcomes as involved in carrying or borrowing (Hitch, 1978).

Although the existence of a link between executive control and arithmetic performance is clear, the literature on executive functions typically considers it a very heterogeneous construct comprising several independent subfunctions (e.g., updating, task shifting, monitoring, etc.; Miyake et al., 2000). In the dual-task studies in particular, the applied executive load typically is not further specified and does not permit one to pinpoint precisely which executive functions contribute to efficient mental arithmetic.

At the same time, the specification of mental arithmetic at the processing level also remains rather vague. Processes such as retrieval- and strategy-based calculation are distinguishable, but many questions remain unanswered: How exactly are these processes executed? On which representations do they depend? How do they elapse over time? From which neural resources do they recruit? Finding an answer to such questions is extremely difficult, especially if one takes into account that executive involvement is very much dependent on the nature of the task and can vary with age and/or skill level (Raghubar et al., 2010).

Although the previous discussion sounds somewhat discouraging, it should not be a reason to abandon hope. Indeed, adding computational modeling and neuroimaging to the arsenal of instruments to study executive control has been shown to permit a level of theorizing that allows a detailed explanatory account of an important cognitive control function. Specifically, by imposing computationally derived and biologically plausible constraints, the executive function of performance adaptation has now become largely understood at a neurocognitive mechanistic level. Recently, this knowledge has been applied to the context of mental arithmetic.

Performance adaptation is the capacity to self-regulate with the goal of optimizing behavior. This form of self-regulation is often referred to as cognitive control. Cognitive control can be seen as the operation of top-down functions that configure the components of a cognitive system in such a way that they are adapted to deal efficiently with the task requirements. Cognitive control requires the coupled operation of a monitoring system and an adaptive system, together forming a closed loop system. The monitoring system keeps track of how well the participant is doing, and on the basis of that information the adaptation system imposes top-down control to optimize performance. The neural mechanism that underlies this control loop has been rather well described, with the anterior cingulate cortex (ACC) and the dorsolateral prefrontal cortex (DLPFC) being key structures. The ACC continuously monitors task performance and if problems in performance (a difficult or erroneous response) are detected, then a signal is sent to the DLPFC which, in turn will then impose top-down changes on the neural networks involved in task execution in order to optimize the efficiency of task performance with reduced risk of committing errors (Botvinick, Braver, Barch, Carter, & Cohen, 2001). This mechanism has been mainly investigated with standard conflict tasks such as the Stroop task, the Simon task, or the flanker task, in which relevant stimulus aspects (e.g., ink color) need to be processed in the context of irrelevant stimulus aspects (e.g., color word) that may interfere with the processing of the relevant information. Recently, this framework has begun to be applied to the context of arithmetic performance.

Desmet et al. (2012) investigated the behavioral consequences of errors in mental arithmetic, more precisely in verification of multiplication facts in which participants have to indicate whether a presented answer for a single-digit multiplication problem is correct or not (e.g., $4\times5=20$, or $6\times7=48$). It was observed that accuracy improved and reaction times slowed down on trials that followed errors. This was the case for errors that were committed by the participant, but also for trials in which an erroneous answer was proposed (e.g., $8\times7=48$), provided that the incorrect answer was table-related (i.e., a correct answer to another simple multiplication problem). This finding clearly supports the fact that cognitive control processes are at play when people solve even simple arithmetic problems.

A recent neuroimaging study confirms the involvement of ACC and DLPFC while solving arithmetic problems. Ansari, Grabner, Koschutnig, Reishofer, and Ebner (2011) found stronger activation of ACC and bilateral DLPFC when participants committed an error while solving simple arithmetic problems of any operations. Interestingly, the right DLPFC was found to be modulated by mathematical competence as measured by a standardized mathematical test with highly competent individuals activating this area more on an incorrectly solved trial than less skilled individuals, suggesting efficiency of error detection is related to overall mathematical competence, or that individuals who are sensitive to errors develop stronger mathematical skills.

Rinne and Mazzocco (2014) established a developmental link between cognitive control and proficiency. In a longitudinal study they found that the successful detection of errors contributes to the development of arithmetic proficiency from grades 5 to 8. More specifically, they investigated the extent to which a good alignment or calibration between confidence in judgments of arithmetic performance and the accuracy of those judgments is related to the development of arithmetic skill. Calibration was shown to be related to performance, that it was particularly poor in children with mathematics learning disability, and calibration in grade 5 predicted gains in arithmetic accuracy from grade 5 to grade 8.

It is clear that cognitive control and self-regulation contribute to mathematical skill. It is interesting to see how a neurocognitive framework has proven to be inspirational to start to form an initial picture of the underlying mechanisms. Yet, it is also important to realize that the models that are developed in the context of basic laboratory tasks don't necessarily transfer to the context of mathematical skill. Laboratory tasks are quite simple and often don't require much more than mapping a stimulus to a response, without much variety in solution strategies. In contrast, arithmetic and math often engage tasks that are more complex and can be solved by multiple procedural strategies (Lemaire & Siegler, 1995). An interesting avenue for further research could be to see how quality of performance determines strategy selection (Uittenhove & Lemaire, 2012; for a similar idea in a developmental perspective, see Qin et al., 2014).

DISCUSSION AND CONCLUSIONS

Above I listed a number of neurocognitive components that are meaningfully related to mental arithmetic and mathematical processing and can reasonably be expected to be contributing factors to achievement and performance levels. Certainly, this list is not complete. There have been recent advances in understanding the memory mechanisms that are involved in the learning and retrieval of basic arithmetic facts. An important proposal has been made by De Visscher and Noël (2014b), who specify sensitivity to interference in memory as an important determinant of efficient storage and retrieval of multiplication facts. The nature of the number system results in a high degree of similarity among all arithmetic problems, and this in turn creates the potential for proactive interference that can hinder efficient storage of the arithmetic tables. Children who are particularly sensitive to such interference may experience difficulties memorizing or retrieving basic arithmetic facts (De Visscher & Noël, 2014a), a cardinal feature of mathematical learning disabilities (Geary, 1993). In line with this, there are studies that show that intrusion of related information is common when these children attempt to retrieve addition or multiplication facts (Barrouillet, Fayol, & Lathuliére, 1997; Geary, Hamson, & Hoard, 2000). Another domain where considerable progress is to be expected is the study of the contribution of general learning systems, like the hippocampus for memory formation or the basal ganglia for procedural learning. A recent study describes results that emphasize the fruitfulness of this path. Supekar et al. (2013) found that hippocampal volume and the way the hippocampus is functionally connected to prefrontal areas and to the basal ganglia is predictive of performance gain induced by a math tutoring program in grade 3 children (see Chapter 4).

The picture that emerges from this functional analysis clearly has a broader scope than the idea of a unique explanatory factor determining mathematical competence, specifically the fidelity of number representations in the ANS. Essentially, it states that one can be endowed with a very sharp and efficient ANS, but efficient use of this system is dependent on many other, poorly understood brain and cognitive systems. Of course, my view that a broad spectrum of related neurocognitive components needs to be considered is not a plea to dismiss the importance of a well-functioning system for quantity representation. To the contrary, quantity representation is a vital component of number knowledge, and how it can develop and be learned (see e.g., vanMarle et al., 2014, who suggest that the ANS supports children's initial learning of numerical symbols [e.g., number words] and their meaning [i.e., their cardinal value], but then becomes less important as children become more proficient with formal, symbolic mathematics). Yet, I believe that it should be seen as a component in the context of an ensemble of other components that together determines what level of mathematical proficiency can be achieved. It is the study of these interactions that opens up interesting perspectives. Indeed, the efficiency of some components is likely to have an impact on the limits of representational accuracy that can be achieved with practice or training. For instance, with poorly developed systems to map digits to quantity representations (Noël & Rousselle, 2011), it is quite possible that the quantity representations do not receive the proper input needed to increase the acuity of these representations. Alternatively, it is not unthinkable that inaccurate quantity representations put heavy demands on other cognitive components, such as executive control.

Compared to a single-component view, a multicomponent framework leads to a considerable increase in the degrees of freedom for theory development. Although this can be viewed as a significant disadvantage from a pragmatic point of view, a certain degree of theoretical complexity is the only way to fundamentally increase our understanding of the multifaceted nature of mathematical cognition. With increasing complexity and number of components it becomes progressively more difficult to decide on the validity of one theory over the other. An important way out is to nail down the characteristics of the contributing components. This can be achieved by appealing to something other than the behavioral data that one wants to account for. In this respect a neural specification of the characteristics of the proposed components can provide the necessary constraints to limit the number of theoretical accounts (Anderson, 1978). Here, a lot of work has already been done, in the sense that the neural underpinnings of quantity representation have been described in rather good detail. For the other components we aren't quite as far yet, but I believe that, as I have indicated, some groundwork has been laid. By further refining our understanding of the neuro-functional organization of these other cognitive components, we can increasingly constrain the number of viable theoretical frameworks and ultimately come to a satisfactory understanding of how the synergistic interactions among several cognitive components permit the mastery of mathematical skills in all their facets, including how they develop—both normally and atypically as in dyscalculia or disability.

REFERENCES

Abrahamse, E., van Dijck, J.-P., Majerus, S., & Fias, W. (2014). Finding the answer in space: The mental whiteboard hypothesis on serial order in working memory. *Frontiers in Human Neuroscience*, *8*, 1–12.

Anderson, J. R. (1978). Arguments concerning representations for mental imagery. *Psychological Review*, *85*, 249–277.

Ansari, D., Grabner, R. H., Koschutnig, K., Reishofer, G., & Ebner, F. (2011). Individual differences in mathematical competence modulate brain responses to arithmetic errors: An fMRI study. *Learning and Individual Differences*, *21*(6), 636–643.

Arsalidou, M., & Taylor, M. J. (2011). Is 2 + 2 = 4? Meta-analyses of brain areas needed for numbers and calculations. *NeuroImage*, *54*(3), 2382–2393.

Attout, L., Fias, W., Salmon, E., & Majerus, S. (2014). Common neural substrates for ordinal representation in short-term memory, numerical and alphabetical cognition. *PloS One*, *9*(3), e92049.

Bachot, J., Gevers, W., Fias, W., & Roeyers, H. (2005). Number sense in children with visuospatial disabilities: Orientation of the mental number line. *Psychology Science*, *47*(1), 172–183.

Bächtold, D., Baumüller, M., & Brugger, P. (1998). Stimulus response compatibility in representational space. *Neuropsychologia*, *36*(8), 731–735.

Baddeley, A. D. (1986). *Working memory*. New York: Oxford University Press.

Barrouillet, P., Fayol, M., & Lathuliére, E. (1997). Selecting between competitors in multiplication tasks: An explanation of the errors produced by adolescents with learning disabilities. *International Journal of Behavioral Development*, *21*, 253–275.

Botvinick, M. M., Braver, T. S., Barch, D. M., Carter, C. S., & Cohen, J. D. (2001). Conflict monitoring and cognitive control. *Psychological Review*, *108*(3), 624–652.

Botvinick, M., & Watanabe, T. (2007). From numerosity to ordinal rank: A gain-field model of serial order representation in cortical working memory. *Journal of Neuroscience*, *27*(32), 8636–8642.

Bull, R., Johnston, R. S., & Roy, J. A. (1999). Exploring the roles of the visual—Spatial sketch pad and central executive in children's arithmetical skills: Views from cognition and developmental neuropsychology. *Developmental Neuropsychology*, *15*(3), 421–442.

Bull, R., & Lee, K. (2014). Executive functioning and mathematics achievement. *Child Development Perspectives*, *8*(1), 36–41.

Butterworth, B., Varma, S., & Laurillard, D. (2011). Dyscalculia: From brain to education. *Science*, *332*(6033), 1049–1053.

Cantlon, J. F., Brannon, E. M., Carter, E. J., & Pelphrey, K.a. (2006). Functional imaging of numerical processing in adults and 4-y-old children. *PLoS Biology*, *4*(5), e125.

Chen, Q., & Li, J. (2014). Association between individual differences in nonsymbolic number acuity and math performance: A meta-analysis. *Acta Psychologica*, *148*, 163–172.

Cowan, N. (1999). An embedded-processes model of working memory. In A. Miyake & P. Shah (Eds.), *Models of working memory: Mechanisms of active maintenance and executive control* (pp. 62–101). New York, NY: Cambridge University Press.

D'Esposito, M., & Postle, B. R. (2015). The Cognitive neuroscience of working memory. *Annual Review of Psychology*, *66*, 115–142.

De Rammelaere, S., Stuyven, E., & Vandierendonck, A. (2001). Verifying simple arithmetic sums and products: Are the phonological loop and the central executive involved? *Memory & Cognition*, *29*, 267–273.

De Smedt, B., Noël, M. P., Gilmore, C., & Ansari, D. (2013). The relationship between symbolic and nonsymbolic numerical magnitude processing and the typical and atypical development of mathematics: A review of evidence from brain and behavior. *Trends in Neuroscience and Education, 2*, 48–55.

De Visscher, A., & Noël, M.-P. (2014a). Arithmetic facts storage deficit: The hypersensitivity-to-interference in memory hypothesis. *Developmental Science, 17*(3), 434–442.

De Visscher, A., & Noël, M. (2014b). The detrimental effect of interference in multiplication facts storing: Typical development and individual differences. *Journal of Experimental Psychology General, 143*(6), 2380–2400.

Dehaene, S. (1992). Varieties of numerical abilities. *Cognition, 44*, 1–43.

Dehaene, S., Bossini, S., & Giraux, P. (1993). The mental representation of parity and number magnitude. *Journal of Experimental Psychology General, 122*(3), 371–396.

Dehaene, S., Molko, N., Cohen, L., & Wilson, A. J. (2004). Arithmetic and the brain. *Current Opinion in Neurobiology, 14*(2), 218–224.

Desmet, C., Imbo, I., De Brauwer, J., Brass, M., Fias, W., & Notebaert, W. (2012). Error adaptation in mental arithmetic. *Quarterly Journal of Experimental Psychology, 65*, 1059–1067.

Doricchi, F., Guariglia, P., Gasparini, M., & Tomaiuolo, F. (2005). Dissociation between physical and mental number line bisection in right hemisphere brain damage. *Nature Neuroscience, 8*(12), 1663–1665.

Dumontheil, I., & Klingberg, T. (2012). Brain activity during a visuospatial working memory task predicts arithmetical performance 2 years later. *Cerebral Cortex, 22*(5), 1078–1085.

Fazio, L. K., Bailey, D. H., Thompson, C.a, & Siegler, R. S. (2014). Relations of different types of numerical magnitude representations to each other and to mathematics achievement. *Journal of Experimental Child Psychology, 123*, 53–72.

Fias, W., Brysbaert, M., Geypens, F., & Ydewalle, G. (1996). The importance of magnitude information in numerical processing: Evidence from the SNARC effect. *Mathematical Cognition, 2*(1), 95–110.

Fias, W., Lammertyn, J., Caessens, B., & Orban, G. A. (2007). Processing of abstract ordinal knowledge in the horizontal segment of the intraparietal sulcus. *Journal of Neuroscience, 27*(33), 8952–8956.

Fias, W., Lauwereyns, J., & Lammertyn, J. (2001). Irrelevant digits affect feature-based attention depending on the overlap of neural circuits. *Cognitive Brain Research, 12*, 415–423.

Fias, W., Menon, V., & Szucs, D. (2013). Multiple components of developmental dyscalculia. *Trends in Neuroscience and Education, 2*(2), 43–47.

Fias, W., van Dijck, J.-P., & Gevers, W. (2011). How is number associated with space? The role of working memory. In S. Dehaene & E. M. Brannon (Eds.), *Space, time and number in the brain: Searching for the foundations of mathematical thought* (pp. 133–148). Amsterdam: Elsevier.

Fischer, M. H., Castel, A. D., Dodd, M. D., & Pratt, J. (2003). Perceiving numbers causes spatial shifts of attention. *Nature Neuroscience, 6*(6), 555–556.

Geary, D. C. (1993). Mathematical disabilities: Cognitive, neuropsychological and genetic components. *Psychological Bulletin, 114*(2), 345–362.

Geary, D. C., Berch, D. B., & Mann Koepke, K. (Eds.), (2015). *Evolutionary origins and early development of number processing*. San Diego, CA: Academic Press.

Geary, D. C., Hamson, C. O., & Hoard, M. K. (2000). Numerical and arithmetical cognition: A longitudinal study of process and concept deficits in children with learning disability. *Journal of Experimental Child Psychology, 77*, 236–263.

Gevers, W., Reynvoet, B., & Fias, W. (2003). The mental representation of ordinal sequences is spatially organised. *Cognition, 87*, B87–B95.

Gevers, W., Reynvoet, B., & Fias, W. (2004). The mental representation of ordinal sequences is spatially organized: Evidence from days of the week. *Cortex*, *40*, 171–172.

Halberda, J., Mazzocco, M. M. M., & Feigenson, L. (2008). Individual differences in nonverbal number acuity correlate with maths achievement. *Nature*, *455*(7213), 665–668.

Herrera, A., Macizo, P., & Semenza, C. (2008). The role of working memory in the association between number magnitude and space. *Acta Psychologica*, *128*(2), 225–237.

Hitch, G. J. (1978). The role of short-term working memory in mental arithmetic. *Cognitive Psychology*, *10*, 302–323.

Hoffmann, D., Mussolin, C., Martin, R., & Schiltz, C. (2014). The impact of mathematical proficiency on the number-space association. *PloS One*, *9*(1), e85048.

Hoffmann, D., Pigat, D., & Schiltz, C. (2014). The impact of inhibition capacities and age on number-space associations. *Cognitive Processing*, *15*(3), 329–342.

Hubbard, E. M., Piazza, M., Pinel, P., & Dehaene, S. (2005). Interactions between number and space in parietal cortex. *Nature Reviews Neuroscience*, *6*(6), 435–448.

Hyde, D. C., Boas, D. A., Blair, C., & Carey, S. (2010). Near-infrared spectroscopy shows right parietal specialization for number in pre-verbal infants. *NeuroImage*, *53*, 647–665.

Imbo, I., Vandierendonck, A., & De Rammelaere, S. (2007). The role of working memory in the carry operation in mental arithmetic: Number and value of the carry. *Quarterly Journal of Experimental Psychology*, *60*, 708–731.

Imbo, I., Vandierendonck, A., & Vergauwe, E. (2007). The role of working memory in carrying and borrowing. *Psychological Research*, *71*, 467–483.

Izard, V., Dehaene-Lambertz, G., & Dehaene, S. (2008). Distinct cerebral pathways for object identity and number in human infants. *PLoS Biology*, *6*(2), 0275–0285.

Knops, A., Piazza, M., Sengupta, R., Eger, E., & Melcher, D. (2014). A shared, flexible neural map architecture reflects capacity limits in both visual short-term memory and enumeration. *Journal of Neuroscience*, *34*(30), 9857–9866.

Knops, A., Thirion, B., Hubbard, E. M., Michel, V., & Dehaene, S. (2009). Recruitment of an area involved in eye movements during mental arithmetic. *Science*, *324*(5934), 1583–1585.

Lemaire, P., & Siegler, R. S. (1995). Four aspects of strategic change: Contributions to children's learning multiplication. *Journal of Experimental Psychology General*, *124*, 83–97.

Majerus, S., D'Argembeau, A., Martinez Perez, T., Belayachi, S., Van der Linden, M., Collette, F., et al. (2010). The commonality of neural networks for verbal and visual short-term memory. *Journal of Cognitive Neuroscience*, *22*(11), 2570–2593.

Majerus, S., Poncelet, M., Greffe, C., & Van der Linden, M. (2006). Relations between vocabulary development and verbal short-term memory: The relative importance of short-term memory for serial order and item information. *Journal of Experimental Child Psychology*, *93*, 95–119.

Marshuetz, C. (2005). Order information in working memory: An integrative review of evidence from brain and behavior. *Psychological Bulletin*, *131*, 323–339.

Marshuetz, C., Smith, E. E., Jonides, J., DeGutis, J., & Chenevert, T. L. (2000). Order information in working memory: fMRI evidence for parietal and prefrontal mechanisms. *Journal of Cognitive Neuroscience*, *12*, 130–144.

Miyake, a., Friedman, N. P., Emerson, M. J., Witzki, a.H., Howerter, a, & Wager, T. D. (2000). The unity and diversity of executive functions and their contributions to complex "Frontal Lobe" tasks: A latent variable analysis. *Cognitive Psychology*, *41*, 49–100.

Molko, N., Cachia, A., Rivière, D., Mangin, J. F., Bruandet, M., Le Bihan, D., et al. (2003). Functional and structural alterations of the intraparietal sulcus in a developmental dyscalculia of genetic origin. *Neuron*, *40*(4), 847–858.

Monsell, S. (2003). Task switching. *Trends in Cognitive Sciences*, *7*(3), 134–140.

Nieder, A., & Miller, E. K. (2004). A parieto-frontal network for visual numerical information in the monkey. *Proceedings of the National Academy of Sciences of the United States of America, 101*(19), 7457–7462.

Noël, M. P. (2009). Counting on working memory when learning to count and to add: A preschool study. *Developmental Psychology, 45*(6), 1630–1643.

Noël, M.-P., & Rousselle, L. (2011). Developmental changes in the profiles of dyscalculia: An explanation based on a double exact-and-approximate number representation model. *Frontiers in Human Neuroscience, 5*, 165.

Oberauer. (2009). Design for a working memory. *Psychology of Learning and Motivation, 51*, 45–100.

Piazza, M., Izard, V., Pinel, P., Le Bihan, D., & Dehaene, S. (2004). Tuning curves for approximate numerosity in the human intraparietal sulcus. *Neuron, 44*(3), 547–555.

Piazza, M., Pinel, P., Le Bihan, D., & Dehaene, S. (2007). A magnitude code common to numerosities and number symbols in human intraparietal cortex. *Neuron, 53*(2), 293–305.

Pinel, P., & Dehaene, S. (2013). Genetic and environmental contributions to brain activation during calculation. *NeuroImage, 81*, 306–316.

Qin, S., Cho, S., Chen, T., Rosenberg-Lee, M., Geary, D. C., & Menon, V. (2014). Hippocampal-neocortical functional reorganization underlies children's cognitive development. *Nature Neuroscience, 17*(9), 1263–1269.

Raghubar, K. P., Barnes, M. A., & Hecht, S. A. (2010). Working memory and mathematics: A review of developmental, individual difference, and cognitive approaches. *Learning and Individual Differences, 20*(2), 110–122.

Rinne, L. F., & Mazzocco, M. M. M. (2014). Knowing right from wrong in mental arithmetic judgments: Calibration of confidence predicts the development of accuracy. *PloS One, 9*(7), e98663.

Roggeman, C., Fias, W., & Verguts, T. (2010). Salience maps in parietal cortex: Imaging and computational modeling. *NeuroImage, 52*(3), 1005–1014.

Roggeman, C., Santens, S., Fias, W., & Verguts, T. (2011). Stages of nonsymbolic number processing in occipitoparietal cortex disentangled by FMRI adaptation. *Journal of Neuroscience, 31*(19), 7168–7173.

Roitman, J. D., Brannon, E. M., & Platt, M. L. (2007). Monotonic coding of numerosity in macaque lateral intraparietal area. *PLoS Biology, 5*(8), e208.

Rossetti, Y., Jacquin-Courtois, S., Aiello, M., Ishihara, M., Brozzoli, C., & Doricchi, F. (2011). Neglect "around the clock": Dissociating number and spatial neglect in right brain damage. In S. Dehaene, & E. M. Brannon (Eds.), *Space, time and number in the brain: Searching for the foundations of mathematical thought* (pp. 149–173). Burlington, MA: Elsevier/Academic Press.

Rousselle, L., & Noël, M.-P. (2007). Basic numerical skills in children with mathematics learning disabilities: A comparison of symbolic vs nonsymbolic number magnitude. *Cognition, 102*, 361–395.

Rumelhart, D. E., McClelland, J. L., the PDP research group. (1986). *Parallel distributed processing: Explorations in the microstructure of cognition: Vol. I.* Cambridge, MA: MIT Press.

Santens, S., Roggeman, C., Fias, W., & Verguts, T. (2010). Number processing pathways in human parietal cortex. *Cerebral Cortex, 20*(1), 77–88.

Shaki, S., & Fischer, M. H. (2008). Reading space into numbers—A cross-linguistic comparison of the SNARC effect. *Cognition, 108*(2), 590–599.

Shalev, R. S., Auerbach, J., Manor, O., & Gross-Tsur, V. (2000). Developmental dyscalculia: Prevalence and prognosis. *European Child & Adolescent Psychiatry, 9*(Suppl. 2(Dc)), II58–II64.

Shum, J., Hermes, D., Foster, B. L., Dastjerdi, M., Rangarajan, V., Winawer, J., et al. (2013). A brain area for visual numerals. *Journal of Neuroscience, 33*(16), 6709–6715.

Storer, L., & Demeyere, N. (2014). Disruptions to number bisection after brain injury: Neglecting parts of the Mental Number Line or working memory impairments? *Brain and Cognition*, *86*, 116–123.

Supekar, K., Swigart, A. G., Tenison, C., Jolles, D. D., Rosenberg-lee, M., Fuchs, L., et al. (2013). Neural predictors of individual differences in response to math tutoring in primary-grade school children. *Proceedings of the National Academy of Sciences*, *110*(20), 8230–8235.

Todd, J. J., & Marois, R. (2004). Capacity limit of visual short-term memory in human posterior parietal cortex. *Nature*, *428*(6984), 751–754.

Uittenhove, K., & Lemaire, P. (2012). Sequential difficulty effects during strategy execution: A study in arithmetic. *Experimental Psychology*, *59*(5), 295–301.

van Dijck, J.-P., Abrahamse, E. L., Acar, F., Ketels, B., & Fias, W. (2014). A working memory account of the interaction between numbers and spatial attention. *Quarterly Journal of Experimental Psychology*, *67*(8), 1500–1513.

van Dijck, J.-P., Abrahamse, E. L., Majerus, S., & Fias, W. (2013). Spatial attention interacts with serial-order retrieval from verbal working memory. *Psychological Science*, *24*(9), 1854–1859.

van Dijck, J.-P., & Fias, W. (2011). A working memory account for spatial-numerical associations. *Cognition*, *119*(1), 114–119.

van Dijck, J.-P., Gevers, W., & Fias, W. (2009). Numbers are associated with different types of spatial information depending on the task. *Cognition*, *113*(2), 248–253.

van Dijck, J.-P., Gevers, W., Lafosse, C., Doricchi, F., & Fias, W. (2011). Nonspatial neglect for the mental number line. *Neuropsychologia*, *49*, 2570–2583.

van Dijck, J.-P., Gevers, W., Lafosse, C., & Fias, W. (2012). The heterogeneous nature of number–space interactions. *Frontiers in Human Neuroscience*, *5*, 1–13.

vanMarle, K., Chu, F. W., Li, Y., & Geary, D. C. (2014). Acuity of the approximate number system and preschoolers' quantitative development. *Developmental Science*, *17*, 492–505.

Verguts, T., & Fias, W. (2004). Representation of number in animals and humans: A neural model. *Journal of Cognitive Neuroscience*, *16*(9), 1493–1504.

Zanolie, K., & Pecher, D. (2014). Number-induced shifts in spatial attention: A replication study. *Frontiers in Psychology*, *5*, 987.

Zebian, S. (2005). Linkages between number concepts, spatial thinking, and directionality of writing: The SNARC effect and the reverse SNARC effect in English and Arabic monoliterates, biliterates, and illiterate Arabic speakers. *Journal of Cognition and Culture*, *5*, 165–190.

Zorzi, M., Priftis, K., & Umiltà, C. (2002). Neglect disrupts the mental number line. *Nature*, *417*, 138–139.

Chapter 9

Individual Differences in Arithmetic Fact Retrieval

Bert De Smedt
Faculty of Psychology and Educational Sciences, University of Leuven, Leuven, Belgium

INTRODUCTION

The ability to retrieve arithmetic facts is a major building block in children's mathematical development (e.g., Kilpatrick, Swafford, & Findell, 2001). It represents a core element of the mathematics curriculum in (early) primary school (e.g., National Mathematics Advisory Panel, 2008) and constitutes a prerequisite for the learning of more complex calculations and algebra. Alternatively, the difficulty in retrieving arithmetic facts is considered the hallmark of children with mathematical learning disabilities or dyscalculia, who experience persistent deficits in acquiring basic mathematical competencies (American Psychiatric Association, 2013; Geary, 2004). Similarly, in the typically achieving population, there are large individual differences in the development and acquisition of these arithmetic facts (Dowker, 2005; Jordan, Hanich, & Kaplan, 2003); such differences can also be observed at the neural level (De Smedt, Holloway, & Ansari, 2011; Grabner et al., 2007; Menon et al., 2000). Understanding the origins of these individual differences is important because it provides the crucial basis for designing and developing educational interventions that are tailored to the needs of the individual.

In this chapter, I will summarize the available evidence on individual differences in arithmetic fact retrieval as viewed through the lens of educational neuroscience (e.g., De Smedt, 2014). Specifically, I will highlight two ways in which educational neuroscience research has improved our understanding of individual differences in arithmetic fact retrieval. First, one indirect way has been by generating predictions of underlying processes based on (often adult) brain imaging data to be tested in cognitive developmental behavioral studies. These studies have pointed to the importance of numerical magnitude processing (see below) and phonological processing (see below) as cognitive factors that contribute to individual differences in arithmetic fact retrieval. The second contribution is direct, by investigating the neural underpinnings of school-taught

Development of Mathematical Cognition. http://dx.doi.org/10.1016/B978-0-12-801871-2.00009-5

arithmetic, in this case arithmetic fact retrieval. Although many brain imaging studies have been carried out with adults, a small but steadily increasing number have been conducted with children.

The remainder of this chapter consists of four sections. In the first, I describe how children develop memory representations of arithmetic facts and how researchers typically have assessed arithmetic fact retrieval. In the second section, I focus on two neurocognitive processes that have been proposed as potential origins of individual differences in arithmetic fact retrieval: numerical magnitude processing and phonological skills. In the third section, I move to the biological level and review neuroimaging studies of arithmetic fact retrieval, and in the final section, I highlight important issues that should be resolved in future studies on this topic.

DEVELOPMENT AND MEASUREMENT OF ARITHMETIC FACT RETRIEVAL

Evidence from various studies has provided a reasonably clear account of how children learn to solve single-digit arithmetic problems and how this learning process is characterized by the development of different arithmetic strategies (e.g., Geary, 2004; Jordan et al., 2003; Siegler, 1996). Initially, children use counting procedures to solve answers to basic arithmetic problems. These counting strategies are often executed with the use of fingers and eventually without them (verbal counting). Whether or not they use their fingers to count, children initially count both addends in their entirety (counting-all, 3+6=1,2,3 and 1,2,3,4,5,6 so 1,2,3,4,5,6,7,8,9), after which they start to count from the first number (counting-on, 3+6=4,5,6,7,8,9) and eventually, they start from the larger number (counting-on-larger, 3+6=7,8,9) (Geary, Bow-Thomas, & Yao, 1992). Through the repeated use of these counting procedures, children build problem-answer associations or arithmetic facts that are stored in their long-term memory (Siegler & Shrager, 1984). The development of strong problem-answer associations is important because these enable direct retrieval of arithmetic facts, which is more efficient (i.e., faster and more accurate) than using time-consuming and more error-prone counting strategies. On top of direct retrieval, knowing some basic facts can facilitate the use of a strategy wherein problems are decomposed into smaller problems, as in the case of procedural strategies such as transformation or decomposition—for example 6+8=14, 6+4=10+4=14.

It is important to point out that this development in strategy use is not an abrupt change from one stage to the other (e.g., from using counting only to using retrieving strategies). Rather, there is a change in the distributions of strategies children use, so-called *overlapping waves* (Siegler, 1996), with a decreased reliance on procedural strategies and increased reliance on fact-retrieval strategies (Geary, Hoard, Byrd-Craven, & Desoto, 2004; Jordan et al., 2003; Siegler, 1996), with the use of these several strategies continuing throughout adulthood. Such developmental changes in strategic shifts from error-prone and slower strategies to more efficient direct strategies can be observed in other academic domains, such as spelling or reasoning (Siegler, 1996) and reading, where development is characterized by decreased reliance on time-consuming phonological decoding to efficient orthographic direct recognition (e.g., Schlaggar & Mccandliss, 2007). In all, there are large individual differences in the development of arithmetic strategies (e.g., Dowker, 2005; Jordan et al., 2003) and one important goal of developmental science is to understand the neurocognitive underpinnings of these individual differences.

How do researchers measure the strategies that children and adults use to solve arithmetic problems? In cognitive (behavioral) studies, arithmetic strategy use is commonly assessed by means of verbal report data (e.g., Campbell & Xue, 2001; Geary et al., 2004; Imbo & Vandierendonck, 2008; Lefevre, Sadesky, & Bisanz, 1996; Siegler, 1996; Torbeyns, Verschaffel, & Ghesquière, 2004). In a typical study, participants are presented with an arithmetic problem (e.g., $6+3$) and they have to solve it as quickly and as accurately as possible without instruction as to which strategy should be used. Immediately after stating the solution to a problem, participants are asked to verbally report the strategy they used to solve that problem by answering the question "How did you solve the problem?" These trial-by-trial verbal reports, sometimes accompanied by experimenter observations, can be reliably classified into different categories: *retrieval strategy* (i.e., the individual immediately knew the answer and there were no overt signs of calculating) or *procedural strategy* (i.e., the individual reported a counting strategy or decomposed the problem into smaller subproblems; some studies distinguish different types of counting strategies, often depending on the age of the participants under investigation).

It is important to point out that use of such verbal reports is not without limitations (e.g., Kirk & Ashcraft, 2001; Thevenot, Fanget, & Fayol, 2007). The mental processes that underlie the solution process may not always be accurately reflected in verbal reports because some strategies can be executed very quickly and automatically, and participants may not always be aware of the different steps they executed, as is the case in very fast counting strategies (Thevenot et al., 2007). It is also possible that some individuals, particularly young children or children with dyscalculia, only report strategies for which they have descriptors, yet actually use more strategies than their language skills allow them to report. For this reason, researchers have sometimes used other approaches that do not rely on verbal reports to investigate strategy use. Examples include the statistical analysis of chronometric data (i.e., reaction times; e.g., Geary & Wiley, 1991; Groen & Parkman, 1972) or, as I describe below, the comparison of sets of problems on which the use of a particular strategy is derived from specific characteristics of the arithmetic problem (e.g., its size or the operation).

In the problem-size approach, arithmetic problems are divided on the basis of the size of the operands into so-called small problems (sums ≤ 10) and large problems (sums > 10). Researchers assume that retrieval strategies are typically used more often to solve small problems and procedural strategies more frequently to solve large problems, an assumption that has been confirmed by empirical data in both adults (e.g., Campbell & Xue, 2001; Lefevre et al., 1996) and children (e.g., Barrouillet, Mignon, & Thevenot, 2008; Imbo & Vandierendonck, 2008). In the operation approach, researchers have contrasted arithmetic operations that are assumed to elicit either more retrieval or procedural strategies. One example is the comparison of subtraction and multiplication, because multiplication problems are more often solved by means of retrieval than subtraction (e.g., Berteletti, Prado, & Booth, 2014; Campbell & Xue, 2001). These problem size (e.g., De Smedt et al., 2011; Grabner et al., 2009) and operation approaches (e.g., Berteletti et al., 2014; Prado, Mutreja, & Booth, 2014) have been typically used in functional magnetic resonance imaging (fMRI) studies that aimed to investigate the brain networks underlying retrieval and procedural strategies as well as individual differences in the recruitment of these networks (e.g., De Smedt et al., 2011; Grabner et al., 2009; also see Menon, this volume). Such an approach is taken because collecting verbal reports during fMRI data acquisition is impractical due to the noise of the scanning environment (but see Andres, Pelgrims, Michaux, Olivier, & Pesenti, 2011).

It is, however, important to emphasize that these problem-size and operation approaches are rather limited because not all problems of a particular size or operation are solved by means of the same strategy, which might lead to misleading conclusions about how people solve these problems (Siegler, 1987). This is particularly problematic in the context of developmental research because the strategies that children use to solve particular types of problems change over time (i.e., with education and practice). Limitations of these assumptions also apply to the comparison of typically and atypically developing groups of participants, who might differ in the strategies they use for a specific problem type or operation. Therefore, verbal self-reports might be the most appropriate way to capture arithmetic strategy use (see also Grabner & De Smedt, 2011, for an electrophysiological validation of these verbal reports).

NEUROCOGNITIVE DETERMINANTS OF INDIVIDUAL DIFFERENCES IN ARITHMETIC FACT RETRIEVAL

Various studies have highlighted that there are large individual differences in the acquisition and use of arithmetic facts (e.g., Bailey, Littlefield, & Geary, 2012; Barrouillet et al., 2008; Dowker, 2005; Jordan et al., 2003). Originally, many studies focused on children with atypical mathematical development, in particular, children with dyscalculia, who have persistent deficits in the acquisition of these arithmetic facts (e.g., Geary, 2004) although individual differences can also be observed in the general population. For example, Jordan et al. (2003) distinguished subgroups of arithmetic fact mastery based on children's performance on a retrieval-only task, where children are asked to quickly solve addition problems using only memory retrieval. More recently, Vanbinst, Ceulemans, Ghesquière, and De Smedt (2015) used a model-based clustering approach, where an automatic numerical grouping procedure (based on a probability model) is applied to different parameters of performance to assign individuals to cohesive clusters or profiles of arithmetic fact development, investigating whether these distinct profiles differed in neurocognitive skills that have previously been associated with individual differences in arithmetic.

What might be important neurocognitive determinants that underlie these individual differences? As previously highlighted, findings from brain imaging studies on arithmetic (see Menon, 2015 for a review, as well as Menon, this volume) have been used to generate predictions about the cognitive processes that underlie arithmetic fact-retrieval abilities and individual differences, and these hypotheses have been tested in cognitive developmental behavioral studies. Specifically, brain imaging studies have indicated the involvement of a widespread fronto-parietal network during arithmetic problem solving (e.g., Menon, 2015, for a review). In this chapter, I will specifically focus on areas in the inferior parietal cortex, which includes the intraparietal sulci (IPS) as well as supramarginal and angular gyri (Figure 1, left panel). These areas are known to play a role in numerical magnitude processing (e.g., Ansari, 2008) and phonological processing (e.g., Dehaene, Piazza, Pinel, & Cohen, 2003). Against this background, these cognitive factors have been highlighted as potential correlates of individual differences in arithmetic fact retrieval as discussed in the following. It is important to recognize that other domain-general cognitive factors also play a role in arithmetic fact retrieval, specifically those functions related to dorsolateral prefrontal cortex—working memory and executive

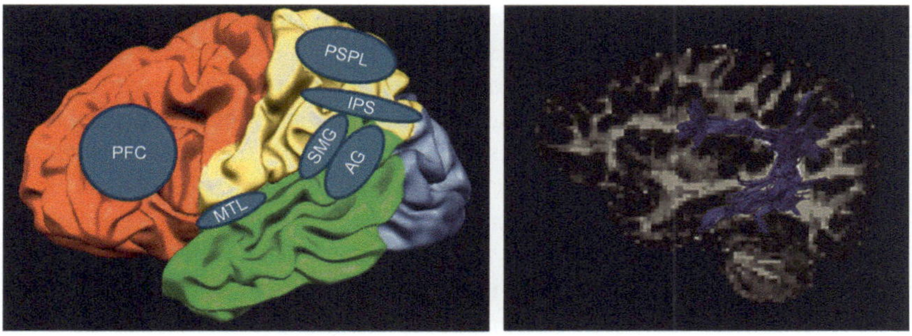

FIGURE 1 Left panel: Brain areas involved in arithmetic fact retrieval. The left hemisphere is depicted and the four lobes of the brain are shown in different colors: frontal lobe = red, parietal lobe = yellow, temporal lobe = green, occipital lobe = blue. PFC = prefrontal cortex. MTL = medial temporal lobe. IPS = intrparietal sulcus. SMG = supramarginal gyrus. AG = angular gyrus. PSPL = posterior superior parietal lobe. Right panel: the arcuate fasciculus in a child on sagittal fractional anisotrophy data depicted in blue. The white matter integrity of the anterior part of this tract, which connects the frontal and parietal lobe, has been related to individual differences in arithmetic fact retrieval (Van Beek, Ghesquière, Lagae, & De Smedt, 2014).

function—that have more generic roles in the acquisition of academic competencies (e.g., Alloway & Alloway, 2010). These will not be discussed in this chapter; the interested reader is referred to the reviews of Raghubar, Barnes, and Hecht (2010) and Friso-Van Den Bos, Van Der Ven, Kroesbergen, and Van Luit (2013) on the role of working memory in mathematics achievement.

Numerical Magnitude Processing

Various neuroimaging studies have shown that the IPS is consistently active whenever people have to perform arithmetic tasks, suggesting that the activity in these areas reflects the processing of numerical magnitudes during calculation (Ansari, 2008; Arsalidou & Taylor, 2011; Kaufmann, Wood, Rubinsten, & Henik, 2011; Menon, 2015). Relatedly, brain imaging studies in dyscalculia have pointed to structural and functional abnormalities in the IPS in these children that coincide with poor arithmetical performance (see Price & Ansari, 2013, for a review). These findings suggest that the processing of numerical magnitudes is key to successful mathematical development, and this suggestion has fueled a large number of psychological and educational studies that have confirmed this hypothesis at the behavioral level (e.g., De Smedt, Noël, Gilmore, & Ansari, 2013; see also Geary, Berch, & Mann Koepke, 2015, for reviews).

The ability to process numerical magnitudes has typically been investigated by means of a numerical magnitude comparison task (see De Smedt et al., 2013; Geary et al., 2015, for reviews). In this task, participants have to indicate the larger of two presented numerical magnitudes. Stimuli can be presented in symbolic (digits) or nonsymbolic (sets of dots) formats. Performance with both types of formats has been related to individual differences in children's general mathematics achievement (e.g., De Smedt et al., 2013, for a review), indicating that children who perform faster and more accurately on these comparison tasks show higher mathematics achievement. However, there exists considerable debate as to whether the processing of nonsymbolic magnitude representations (dots), symbolic magnitude representations (digits) or both are relevant for learning more advanced mathematical competencies. Conflicting results have been reported, in particular for nonsymbolic measures of magnitude processing (De Smedt et al., 2013, for a narrative review), but meta-analytic data, focusing only on the association between nonsymbolic measures and mathematics achievement, suggest that this association tends to be moderate in size (Chen & Li, 2014; Fazio et al., 2014). A recent meta-analysis by Schneider et al. (submitted) indicates that the effect size (a quantitative measure of the strength of an outcome) for symbolic measures is significantly higher than for nonsymbolic measures. Most importantly, these authors also showed that the type of mathematical competency moderates even more strongly the size of the association between numerical magnitude comparison and mathematical development. It is therefore particularly important to investigate how numerical magnitude processing is related to any specific type of mathematical competence—in this case arithmetic fact retrieval.

There are various theoretical reasons for speculating about a connection between individual differences in arithmetic fact mastery and numerical magnitude processing. As indicated above, children gradually progress in their arithmetic development from counting-all to counting-on-from-larger (Geary et al., 1992). This latter advanced counting strategy requires a determination on the larger number and therefore draws on the understanding of numerical magnitudes. Proficient numerical magnitude processing skills might facilitate children's learning of the cardinal values (i.e., number or size of a set) associated with Arabic numerals and number words (vanMarle, Chu, Li, & Geary, 2014), speeding the transition to the counting-on-larger strategy. The use of counting-on in turn might foster the memorization of problem-answer associations in long-term memory, that is, the development of arithmetic facts. Interestingly, it also appears that arithmetic facts are stored in long-term memory in a meaningful way—according to their magnitude (Butterworth, Zorzi, Girelli, & Jonckheere, 2001; Campbell, 1995; Robinson, Menchetti, & Torgesen, 2002).

Against this background, Vanbinst, Ghesquière, and De Smedt (2012) investigated the association between numerical magnitude processing, as indexed by means of symbolic and nonsymbolic comparison tasks, and arithmetic fact retrieval in typically developing third graders. In this study, children had to solve single-digit arithmetic additions and subtractions, and their strategy use was assessed by means of verbal report on a trial-by-trial basis. Their data revealed a specific and unique association between symbolic but not nonsymbolic magnitude comparison skills and arithmetic fact retrieval: Children who were faster in comparing Arabic digits more frequently used fact retrieval and were more efficient when doing so (i.e., they were faster and more accurate). This association remained significant even when individual differences in intelligence, general mathematics achievement, or processing speed were controlled for.

Although these findings are a first step in showing an association between (symbolic) numerical magnitude processing and arithmetic fact retrieval, they are limited for several reasons. As these data were collected at the same point in time, they do not allow us to determine whether numerical magnitude processing skills are precursors or consequences of individual differences in arithmetic fact retrieval. Furthermore, these correlations do not capture developmental processes in the acquisition of fact-retrieval skills (the change in strategy mix from procedural to retrieval strategies), leaving it unresolved whether numerical magnitude processing plays a role in the acquisition of arithmetic facts. To assess the mechanisms that are driving this cognitive change, one should focus on the age where the rate of change is the steepest—in this case, the early grades of primary school.

In an attempt to address these issues, Vanbinst, Ghesquière, and De Smedt (2015) designed a prospective longitudinal study. They selected a sample of typically developing children at the onset of formal schooling (the beginning of first grade), because these children have not yet received formal mathematics instruction. During the initial weeks of first grade, all children completed nonsymbolic and symbolic magnitude comparison tasks. Arithmetic skills and strategy use were subsequently evaluated in the middle of first grade and at the start of second grade. To carefully examine the unique predictive value of numerical magnitude processing for future individual differences in arithmetic, the authors took into account various potential confounds. For example, many children have acquired a good deal of mathematical knowledge before they receive formal mathematics instruction (e.g., Resnick, 1989) and this early acquired knowledge predicts children's subsequent growth in mathematics (e.g., Duncan et al., 2007). To control for this, all children completed a general curriculum-based standardized mathematics achievement test at primary school entrance.

Moreover, nonnumerical skills that have been associated with individual differences in fact-retrieval development, such as working memory (e.g., Friso-Van Den Bos et al., 2013), speed of number word articulation (Hecht, Torgesen, Wagner, & Rashotte, 2001), and intelligence were controlled. Vanbinst, Ghesquière, and De Smedt (2015) showed that symbolic numerical magnitude processing assessed at the start of first grade uniquely predicted single-digit arithmetic accuracy, reaction time, and frequency of use of the fact-retrieval strategy in the middle of first grade as well as arithmetic reaction time and fact-retrieval frequency at the start of second grade. Interestingly, symbolic numerical magnitude processing also uniquely predicted the developmental change from first to second grade, that is, the decrease in reaction time and the increase in fact-retrieval frequency. Associations with measures of nonsymbolic processing were either weak or nonsignificant. These findings suggest that symbolic numerical magnitude processing may play a unique role in the individual differences and developmental changes in children's early arithmetic fact acquisition and suggest that proficient symbolic magnitude processing provides an important scaffold for developing arithmetical facts.

Does numerical processing also play a role at later stages of arithmetic facts mastery? To answer this question, Vanbinst, Ceulemans, et al. (2015) carried out a three-year longitudinal study starting in third grade. They investigated arithmetic fact mastery in a random sample that was tested at the start of each of three successive school years, that is third, fourth, and fifth grade. At each time point, children were asked to solve sets of single-digit addition and subtraction problems, and their strategy use was evaluated by means of trial-by-trial verbal reports. They also completed symbolic (i.e., comparing the magnitudes of Arabic numerals) and nonsymbolic magnitude comparison tasks at each time point. The authors used a model-based clustering approach (Banfield & Raftery, 1993) on different parameters of arithmetic fact mastery at the start of each grade, where frequency of fact retrieval, mean reaction time, and variance in reaction time were used to delineate clusters or profiles of arithmetic fact development.

The analysis of Vanbinst and colleagues revealed three profiles, "*slow and variable*," "*average*," and "*efficient*," which differed in their arithmetic facts development from third to fifth grade. It is important to note that the children comprising each of these profiles did not differ in age, sex, socioeconomic status, intelligence, or working memory. Children with a *slow and variable* profile relied less frequently on fact retrieval and were slower and more variable in reaction times. Although these children improved over time, they continued to perform more poorly than the *average* and *efficient* profiles, particularly in reaction time and variability of reaction time. *Average* children were significantly slower at each time point than the *efficient* children, but these group differences in reaction time variability diminished over time. Importantly, the three profiles differed in their symbolic magnitude processing skills: at each time point, children with the *slow and variable* profile were significantly slower than children with the *average* profile, who in turn were significantly slower than the children with the *efficient* profile. No such differences in nonsymbolic processing were observed. These data clearly suggest that children's symbolic magnitude processing may play a particularly important role in children's arithmetic fact development.

Deficits in arithmetic fact retrieval have been put forward as the hallmark feature of children with dyscalculia (American Psychiatric Association, 2013; Geary, 1993). Do these arithmetic fact retrieval deficits coincide with impairments in numerical magnitude processing? To answer these questions, Vanbinst, Ghesquière, et al. (2014) conducted a two-year longitudinal study with three measurement points in children diagnosed with dyscalculia based on persistent poor performance on a general standardized achievement test (American Psychiatric Association, 2013). Their findings revealed that the children with dyscalculia had deficits in arithmetic fact retrieval at each time point as compared to a group of typically developing children matched on age, intelligence, reading ability, and working memory. Crucially, the children with dyscalculia performed more poorly on symbolic numerical magnitude comparisons at each time point, whereas no group differences in nonsymbolic numerical magnitude comparisons were observed. Interestingly, children with dyscalculia also performed more poorly on measures of phonological processing (see the following). Importantly, these group differences could not be explained by poor reading skills, because the group of children with mathematical learning disability (MLD) did not differ from matched controls in reading ability, and all scored close to the population average, indicating that the MLD group did not have a co-occurring reading disability. Taken together, these data suggest that persistent deficits in arithmetic fact retrieval coincide with persistent impairments in symbolic numerical magnitude processing.

Phonological Processing

One of the most influential neurocognitive models of processing number, the triple-code model (Dehaene, 1992), originally postulated that arithmetic fact retrieval might depend on phonological codes. Specifically, the model stated that numerals can be represented in a verbal-phonological code and that this code is probably used in verbally mediated arithmetic tasks, such as retrieval of arithmetic facts from memory, particularly in multiplication. Findings from cognitive neuroscience studies also point to shared neural correlates for reading and arithmetic in the left temporo-parietal cortex, leading to the suggestion that there exists an association between phonological abilities and arithmetic (e.g., De Smedt, Taylor, Archibald, & Ansari, 2010; see also Geary, 1993; Geary, 2011). More specifically, developmental neuroimaging studies of reading have revealed that the left temporo-parietal cortex is consistently activated during phonological processing tasks, such as rhyme judgment tasks and phonological reading strategies—reading strategies that require children to convert letters (graphemes) into sounds (phonemes), as often happens in the initial stages of learning to read and in reading nonwords (e.g., Schlaggar & Mccandliss, 2007, for a review).

At the same time, studies in arithmetic have shown that the left temporo-parietal cortex, including the angular gyrus, is particularly active whenever the answer to a problem is directly retrieved from long-term memory (Grabner et al., 2009), especially in multiplication (Dehaene et al., 2003; Grabner et al., 2007) and problems that have been highly practiced (Delazer et al., 2003). These overlapping neural processes between reading and arithmetic may therefore predict an association between phonological processing and arithmetic fact retrieval, due to their common reliance on phonological codes (see also Simmons & Singleton, 2008).

Various behavioral studies have examined associations between phonological processing and individual differences in general mathematical competence (Fuchs et al., 2006; Hecht et al., 2001; Lefevre et al., 2010; Vukovic & Lesaux, 2013), but none of them had specifically focused on arithmetic fact retrieval. Testing the prediction derived from the above-mentioned cognitive neuroscience studies, De Smedt et al. (2010) investigated the specific association between phonological processing and arithmetic fact retrieval in a behavioral study in typically developing fourth and fifth graders. In this study, children were asked to solve single-digit arithmetic additions, subtractions, and multiplications. The authors used a problem-size approach to determine arithmetic strategy use and categorized their items into retrieval problems, that is, problems that are likely to be solved by means of fact retrieval by children of this age (i.e., small additions, small subtractions and multiplications), and procedural problems, those that are likely to be solved using a procedural strategy by children of this age (i.e., large additions and large subtractions; see Barrouillet et al., 2008; Imbo & Vandierendonck, 2008). Phonological processing was investigated by means of a phoneme deletion task, in which children were asked to indicate what a word would be if a specified phoneme in the word were to be deleted (e.g., saying cup without the phoneme/k/). As predicted from the aforementioned neuroimaging studies, De Smedt et al. (2010) found that individual differences in phonological processing were uniquely related to problems that children are likely to retrieve from long-term memory, such as small additions, small subtractions (e.g., 3 − 2), and multiplications, but unrelated to problems that are likely to be solved by means of procedural strategies (e.g., 8 + 9 or 15 − 7). Consequently, these authors concluded that phonological processing abilities are an important determinant of individual differences in arithmetic fact retrieval.

If a reliable association between phonological processing and arithmetic fact retrieval exists, then individuals with deficits in phonological processing are expected to also show difficulties with arithmetic fact retrieval. One group of individuals with persistent phonological processing deficits are individuals with developmental dyslexia, a specific disorder in learning to read and to write, despite normal intelligence and adequate instruction, which is caused by impairments in phonological processing (Ramus et al., 2003; Snowling, 2000). Several reports have indicated difficulties in multiplication in children with dyslexia (Simmons & Singleton, 2008), but none of them systematically investigated the strategies that children use when solving arithmetic problems.

De Smedt and Boets (2010) investigated this issue in adults and used verbal reports during the solution of single-digit subtractions and multiplications to directly test whether individuals with dyslexia retrieved fewer arithmetic facts compared to controls (matched on age, education, and intelligence), and whether these arithmetic fact-retrieval difficulties could be explained by poorer phonological processing skills. The authors selected a specific group of individuals with dyslexia, namely those without dyscalculia. This was done because dyslexia and dyscalculia are known to co-occur frequently (e.g., Landerl & Moll, 2010), and because in individuals with both dyslexia and dyscalculia, arithmetic fact-retrieval difficulties might also be explained by their impaired numerical magnitude processing deficits, which are common in dyscalculia (e.g., De Smedt et al., 2013; Noel & Rousselle, 2011 for reviews). To exclude this possibility, none of the individuals with dyslexia in the De Smedt and Boets (2010) study differed from their matched controls on a numerical magnitude processing task. The results of this study showed that individuals with developmental dyslexia retrieved fewer arithmetic facts from memory, as reflected in their significantly lowered frequencies of retrieval use. They also were less efficient than controls in executing these retrieval strategies. These difficulties were observed in both subtraction and multiplication. As these adults with dyslexia had weaker phonological skills than controls, their phonological impairments coincided with less frequent fact retrieval from memory. Moreover, phonological processing skills were significantly associated with the frequency of fact retrieval for both control and dyslexia groups, even when individual differences in reading ability were controlled for. In all, these data from individuals with a specific learning disorder confirm the hypothesis of an association between phonological processing and arithmetic fact retrieval (see also Geary, 1993, for a discussion).

Although the aforementioned studies point to a behavioral association between phonological processing and arithmetic fact retrieval, several issues remain that should be tested in subsequent studies. For example, longitudinal data are needed to further determine whether phonological skills can predict subsequent arithmetic fact development. Likewise, what remains to be established is the point in the developmental trajectory of arithmetic fact learning where these phonological skills play the most prominent role. Furthermore, in addition to phonological processing, symbolic numerical magnitude processing can affect the development of arithmetic fact retrieval (see above) and research is needed to determine how each of these cognitive processes uniquely contributes to growth in arithmetic fact retrieval. It is also clear that different pathways can contribute to individual differences in fact development (see Lefevre et al., 2010, for a similar rationale) and it might be that poor numerical processing skills and phonological skills constitute separate risk factors for developing subsequent arithmetic fact-retrieval difficulties.

NEURAL CORRELATES OF ARITHMETIC FACT RETRIEVAL

Arithmetic Fact Retrieval in the (Developing) Brain

Another important goal in educational neuroscience is to characterize the brain networks of complex mathematical skills that are formally taught in school, such as the retrieval of arithmetic facts (e.g., De Smedt et al., 2011). The majority of this work has been carried out with adults by means of fMRI in studies comparing brain activity during the solving of different types of arithmetic problems, for example, contrasting small versus large problems or contrasting different operations, such as addition and subtraction (Menon, 2015, for a review; Arsalidou & Taylor, 2011 for a meta-analysis).

Other studies have attempted to simulate the developmental changes leading to the use of arithmetic fact retrieval by training adults to learn to retrieve the answers to more complex problems (e.g., 23×8) and then examining how brain activity changes as a function of this learning process (see Zamarian & Delazer, 2015, for a review). These studies have revealed that a widespread fronto-parietal network (Figure 1, left panel) is active during calculation in the adult brain, and that the activity in this network is modulated by problem size and operation, or more broadly by the strategy that is used to solve arithmetic problems (Menon, 2015; Zamarian & Delazer, 2015). Specifically, the (pre) frontal cortex has an auxiliary role during calculation and reflects the allocation of working memory and attentional resources. This part of the network is particularly active during the solution of large problems, the initial stages of arithmetic learning, and the execution of procedural strategies.

Two areas of the inferior parietal cortex are important, both of which appear to have distinct roles in calculation. Activity in the IPS is thought to reflect the involvement of quantity-based processing that has been assumed to be more activated during the application of procedural strategies—specifically during the solution of large problems and/or subtractions. The angular gyrus appears to be particularly active when solving multiplications and small problems and specifically during the retrieval of arithmetic facts from semantic memory (e.g., Grabner et al., 2009). Although earlier studies have suggested that this activity reflects the involvement of phonological processes (Dehaene et al., 2003), more recent work has questioned this functional interpretation in calculation (e.g., Menon, 2015, for a discussion).

It is important to point out that these types of findings from adult neuroimaging studies are not readily generalizable to children because these studies reflect the end state of arithmetic development in (often) highly skilled individuals (Ansari, 2010; Menon, 2015). Such data might mistakenly suggest that only a small set of brain areas, particularly in the parietal cortex, are activated during calculation, that these brain networks are the same in children as in adults and that this continues to be the case over development. In other words, adult studies miss important changes in the functional organization of brain networks throughout development and schooling (see also Qin et al., 2014, as discussed in the following). To understand this emergence of brain networks, a developmental perspective is crucial (Menon, 2015 and Menon, this volume), even though neuroimaging research in children is more practically challenging than in adults.

Only but a few developmental imaging studies have examined brain activity when children of different ages solve arithmetic problems. Rivera, Reiss, Eckert, and Menon (2005) were the first to investigate the neural correlates of arithmetic skill development in 8- to 19-year-old children. Specifically, they investigated how the activation of the network underlying small addition and subtractions changes as a function of years in school. Their participants had to verify the correctness of single-digit equations, such as $2+3=6$, which can be assumed to be increasingly solved by fact retrieval across grades. As such, their brain imaging data probably reflect the neural changes that are associated with an increasing reliance on fact retrieval. These authors observed an increased specialization of the left-parietal cortex (i.e., supramarginal gyrus, anterior angular gyrus, and adjoining intraparietal sulcus) and inferior temporal cortex with age, coupled with a decreasing activity in prefrontal cortices, presumably reflecting a decrease in attentional resources and working memory needed to solve these problems using counting procedures. Although these data were a first important step toward understanding the brain networks underlying arithmetic fact retrieval, the sample included a very wide age range and therefore may have missed very important changes in specific developmental periods.

To overcome this issue, more recent studies have focused on samples of more narrow age ranges, as is typically the case in behavioral studies (see section "Neurocognitive Determinants of Individual Differences in Arithmetic Fact Retrieval"). For example, De Smedt et al. (2011) investigated brain activity in children aged 10-12 years during single-digit addition and subtraction. Specifically, the authors aimed to examine how brain activity is modulated by problem size and operation. Commensurate with adult data, De Smedt et al. observed that large problems and subtractions activated a wide fronto-parietal network, including the IPS and prefrontal cortices. This network corresponds to the adult brain areas that are more active during self-reported procedural strategies than during arithmetic fact retrieval (Grabner et al., 2009) and those that are more active during untrained than during trained problems (Zamarian & Delazer, 2015). However, brain activity in these children during the solution of small problems and additions, that is those problems that are assumed to be solved by means of fact retrieval revealed a different pattern from what is observed in adults. During the solution of these problems, the angular gyrus did not show increases in activity, as is typically observed in adults (Dehaene et al., 2003; Grabner et al., 2007). Furthermore, the medial temporal lobe, specifically the (left) hippocampus, was particularly active during the solution of those problems, suggesting a specific role of the hippocampus in children's arithmetic fact retrieval (see Menon, this volume). It has been suggested that the hippocampus has a time-limited role in semantic memory, particularly in the early stages of learning and retrieving recently learned facts (Smith & Squire, 2009). One potential hypothesis is that the hippocampus plays an important role in the initial consolidation of arithmetic facts in long term memory, whereas over development the angular gyrus becomes more involved when highly automatized facts are retrieved.

More recent developmental imaging studies have confirmed the role of the hippocampus in learning arithmetic facts. Cho, Ryali, Geary, and Menon (2011) compared the brain activity of two groups of 7- to 9-year-old children who primarily relied on either retrieval (i.e., retrievers) or procedural strategies (i.e., counters) during small additions. The authors used specific fine-grained multivariate analysis techniques, so-called MultiVoxel Pattern Analyis (e.g., Norman, Polyn, Detre, & Haxby, 2006) to analyze their data. In contrast to traditional univariate fMRI analyses, which only compare the overall mean activity in different brain regions between the two groups, MVPA allows one to investigate, on a much finer scale, spatial patterns of activity (in a particular region) and how these spatial patterns differ between groups. For example, brain regions might be recruited to a similar extent in two groups, revealing no differences in univariate comparisons, yet the spatial pattern of activity in these regions might be different, pointing to differences in processing between groups. These spatial patterns can be analyzed with MVPA and as a result, these methods are particularly promising for detecting fine-grained differences between conditions (e.g., use of retrieval vs. counting for problem solving) and groups of individuals.

Comparing the retrievers with the counters, Cho et al. (2011) revealed that the two groups showed distinct brain activity patterns in the bilateral hippocampus. In a follow-up study, Cho et al. (2012) replicated and extended this finding in children of a similar age range by investigating the correlation between retrieval fluency, the number of facts retrieved from memory, and brain activity during addition. Their data showed that higher frequencies of retrieval use were correlated with increased activity in the bilateral hippocampus.

Recently, Qin et al. (2014) collected longitudinal data on developmental changes leading to the use of retrieval during a one-year interval, with the same paradigm in 7- to 9-year-old children. Behaviorally, children showed an increased reliance on fact-retrieval strategies over one year's time. At the neural level, there was an increase in bilateral hippocampal activity as well as a decrease in activity in the bilateral dorsolateral prefrontal cortex, left superior parietal cortex, and right posterior parietal-occipital cortex—areas that have been implicated in working memory, executive control, and use of effortful counting strategies.

To conclude, these data clearly show that results derived from adult neuroimaging studies are not always adequate for characterizing functional brain organization in children (Ansari, 2010); in particular, the hippocampus is critical for the early learning of arithmetic facts, but not their retrieval in adulthood. It also highlights that from a developmental perspective, it is inappropriate to focus only on parietal circuits when investigating arithmetic in children. It is not clear at which age or level of expertise the angular gyrus starts to play a role in fact retrieval, and this issue should be addressed in subsequent longitudinal, developmental imaging studies.

Against the background of findings by Geary, Bow-Thomas, Liu, and Siegler (1996), who showed at the behavioral level a substantial advantage in retrieval learning in Chinese compared to U.S. elementary schoolchildren, cross-cultural studies comparing Chinese children's fact retrieval with same-age U.S. children might help to separate age-related changes in brain maturation from the effects of mathematics education. It can be speculated that retrieval is a graded phenomenon, with different levels in the automaticity of fact retrieval being associated with different brain regions: The hippocampus might be important for the initial encoding and retrieval of arithmetic facts (Qin et al., 2014), after which its role might diminish and the angular gyrus might take over for the retrieval of fully consolidated facts (see Menon, this volume). These differences might be difficult to capture by means of behavioral data alone (i.e., verbal reports and reaction times). In this context, neuroimaging data might provide a level of analysis and measurement that cannot be accessed by behavioral studies alone, adding new insights to theories of arithmetic fact-retrieval development in children.

Individual Differences in Brain Activity During Fact Retrieval

How is activity in this arithmetic network modulated by individual differences in arithmetic fact-retrieval skills? In their study, De Smedt et al. (2011) compared the brain activity during single-digit addition and subtraction in 10- to 12-year-old children with low and average levels of arithmetic fluency, as determined by performance on a standardized Woodcock-Johnson test of math fluency (Woodcock, McGrew, & Mather, 2003). In this timed test, children had to answer as many single-digit addition, subtraction, and multiplication problems as possible within a 3-min period. Performance on this test is highly dependent on one's retrieval ability, because the use of a retrieval strategy is always faster and less error prone than the use of a procedural strategy. As a result, these two groups of children can be considered poor versus average retrievers. Comparison of both groups revealed a group by problem-size interaction in the right IPS. The average retrievers showed a typical problem-size effect in this area, with more activity during the solution of large as compared to small problems. In contrast, the poor retrievers did not show such a problem-size effect and recruited the right IPS to the same extent during both problem sets. In other words, the poor retrievers showed higher activation in the right IPS during the solution of small problems, which might represent a neural correlate of their fact-retrieval impairments.

It has been suggested that the right IPS shows a process of developmental specialization (i.e., an increase in focused activity over developmental time) for the representation of magnitude, highlighting the crucial importance of this area for arithmetic development (Holloway, Price, & Ansari, 2010). In contrast, structural abnormalities (e.g., Rotzer et al., 2008) and functional alterations (e.g., Price, Holloway, Rasanen, Vesterinen, & Ansari, 2007) during number processing have been observed in children with dyscalculia. It could be that these children rely on an impaired system of magnitude representations which prevented them from developing the use of a fact-retrieval strategy. Alternatively, these findings might reflect their persistent reliance on the use of procedural quantity-based strategies (such as counting) in the poor retrievers, which could stem from atypical functioning or development of brain regions involved in fact retrieval.

Connections Between Areas of the Arithmetic Fact-Retrieval Network

It is important to point out that the different cortical areas of the arithmetic network described earlier are physically distant from each other. Although the majority of the brain imaging studies on arithmetic fact retrieval focused on the localization of activity in specific areas, there is an increasing awareness that the activity in these areas depends on interactions with other regions of the arithmetic network (Menon, 2015). As a result, the analysis of the structural and functional connections between the different areas of the arithmetic network is important. Structural connections can be investigated by means of diffusion tensor imaging (DTI) (e.g., Matejko & Ansari, 2015), which analyzes the quality of the white matter pathways between different areas in the brain and allows one to investigate the association between the integrity of these white matter tracts and behavioral performance, such as reading (e.g., Vandermosten, Boets, Wouters, & Ghesquière, 2012) or arithmetic (e.g., Matejko & Ansari, 2015). Functional connections, on the other hand, are typically investigated in fMRI studies by examining the temporal correlations in activity between remote brain areas (e.g., Friston, 1994).

Using DTI, Van Beek et al. (2014) investigated the association between different arithmetic operations and white matter connections between the frontal and parietal cortex in typically developing 12-year-olds. All children completed standardized timed measures of addition, subtraction, multiplication, and division. The authors used DTI tractography and delineated for each child the arcuate fasciculus (Figure 1, right panel), a bundle of pathways that connects frontal and temporo-parietal regions that are commonly active during calculation tasks. For each individual, the integrity of different segments of this tract was determined by calculating the fractional anisotropy. Their data revealed that the quality of the anterior part of the arcuate fasciculus was positively correlated with performance in addition and multiplication, but not with subtraction and division, suggesting a specific role of this left anterior segment in problems that are typically solved by fact retrieval. Importantly, the association was not explained by age, intelligence, or working memory.

It should also be pointed out that the arcuate fasciculus is typically associated with individual differences in language and reading abilities (Vandermosten et al., 2012). Van Beek et al. (2014) therefore ran a series of follow-up analyses to further investigate whether the association between the arcuate fasciculus and arithmetic fact retrieval could be explained by reading ability. Interestingly, the authors controlled for the ability to read words as well as pseudowords, with the latter depending more heavily on phonological processing. The analyses revealed that the associations between the arcuate fasciculus and fact retrieval remained when controlling for word reading but disappeared when controlling for pseudoword reading. These findings indicate common processes in arithmetic fact retrieval and pseudoword reading and further confirm, at the neural level, an association between phonological processing and arithmetic fact retrieval. They also suggest shared neural networks between arithmetic fact retrieval and phonological processing in the left temporo-parietal cortex, potentially related to the formation of problem-answer associations in long-term memory as already suggested at the behavioral level by Siegler and Shrager (1984) and Geary (1993).

In their recent longitudinal study on arithmetic fact retrieval in children, Qin et al. (2014) investigated functional connectivity of the hippocampus with cortical areas. Their data indicated that this hippocampal-neocortical connectivity changes over this one-year period of arithmetic (fact) learning. Most interestingly, their data showed that the increase in frequency of retrieval use was significantly associated with the strength of hippocampal connectivity to various prefrontal and parietal areas, including the IPS. Furthermore, it was the change in the hippocampal-cortical functional connectivity rather than the change in activity in the hippocampus itself that predicted the development of fact retrieval. Cross-sectional analyses that also included adolescents and adults further revealed that the continued improvement in retrieval efficiency appeared to be related to the functional integration of prefrontal and hippocampal regions.

It is important to point out the number of studies on structural and functional connectivity of the arithmetic network in children is very limited. The current findings are therefore preliminary and require replication with larger samples. Nonetheless, these studies highlight that the investigation of connections between areas of the arithmetic network and how such connections change over developmental time represent an important agenda for future research.

CONCLUSIONS AND FUTURE DIRECTIONS

In this chapter I have summarized the research on individual differences in arithmetic fact retrieval through the lens of educational neuroscience (e.g., De Smedt, 2014). One the one hand, this field improves our understanding of individual differences in arithmetic fact retrieval by generating predictions concerning the underlying cognitive processes on the basis of (often adult) brain imaging data that are tested in cognitive developmental behavioral studies. On the other hand, this field specifically investigates the neural underpinnings of school-taught abilities, in this case arithmetic fact retrieval.

The findings reviewed indicate that particularly symbolic numerical magnitude processing plays a unique role in individual differences and developmental changes in children's early arithmetic facts acquisition. It also suggests that proficient symbolic magnitude processing provides a major scaffold for developing arithmetical facts. It is important to emphasize that the studies reviewed earlier are all correlational in nature and do not allow us to draw causal conclusions on the association between symbolic numerical magnitude processing and arithmetic fact retrieval. To establish such a causal connection, intervention studies, in which symbolic numerical magnitude processing is trained, should be carried out and the effects of these interventions on arithmetic fact retrieval should be investigated (see, for instance, Fuchs et al., 2013; Supekar et al., 2013). Preferably, these studies need to be conducted at relatively young ages during which the development of fact retrieval is the greatest. That being said, the longitudinal findings of demonstrated that symbolic numerical processing precedes the development of arithmetic fact retrieval.

Against this background, measures of symbolic numerical magnitude processing can be used for screening and early identification of children who are at risk for developing difficulties in arithmetic fact retrieval, and such diagnostic instruments have recently been developed (Geary, Bailey, & Hoard, 2009; Nosworthy, Bugden, Archibald, Evans, & Ansari, 2013). Finally, it should be noted that the existing studies on the association between symbolic magnitude processing and fact retrieval have included only addition and subtraction as operations in their study design, and it remains to be determined the extent to which these findings can also be generalized to the retrieval of facts of other operations, such as multiplication and division.

The studies reviewed in this chapter also pointed to the role of phonological processing in arithmetic fact retrieval, although these findings should be interpreted with caution given that none of the evidence to date is longitudinal (except for Hecht et al., 2001). As the ages of the participants in the reported studies vary considerably, it also remains to be determined at which point in the developmental trajectory of arithmetic fact learning these phonological skills play the most prominent role. Future studies should therefore collect longitudinal data and precisely determine the unique role of phonological processing in predicting arithmetic fact development, thereby controlling for other cognitive factors that have been associated with this development. Deriving educational implications from a link between phonological processing and arithmetic fact retrieval is challenging. One possibility is to explore whether instruction in reading, that emphasizes phonological strategies, improves arithmetic fact retrieval. Still, difficulties in phonological processing constitute a risk factor for developing arithmetic fact deficits, and therefore, such deficits should be evaluated in the diagnosis of reading difficulties or dyslexia.

Future research on the correlates of individual differences in arithmetic fact retrieval should also consider other cognitive factors that might contribute to these differences. For example, Geary, Hoard, and Bailey (2012) suggested that fact-retrieval deficits in children with dyscalculia might be due to the poor inhibition of relevant information during the fact-retrieval process, a possibility that needs further empirical testing. Recently, De Visscher and Noel (2014a, b) proposed an interesting hypothesis that might explain these fact-retrieval deficits. Specifically, they argued that hypersensitivity to interference could play a prominent role in these deficits. Starting from interference-based forgetting models of memory, De Visscher and Noël suggested that the feature overlap between arithmetic facts might provoke interference that hampers the storage of arithmetic facts when learners are sensitive to this interference. In a recent developmental study, they showed that fourth-grade children with poor arithmetic fact fluency differed from matched controls in an associative memory task, but only when learning high-interfering associations. As a result, hypersensitivity to interference might be an important factor to consider in future studies.

Brain imaging studies that have investigated the neural correlates of arithmetic fact retrieval have revealed a network that includes areas in the prefrontal cortex, inferior parietal cortex (i.e., intraparietal sulcus, angular and supramarginal gyri), and the medial temporal lobe (i.e., hippocampus), that is active during the retrieval of arithmetic facts. It is important to emphasize that as the majority of this work has been done with adults, caution is needed when interpreting these findings. As indicated earlier, these adult findings reflect the end state of arithmetic development and might mistakenly suggest that only a small static set of brain areas are activated during arithmetic fact retrieval. On the contrary, there is a surprisingly small but increasing number of studies in children indicating that activity in this network changes over developmental time, with some of these areas, such as the hippocampus (Qin et al., 2014), playing only a time-limited role in this development. In all, these developmental studies suggest that it is not appropriate to focus only on the parietal cortex when investigating arithmetic. Furthermore, the reviewed studies on structural and functional connectivity suggest that it is crucial for future studies to investigate the connections between distant areas in the arithmetic network and how they develop over time. As highlighted above, it is particularly important that such studies focus on specific age ranges in which the learning of arithmetic facts shows the steepest development.

It is important to point out that the development of arithmetic fact retrieval, rather than occurring in isolation, is highly dependent on the educational environment in general and the degree to which a mathematics curriculum emphasizes the acquisition of this skill in particular. For example, Campbell and Xue (2001) showed that retrieval frequencies of university students were significantly higher in Asian Chinese students (85%) than in Non-Asian Canadian students (72%), and a similar pattern of findings has been reported when comparing Chinese and U.S. elementary schoolchildren (Geary et al., 1996). Except for this comparison between Chinese and U.S. children, there are, to the best of my knowledge, no other cross-cultural studies that have compared these retrieval frequencies in children.

Many of the studies reviewed in this chapter involved Belgian (e.g., De Smedt & Boets, 2010; Van Beek et al., 2014; Vanbinst, Ceulemans, et al., 2015) as well as North American (e.g., De Smedt et al., 2010; Qin et al., 2014) children. A comparison of third-graders' retrieval frequencies in single-digit arithmetic addition and subtraction between U.S. children (Geary et al., 2004) and Belgian children (Torbeyns et al., 2004) revealed that the relative frequency of retrieval was about 38% in North American children and about 88% in Belgian children. Relatedly, frequencies of counting were approximately 30% in U.S. children while such strategies were very rare in Belgian children (4%). These differences are probably due to differences in the extent to which curricular programs place an emphasis on arithmetic fact retrieval. It would be interesting to compare mathematics curricula that differ in their emphasis on arithmetic fact retrieval and investigate how neurocognitive determinants of arithmetic fact retrieval at both cognitive and neural levels differ as a function of the type of mathematics education. At the very least, future studies in educational neuroscience on the development of fact retrieval should carefully document the characteristics of the mathematics curriculum and take these into account when interpreting their findings.

REFERENCES

Alloway, T. P., & Alloway, R. G. (2010). Investigating the predictive roles of working memory and IQ in academic attainment. *Journal of Experimental Child Psychology*, *106*, 20–29.

American Psychiatric Association. (2013). *Diagnostic and statistical manual of mental disorders* (5th ed.). Washington, DC: American Psychiatric Association.

Andres, M., Pelgrims, B., Michaux, N., Olivier, E., & Pesenti, M. (2011). Role of distinct parietal areas in arithmetic: An fMRI-guided TMS study. *NeuroImage*, *54*, 3048–3056.

Ansari, D. (2008). Effects of development and enculturation on number representation in the brain. *Nature Reviews Neuroscience*, *9*, 278–291.

Ansari, D. (2010). Neurocognitive approaches to developmental disorders of numerical and mathematical cognition: The perils of neglecting the role of development. *Learning and Individual Differences*, *20*, 123–129.

Arsalidou, M., & Taylor, M. J. (2011). Is 2 + 2 = 4? Meta-analyses of brain areas needed for numbers and calculations. *NeuroImage*, *54*, 2382–2393.

Bailey, D. H., Littlefield, A., & Geary, D. C. (2012). The co-development of skill at and preference for use of retrieval-based processes for solving addition problems: Individual and sex differences from first to sixth grades. *Journal of Experimental Child Psychology*, *113*, 78–92.

Banfield, J. D., & Raftery, A. E. (1993). Model-based Gaussian and nonGaussian clustering. *Biometrics*, *49*, 803–821.

Barrouillet, P., Mignon, M., & Thevenot, C. (2008). Strategies in subtraction problem solving in children. *Journal of Experimental Child Psychology*, *99*, 233–251.

Berteletti, I., Prado, J., & Booth, J. R. (2014). Children with mathematical learning disability fail in recruiting verbal and numerical brain regions when solving simple multiplication problems. *Cortex*, *57*, 143–155.

Butterworth, B., Zorzi, M., Girelli, L., & Jonckheere, A. R. (2001). Storage and retrieval of addition facts: The role of number comparison. *Quarterly Journal of Experimental Psychology Section A: Human Experimental Psychology*, *54*, 1005–1029.

Campbell, J. I. D. (1995). Mechanisms of simple addition and multiplication: A modified network-interference theory and simulation. *Mathematical Cognition*, *1*, 121–164.

Campbell, J. I. D., & Xue, Q. (2001). Cognitive arithmetic across cultures. *Journal of Experimental Psychology: General*, *130*, 299–315.

Chen, Q., & Li, J. (2014). Association between individual differences in non-symbolic number acuity and math performance: A meta-analysis. *Acta Psychologica*, *148*, 163–172.

Cho, S., Metcalfe, A. W. S., Young, C. B., Ryali, S., Geary, D. C., & Menon, V. (2012). Hippocampal-prefrontal engagement and dynamic causal interactions in the maturation of children's fact retrieval. *Journal of Cognitive Neuroscience*, *24*, 1849–1866.

Cho, S., Ryali, S., Geary, D. C., & Menon, V. (2011). How does a child solve 7 + 8? Decoding brain activity patterns associated with counting and retrieval strategies. *Developmental Science*, *14*, 989–1001.

De Smedt, B. (2014). Advances in the use of neuroscience methods in research on learning and instruction. *Frontline Learning Research*, *2*, 7–14.

De Smedt, B., & Boets, B. (2010). Phonological processing and fact retrieval: Evidence from developmental dyslexia. *Neuropsychologia*, *48*, 3973–3981.

De Smedt, B., Holloway, I. D., & Ansari, D. (2011). Effects of problem size and arithmetic operation on brain activation during calculation in children with varying levels of arithmetical fluency. *NeuroImage*, *57*, 771–781.

De Smedt, B., Noël, M. P., Gilmore, C., & Ansari, D. (2013). The relationship between symbolic and nonsymbolic numerical magnitude processing skills and the typical and atypical development of mathematics: A review of evidence from brain and behavior. *Trends in Neuroscience and Education*, *2*, 48–55.

De Smedt, B., Taylor, J., Archibald, L., & Ansari, D. (2010). How is phonological processing related to individual differences in children's arithmetic skills. *Developmental Science*, *13*, 508–520.

De Visscher, A., & Noel, M. P. (2014a). Arithmetic facts storage deficit: The hypersensitivity-to-interference in memory hypothesis. *Developmental Science*, *17*, 434–442.

De Visscher, A., & Noel, M. P. (2014b). The detrimental effect of interference in multiplication facts storing: Typical development and individual differences. *Journal of Experimental Psychology: General*, *143*, 2380–2400.

Dehaene, S. (1992). Varieties of numerical abilities. *Cognition*, *44*, 1–42.

Dehaene, S., Piazza, M., Pinel, P., & Cohen, L. (2003). Three parietal circuits for number processing. *Cognitive Neuropsychology*, *20*, 487–506.

Delazer, M., Domahs, F., Bartha, L., Brenneis, C., Lochy, A., Trieb, T., et al. (2003). Learning complex arithmetic—An fMRI study. *Cognitive Brain Research*, *18*, 76–88.

Dowker, A. (2005). *Individual differences in arithmetic: Implications for psychology, neuroscience and education*. Hove: Psychology Press.

Duncan, G. J., Claessens, A. M. Y., Huston, A. C., Pagani, L. S., Engel, M., Sexton, H., et al. (2007). School readiness and later achievement. *Developmental Psychology*, *43*, 1428–1446.

Fazio, L. K., Bailey, D. H., Thompson, C. A., & Siegler, R. S. (2014). Relations of different types of numerical magnitude representations to each other and to mathematics achievement. *Journal of Experimental Child Psychology*, *123*, 53–72.

Friso-Van Den Bos, I., Van Der Ven, S. H. G., Kroesbergen, E. H., & Van Luit, J. E. H. (2013). Working memory and mathematics in primary school children: A meta-analysis. *Educational Research Review*, *10*, 29–44.

Friston, K. (1994). Functional and effective connectivity in neuroimaging: A synthesis. *Human Brain Mapping*, *2*, 56–78.

Fuchs, L. S., Fuchs, D., Compton, D. L., Powell, S. R., Seethaler, P. M., Capizzi, A. M., et al. (2006). The cognitive correlates of third-grade skill in arithmetic, algorithmic computation, and arithmetic word problems. *Journal of Educational Psychology*, *98*, 29–43.

Fuchs, L. S., Geary, D. C., Compton, D. L., Fuchs, D., Schatschneider, C., Hamlett, C. L., et al. (2013). Effects of first-grade number knowledge tutoring with contrasting forms of practice. *Journal of Educational Psychology*, *105*, 58–77.

Geary, D. C. (1993). Mathematical disabilities: Cognitive, neuropsychological, and genetic components. *Psychological Bulletin*, *114*, 345–362.

Geary, D. C. (2004). Mathematics and learning disabilities. *Journal of Learning Disabilities*, *37*, 4–15.

Geary, D. C. (2011). Consequences, characteristics, and causes of poor mathematics achievement and mathematical learning disabilities. *Journal of Developmental and Behavioral Pediatrics*, *32*, 250–263.

Geary, D. C., Bailey, D. H., & Hoard, M. K. (2009). Predicting mathematical achievement and mathematical learning disability with a simple screening tool: The Number Sets Test. *Journal of Psychoeducational Assessment*, *27*, 265–279.

Geary, D. C., Berch, D. B., & Mann Koepke, K. (2015). *Evolutionary origins and early development of number processing*. San Diego, CA: Elsevier.

Geary, D. C., Bow-Thomas, C. C., Liu, F., & Siegler, R. S. (1996). Development of arithmetical competencies in Chinese and American children: Influence of age, language, and schooling. *Child Development*, *67*, 2022–2044.

Geary, D. C., Bow-Thomas, C. C., & Yao, Y. H. (1992). Counting knowledge and skill in cognitive addition—A comparison of normal and mathematically disabled children. *Journal of Experimental Child Psychology*, *54*, 372–391.

Geary, D. C., Hoard, M. K., & Bailey, D. H. (2012). Fact-retrieval deficits in low achieving children and children with mathematical learning disability. *Journal of Learning Disabilities*, *45*, 291–307.

Geary, D. C., Hoard, M. K., Byrd-Craven, J., & Desoto, M. C. (2004). Strategy choices in simple and complex addition: Contributions of working memory and counting knowledge for children with mathematical disability. *Journal of Experimental Child Psychology*, *88*, 121–151.

Geary, D. C., & Wiley, J. G. (1991). Cognitive addition: Strategy choice and speed-of-processing differences in young and elderly adults. *Psychology and Aging*, *6*, 474–483.

Grabner, R. H., Ansari, D., Koschutnig, K., Reishofer, G., Ebner, F., & Neuper, C. (2009). To retrieve or to calculate? Left angular gyrus mediates the retrieval of arithmetic facts during problem solving. *Neuropsychologia*, *47*, 604–608.

Grabner, R. H., Ansari, D., Reishofer, G., Stern, E., Ebner, F., & Neuper, C. (2007). Individual differences in mathematical competence predict parietal brain activation during mental calculation. *NeuroImage*, *38*, 346–356.

Grabner, R. H., & De Smedt, B. (2011). Neurophysiological evidence for the validity of verbal strategy reports in mental arithmetic. *Biological Psychology*, *87*, 128–136.

Groen, G. J., & Parkman, J. M. (1972). Chronometric analysis of simple addition. *Psychological Review*, *79*, 329–343.

Hecht, S. A., Torgesen, J. K., Wagner, R. K., & Rashotte, C. A. (2001). The relations between phonological processing abilities and emerging individual differences in mathematical computation skills: A longitudinal study from second to fifth grades. *Journal of Experimental Child Psychology*, *79*, 192–227.

Holloway, I. D., Price, G. R., & Ansari, D. (2010). Common and segregated neural pathways for the processing of symbolic and nonsymbolic numerical magnitude: An fMRI study. *NeuroImage*, *49*, 1006–1017.

Imbo, I., & Vandierendonck, A. (2008). Effects of problem size, operation, and working-memory span on simple-arithmetic strategies: Differences between children and adults? *Psychological Research-Psychologische Forschung*, *72*, 331–346.

Jordan, N. C., Hanich, L. B., & Kaplan, D. (2003). Arithmetic fact mastery in young children: A longitudinal investigation. *Journal of Experimental Child Psychology*, *85*, 103–119.

Kaufmann, L., Wood, G., Rubinsten, O., & Henik, A. (2011). Meta-analyses of developmental fMRI studies investigating typical and atypical trajectories of number processing and calculation. *Developmental Neuropsychology*, *36*, 763–787.

Kilpatrick, J., Swafford, J., & Findell, B. (2001). *Adding it up. Helping children learn mathematics*. Washington, DC: National Academy Press.

Kirk, E. P., & Ashcraft, M. H. (2001). Telling stories: The perils and promise of using verbal reports to study math strategies. *Journal of Experimental Psychology-Learning Memory and Cognition*, *27*, 157–175.

Landerl, K., & Moll, K. (2010). Comorbidity of learning disorders: Prevalence and familial transmission. *Journal of Child Psychology and Psychiatry*, *51*, 287–294.

Lefevre, J. A., Fast, L., Skwarchuk, S. L., Smith-Chant, B. L., Bisanz, J., Kamawar, D., et al. (2010). Pathways to mathematics: Longitudinal predictors of performance. *Child Development*, *81*, 1753–1767.

Lefevre, J. A., Sadesky, G. S., & Bisanz, J. (1996). Selection of procedures in mental addition: Reassessing the problem size effect in adults. *Journal of Experimental Psychology: Learning Memory and Cognition*, *22*, 216–230.

Matejko, A., & Ansari, D. (2015). Drawing connections between white matter and numerical and mathematical cognition: A literature review. *Neuroscience & Biobehavioral Reviews*, *48*, 35–52.

Menon, V. (2015). Arithmetic in the child and adult brain. In R. Cohen-Kadosh, & A. Dowker (Eds.), The Oxford handbook of mathematical cognition. Oxford: Oxford University Press.

Menon, V., Rivera, S. M., White, C. D., Eliez, S., Glover, G. H., & Reiss, A. L. (2000). Functional optimization of arithmetic processing in perfect performers. *Cognitive Brain Research, 9*, 343–345.

National Mathematics Advisory Panel. (2008). *Foundations for success: The final report of the National Mathematics Advisory Panel*. Washington, DC: US Department of Education.

Noel, M. P., & Rousselle, L. (2011). Developmental changes in the profiles of dyscalculia: An explanation based on a double exact-and-approximate number representation model. *Frontiers in Human Neuroscience, 5*, 1–4.

Norman, K. A., Polyn, S. M., Detre, G. J., & Haxby, J. V. (2006). Beyond mind-reading: Multi-voxel pattern analysis of fMRI data. *Trends in Cognitive Sciences, 10*, 424–430.

Nosworthy, N., Bugden, S., Archibald, L., Evans, B., & Ansari, A. (2013). A two-minute paper-and-pencil test of symbolic and nonsymbolic numerical magnitude processing explains variability in primary school children's arithmetic competence. *PLoS ONE, 8*, e67918.

Prado, J., Mutreja, R., & Booth, J. R. (2014). Developmental dissociation in the neural responses to simple multiplication and subtraction problems. *Developmental Science, 17*, 537–552.

Price, G. R., & Ansari, D. (2013). Dyscalculia. In O. Dulac, & M. Lassonde (Eds.), Handbook of Clinical Neurology (pp. 241–244). London: Elsevier.

Price, G. R., Holloway, I., Rasanen, P., Vesterinen, M., & Ansari, D. (2007). Impaired parietal magnitude processing in developmental dyscalculia. *Current Biology, 17*, R1042–R1043.

Qin, S., Cho, S., Chen, T., Rosenberg-Lee, M., Geary, D. C., & Menon, V. (2014). Hippocampal-neocortical functional reorganization underlies children's cognitive development. *Nature Neuroscience, 17*, 1263–1269.

Raghubar, K. P., Barnes, M. A., & Hecht, S. A. (2010). Working memory and mathematics: A review of developmental, individual difference, and cognitive approaches. *Learning and Individual Differences, 20*, 110–122.

Ramus, F., Rosen, S., Dakin, S. C., Day, B. L., Castellote, J. M., White, S., et al. (2003). Theories of developmental dyslexia: Insights from a multiple case study of dyslexic adults. *Brain, 126*, 841–865.

Resnick, L. B. (1989). Developing mathematical knowledge. *American Psychologist, 44*, 162–169.

Rivera, S. M., Reiss, A. L., Eckert, M. A., & Menon, V. (2005). Developmental changes in mental arithmetic: Evidence for increased functional specialization in the left inferior parietal cortex. *Cerebral Cortex, 15*, 1779–1790.

Robinson, C. S., Menchetti, B. M., & Torgesen, J. K. (2002). Toward a two-factor theory of one type of mathematics disabilities. *Learning Disabilities Research and Practice, 17*, 81–89.

Rotzer, S., Kucian, K., Martin, E., Von Aster, M., Klaver, P., & Loenneker, T. (2008). Optimized voxel-based morphometry in children with developmental dyscalculia. *NeuroImage, 39*, 417–422.

Schlaggar, B. L., & Mccandliss, B. D. (2007). Development of neural systems for reading. *Annual Review of Neuroscience, 30*, 475–503.

Schneider, M., Beeres, K., Coban, L., Merz, S., Schmidt, S.S., Stricker, J., & De Smedt, B. (submitted). Associations of nonsymbolic and symbolic numerical magnitude processing with mathematical competence: A meta-analysis. Manuscript submitted for publication.

Siegler, R. S. (1987). The perils of averaging data over strategies—an example from children's addition. *Journal of Experimental Psychology: General, 116*, 250–264.

Siegler, R. S. (1996). *Emerging minds: The process of change in children's thinking*. New York, NY: Oxford University Press.

Siegler, R. S., & Shrager, J. (1984). Strategy choices in addition and subtraction: How do children know what to do? In C. Sophian (Ed.), *The origins of cognitive skills* (pp. 229–293). Hillsdale, NJ: Erlbaum.

Simmons, F. R., & Singleton, C. (2008). Do weak phonological representations impact on arithmetic development? A review of research into arithmetic and dyslexia. *Dyslexia, 14*, 77–94.

Smith, C. N., & Squire, L. R. (2009). Medial temporal lobe activity during retrieval of semantic memory is related to the age of the memory. *Journal of Neuroscience, 29*, 930–938.

Snowling, M. J. (2000). *Dyslexia* (2nd ed.). Malden, MA: Blackwell Publishers.

Supekar, K., Swigart, A. J., Tenison, C., Jolles, D. D., Rosenberg-Lee, M., Fuchs, L., et al. (2013). Neural predictors of individual differences in response to math tutoring in primary-grade school children. *Proceedings of the National Academy of Sciences, 110*, 8230–8235.

Thevenot, C., Fanget, M., & Fayol, M. (2007). Retrieval or nonretrieval strategies in mental arithmetic? An operand recognition paradigm. *Memory & Cognition, 35*, 1344–1352.

Torbeyns, J., Verschaffel, L., & Ghesquière, P. (2004). Strategic aspects of simple addition and subtraction: The influence of mathematical ability. *Learning and Instruction, 14*, 177–195.

Van Beek, L., Ghesquière, P., Lagae, L., & De Smedt, B. (2014). Left fronto-parietal white matter correlates with individual differences in children's ability to solve additions and multiplications: A tractography study. *NeuroImage, 90*, 117–127.

Vanbinst, K., Ceulemans, E., Ghesquière, P., & De Smedt, B. (2015). Profiles of children's arithmetic fact development: A model-based clustering approach. *Journal of Experimental Child Psychology, 133*, 29–46.

Vanbinst, K., Ghesquière, P., & De Smedt, B. (2012). Numerical magnitude representations and arithmetic strategy use. *Mind, Brain, and Education, 6*, 129–136.

Vanbinst, K., Ghesquière, P., & De Smedt, B. (2014). Arithmetic strategy development and its domain-specific and domain-general cognitive correlates: A longitudinal study in children with persistent mathematical learning difficulties. *Research in Developmental Disabilities, 35*, 3001–3013.

Vanbinst, K., Ghesquière, P., & De Smedt, B. (2015). Does numerical processing uniquely predict first graders' future development of single-digit arithmetic? *Learning & Individual Differences, 37*, 153–160.

Vandermosten, M., Boets, B., Wouters, J., & Ghesquière, P. (2012). A qualitative and quantitative review of diffusion tensor imaging studies in reading and dyslexia. *Neuroscience and Biobehavioral Reviews, 36*, 1532–1552.

vanMarle, K., Chu, F. W., Li, Y., & Geary, D. C. (2014). Acuity of the approximate number system and preschoolers' quantitative development. *Developmental Science, 17*, 492–505.

Vukovic, R. K., & Lesaux, N. K. (2013). The relationship between linguistic skills and arithmetic knowledge. *Learning and Individual Differences, 23*, 87–91.

Woodcock, R. W., McGrew, K. S., & Mather, N. (2003). *Woodcock-Johnson III tests of achievement*. Itasca, IL: Riverside.

Zamarian, L., & Delazer, M. (2015). Arithmetic learning in adults: Evidence from brain imaging. In R. Cohen-Kadosh, & A. Dowker (Eds.), *The Oxford handbook of mathematical cognition*. Oxford: Oxford University Press.

Chapter 10

Transcranial Electrical Stimulation and the Enhancement of Numerical Cognition

Amar Sarkar, Roi Cohen Kadosh
Department of Experimental Psychology, University of Oxford, Oxford, UK

INTRODUCTION

The other chapters in this volume discuss the neuroanatomical, neurophysiological, neurochemical, and genetic substrates of numerical cognition and learning. In this chapter, we expand on some of these discussions by considering the enhancement of numerical cognition and numerical performance through a particular exogenous intervention: mild electrical currents. Transmitting electrical signals into the brain to enhance numerical learning might seem better suited to Hollywood. However, transcranial electrical stimulation (tES) has emerged as a technology that can be used to affect brain function across the psychological spectrum, and, moreover, is easy to use, portable, painless, and most importantly, safe (when following the recommended guidelines). Very recently, tES has captured the imagination of the public, as demonstrated by increasing media coverage. In 2006, fewer than 10 print media articles covered transcranial direct current stimulation (tDCS), the most common form of tES. In 2012, there were nearly 70 such articles (Dubljević, Saigle, & Racine, 2014). The academic analogue of this growth in public interest is an increasing number of experimental studies demonstrating a variety of cognitive enhancement effects. Approximately 25 academic articles concerning tDCS were published in 2006. By 2013, the number was greater than 250 (Dubljević et al., 2014). A substantial proportion of the published research has focused on therapeutic applications of tDCS, although the emphasis on cognitive enhancement in healthy populations has also been growing quite rapidly.

A large number of benefits have been documented in healthy individuals, ranging from perception (e.g., Fertonani, Pirulli, & Miniussi, 2011; Kanai, Chaieb, Antal, Walsh, & Paulus, 2008) and motor learning and performance

Development of Mathematical Cognition. http://dx.doi.org/10.1016/B978-0-12-801871-2.00010-1

(Antal et al., 2008; Waters-Metenier, Husain, Wiestler, & Diedrichsen, 2014; Furuya, Klaus, Nitsche, Paulus, & Altenmüller, 2014) to high-level cognition such as visual short-term memory (Tseng et al., 2012), planning (Dockery, Hueckel-Weng, Birbaumer, & Plewnia, 2009), language learning (Flöel, Rösser, Michka, Knecht, & Breitenstein, 2008; Meinzer et al., 2014), and analogical reasoning (Santarnecchi et al., 2013). Though impressive, these enhancements have been entirely restricted to controlled laboratory settings and carefully defined experimental response variables. There is no evidence as yet of the type of enhancement of popular speculation, such as suddenly acquiring a new language, becoming immensely proficient at mathematics, or expectations surrounding a rags-to-riches theme which assume that tES can be used as some sort of cognitive cheat code or brain hack. Though in fact such extraordinary changes are extremely unlikely, tES may have genuine potential for cognitive enhancement. Laboratory findings in healthy research participants themselves are of scientific interest and value, and should pave the way for future investigations of the real-world benefits of targeted brain stimulation.

tES-induced benefits have also been noted in the area of numerical cognition, and therefore are of particular relevance to the material covered in this book series. The research area is new, and the literature on this subject is relatively small. Consequently, there are as yet no reviews at the intersection of tES and numerical cognition, although there are several on tES and cognition in general (e.g., Cohen Kadosh, 2013; Jacobson, Koslowsky, & Lavidor, 2012; Krause & Cohen Kadosh, 2013; Kuo & Nitsche, 2012).

In this chapter, we review the evidence that tES enhances numerical cognition in humans. The chapter is divided into four sections: (1) a short overview of the different forms of tES, the physiological principles thought to underlie tES, and general principles of designing a tES experiment, as well as some of its limitations; (2) the use of tES for the enhancement of some forms of numerical cognition; (3) the potential of tES for overcoming deficits in numerical cognition, as in the case of developmental dyscalculia and mathematics anxiety; and (4) an agenda for future research. Our goal is to offer a straightforward introduction to tES and a discussion of its uses in the enhancement of numerical cognition. Furthermore, numerical cognition can be treated as a model, and the data and theory discussed here may, in some cases, be meaningfully extended to the enhancement of other cognitive processes through tES, and to the design and analysis of behavioral and cognitive experiments more generally.

A Brief History

Humans have intuited the link between exogenous electrical current and mental states and experiences since classical antiquity, examples including Pliny the Elder and Galen of Pergamum, who described the use of electrical currents generated by torpedo fish for the treatment of headaches (Priori, 2003). Possibly the best known contemporary example of the application of electricity to the brain is electroconvulsive therapy (ECT). Modern ECT has been in use since the 1930s (Gilula & Kirsch, 2005) and is vastly different from tES in principle, mechanism, and effect. Nevertheless, this connection lingers in the popular imagination, fostered through media and film representations, and frequently poses difficulties in recruiting participants. Modern ECT involves the application of very strong electrical currents to induce seizures for the treatment of otherwise intractable psychiatric conditions, including depression and bipolar disorder. For reviews on the history and mechanisms of ECT, see Bolwig (2011) and Fink (1984).

Although the present applications of tES may be perceived as something out of science fiction, they are hardly new. A form of tES was popular during the Victorian and Edwardian periods, and was extensively and easily available for public purchase and consumption, appearing regularly in markets and catalogs across Europe (Elliott, 2014). However, as contemporaneous critics pointed out, the noted benefits of this type of popular electrical stimulation (including reports of euphoria, greater attentional abilities or mental energy) may simply have been placebo effects. The growing skepticism, and perhaps the later emergence of psychopharmacology and ECT caused this earlier generation of tES applications, especially for cognitive enhancement, to fade from the spotlight (Elliott, 2014).

tES Today

tES and its enhancement effects are now experiencing a renaissance. With the convenience of the technology itself, more rigorous experimental controls to rule out alternative explanations and more sensitive response measures that can capture subtle behavioral changes, the claims for tES can now be intensively tested. In its current form, tES involves the delivery of a weak electrical current, usually in the range of 1-2 mA—1/500th of the current necessary to light a 100-W bulb—via two or more electrodes. The electrodes are positioned on the scalp, directly above the cortical region of interest. The passage of the current, from electrode to electrode through the brain, alters neuronal activity when applied for a sufficient duration (Nitsche et al., 2008).

The effect of the stimulation on neuronal activity depends on the shape of the current. As a result, there are several forms of tES that produce different effects, each dependent on the type of current applied. Of these, the most commonly used in the enhancement of numerical cognition is transcranial direction current stimulation (tDCS). Another form of tES, transcranial random noise stimulation (tRNS), is also gaining currency in the brain stimulation community and was the stimulation of choice in some prominent studies on numerical cognition. tDCS is the most well known and the most widely used, whereas tRNS, a newer and less well-known form of tES, provides several experimental advantages that overcome some limitations of tDCS. Moreover, there is a growing stream of published work using tRNS, and it is important that the relative advantages and disadvantages of both techniques be appreciated. Finally, it is worth mentioning a third form of tES that is also being used more and more widely, transcranial alternating current stimulation (tACS), although to the best of our knowledge this technique has not been used in studies of numerical cognition.

The Forms of tES

tDCS

Transcranial direct current stimulation uses a constant current that produces two types of stimulation. Anodal stimulation occurs where the current enters the brain (the anode), and cathodal stimulation occurs where the current leaves the brain and reenters the electrode (the cathode). Animal and human investigations have established that the polarity of the current (anodal or cathodal) produces specific types of changes in neural resting membrane potentials (Bindman, Lippold, & Redfearm, 1964; Márquez-Ruiz et al., 2012; Nitsche & Paulus, 2000). When tested on the motor cortex, anodal stimulation has generally been shown to induce depolarization (i.e., bringing resting membrane potentials closer to activation), thereby increasing excitability (Bindman et al., 1964; Fritsch et al., 2010; Nitsche & Paulus, 2000). Cathodal stimulation has generally shown the opposite effect, resulting in hyperpolarization and decreased excitability (Bindman et al., 1964; Fritsch et al., 2010; Nitsche & Paulus, 2000). From a neurobiological perspective, it is important to understand that anodal and cathodal stimulation, while producing opposite effects, are the logical results of one another (i.e., a direct current entering the brain must exit the brain, and the polarity of these currents are opposite to one another).

A general pattern of polarity-performance relationships has emerged over the numerous studies that have employed tDCS. In particular, the majority of published work shows that anodal stimulation enhances performance, whereas cathodal stimulation impairs performance (Jacobson et al., 2012; Kuo & Nitsche, 2012). It should be noted, however, that this is far from axiomatic. In fact, several studies show the opposite pattern, with anodal stimulation impairing performance and cathodal stimulation enhancing it (e.g., Antal et al., 2004; Terhune, Tai, Cowey, Popescu, & Cohen Kadosh, 2011). Such differences may depend on several factors, such as the region being stimulated or baseline activity in the stimulated region (Krause, Márquez-Ruiz, & Cohen Kadosh, 2013; Sarkar, Dowker, & Cohen Kadosh, 2014).

While the effects during stimulation seem to depend crucially on the polarity of the stimulation itself, they also form the basis for a host of other changes that may have implications for the longer-term effects of stimulation. Experiments on the motor system have shown that changes induced by tDCS in a single session of stimulation may outlive the stimulation itself by ~60-90 min (Nitsche & Paulus, 2001). Evidence suggests that the long-lasting effects of tDCS are protein synthesis-dependent, and their mechanisms have features in common with long-term potentiation and long-term depression (Neves, Cooke, & Bliss, 2008). These include modifications of intracellular cyclic adenosine monophosphate (cAMP) and calcium concentrations (Hattori, Moriwaki, & Hori, 1990), brain-derived neurotrophic factor (BDNF) (Fritsch et al., 2010), and activation of adenosine A1 receptors (Márquez-Ruiz et al., 2012).

Research on humans has shown tDCS-induced changes in local concentrations of gamma-aminobutyric acid (GABA) and glutamate (Clark, Coffman, Trumbo, & Gasparovic, 2011; Stagg, Bachtiar, & Johansen-Berg, 2011; Stagg et al., 2009), neurotransmitters that are crucial determinants of neuronal excitation and inhibition, and are involved in learning and memory (Floyer-Lea, Wylezinska, Kincses, & Matthews, 2006; Neves et al., 2008; Stagg et al., 2011). Combining tDCS with neuroimaging methods such as functional magnetic resonance imaging (*f*MRI) or electroencephalography (EGG) has also shown that it modifies brain activation (e.g., Antal et al., 2012; Holland et al., 2011; Keeser et al., 2011; Stagg et al., 2011; Wirth et al., 2011). While studies have found that the tDCS effects are largely localized to the stimulation target (Holland et al., 2011), more recent work is revealing that they are by no means restricted to that target and also affect distributed networks functionally linked to the stimulated region (Keeser et al., 2011; Meinzer et al., 2012; Meinzer, Lindenberg, Antonenko, Flaisch, & Flöel, 2013).

tRNS

Unlike tDCS, which involves a direct-form current, transcranial random noise stimulation involves the application of alternating currents at random frequencies. tRNS generates "samples" several hundred times per second, which are then randomly assigned current amplitudes and normally distributed with a mean current of 0 mA. The fluctuations of these sample currents between positive and negative amplitudes around zero are random, producing "noise" in the targeted region. The first prominent report of transcranial random noise stimulation (Terney, Chaieb, Moliadze, Antal, & Paulus, 2008) demonstrated that this technique increases cortical excitability at *both* electrodes, a finding which subsequent studies have corroborated.

Terney et al. (2008) found that 10 min of tRNS applied to the primary motor cortex increased cortical excitability by 20-50%. Performance on a serial-reaction time task revealed a parallel improvement in the acquisition and early consolidation of implicit motor learning. In a serial-reaction time paradigm, participants respond to a set of cues or stimuli, each of which elicits a fixed response (e.g., mouse click or key press). The probability of different cues is nonrandom, and a general pattern of their appearance can be inferred while performing the task (although in its implicit form, participants are not told that this pattern is present). Over time, participants learn this pattern, and their reaction times steadily reduce (Nissen & Bullemer, 1987; Robertson, 2007). Further research using tRNS showed that as few as 5 min of stimulation can produce lasting excitability (Chaieb, Paulus, & Antal, 2011). Researchers have investigated the effects of tRNS on performance and learning in a variety of other domains, including perceptual learning (Fertonani et al., 2011), numerosity discrimination (Cappelletti et al., 2013), and arithmetic learning (Snowball et al., 2013).

tRNS has several important features. For example, it appears to exert uniform effects at both electrode sites, due to its oscillatory rather than direct form, which ensures its effects are independent of the polarity/direction of the current. Using tDCS terminology, one might compare it to having two anodes, rather than an anode and a cathode, though such a comparison is only meant for illustration, as the physiological mechanisms are different (Chaieb et al., 2009). Also, compared to anodal tDCS, high-frequency tRNS (e.g., 100-640 Hz) yields a larger learning effect (Fertonani et al., 2011), potentially adding to the effect sizes of experiments. Finally, tRNS has a much higher cutaneous threshold compared to tDCS (Ambrus, Paulus, & Antal, 2010), meaning that perceiving it through physical scalp sensations is more difficult and participants are less likely to detect its application. In contrast, although tDCS is not physically painful, it does frequently produce tingling sensations, which fade quickly as the scalp habituates (Nitsche et al., 2008). These sensations are much weaker in tRNS compared to tDCS (Ambrus et al., 2010). This makes it easier to maintain participant blindness to their experimental condition.

The mechanisms of tRNS action are less well understood than tDCS. For tRNS at higher intensities (e.g., 1 mA), the effects appear to be excitatory at both electrodes. The stochastic resonance hypothesis (see Moss, Ward, & Sannita, 2004) postulates that the introduction of a small amount of exogenous noise raises neuronal sensitivity to subthreshold stimuli against a background of random neuronal activity, such that the firing threshold is reached more easily (Fertonani et al., 2011). The response to stimuli can be enhanced with sufficient additional noise.

Another form of brain stimulation, tACS, stimulates the brain at fixed, rather than random, frequencies. One of the ideas behind using tACS for cognitive enhancement is to deliver the stimulation at a frequency similar to naturally occurring electrophysiological signatures of cognitive performance, thereby "locking in" on ongoing activity (for more detailed discussions, see Antal & Paulus, 2013; Battleday, Muller, Clayton, & Cohen Kadosh, 2014; Helfirch et al., 2014; Herrmann et al., 2013). tACS is gaining increasing attention, and traction, in the brain stimulation community. However, to the best of our knowledge, it has not yet been used in studies of numerical cognition and is therefore not discussed in any detail in subsequent sections of this chapter.

PRINCIPLES AND LIMITATIONS OF tES EXPERIMENTS

As with any method, tES is accompanied by a number of limitations that may compromise the internal validity of tES experiments. These limitations can be overcome with rigorous efforts to rule out alternative explanations. In much of the published work, these limitations are either glossed over or are not mentioned at all. Given the recency of tES's emergence into the research spotlight, we feel it would be more rigorous to introduce the limitations of tES at the outset, so the results, which are interesting and important, may be interpreted in light of the limitations of the method.

Placebo Effects

A central issue is the ability of the experiment to account for the placebo effect. As with many research questions in other areas of psychology, cognitive neuroscience, and medicine, the placebo effect cannot be underestimated. Fortunately for tES researchers, an effective and easily applied placebo condition already exists. This entails applying a current of the same strength and intensity, but reducing its duration to just a few seconds (e.g., 30 s, compared to approximately 5 min, at least, for an effective stimulation). This is referred to as sham stimulation. The duration of sham stimulation is too brief to induce changes in excitability (Fritsch et al., 2010). However, it leads to the same physical sensations on the scalp and is indistinguishable from real stimulation. Because of rapid habituation to the stimulation, the sensations for both real and sham stimulation fade quickly, producing a straightforward method to account for the placebo effect (Gandiga, Hummel, & Cohen, 2006).

Online and Offline Effects: Single Session and Training Studies

Changes in performance during (i.e., online) or immediately after the application of tES may be of great scientific value, especially when they reveal how specific regions of the brain are involved in particular types of cognition or situations. For example, Ruff, Ugazio, and Fehr (2013) used tDCS to understand the role of the right lateral prefrontal cortex in aspects of social norm compliance (broadly, the adherence of behavior to wider social rules and expectations). In such cases, tES becomes a tool for understanding neural mechanisms of specialized brain functions. However, when tES is used for cognitive enhancement, where the central theme of the "enhancement" lies in improving some aspect of cognitive processing appreciable in a real-life setting, the offline effects of stimulation should also be assessed. That is, the researcher has to assess whether the stimulation has produced changes that are sustainable beyond the stimulation duration itself, and are sufficiently positive and long-lived to be described as a long-term gain. In this regard, training studies seem to hold particular promise.

Cognitive training has already been shown to be highly effective in modulating performance in a wide variety of populations (Anguera et al., 2013; Ball et al., 2002; Holmes, Gathercole, & Dunning, 2009; Jaeggi, Buschkuehl, Jonides, & Perrig, 2008; Jaeggi, Buschkuehl, Jonides, & Shah, 2011; Klingberg, 2010; Slagter et al., 2007). Task performance engages activity in specific neural networks and regions, and training induces specific neuroanatomical and neurophysiological changes (Boyke, Driemeyer, Gaser, Büchel, & May, 2008;

Draganski et al., 2004; Klingberg, 2010; Slagter et al., 2007). tES may be added to cognitive training to facilitate these ongoing neural engagements and long-term cortical changes. As mentioned elsewhere (Cohen Kadosh, Levy, O'Shea, Shea, & Savulescu, 2012; Reis, Prichard, & Fritsch, 2014), tES sensitizes the neuronal environment to cognitive engagement, potentially making it more receptive to the effects of training, and thereby improves the outcome of any training to a greater extent than training alone. This hypothesis has now been confirmed in several studies that have compared training alone to training combined with tES (e.g., Cappelletti et al., 2013; Cohen Kadosh, Soskic, Iuculano, Kanai, & Walsh, 2010; Ditye, Jacobson, Walsh, & Lavidor, 2012; Reis et al., 2009; Snowball et al., 2013).

Depth of Stimulation

The regions that can be directly stimulated are the areas of the cortex beneath the scalp. While subcortical regions are beyond the direct reach of stimulation, studies in both animals and humans have indicated that activity in subcortical structures can be reliably modulated by targeting cortical brain regions that are anatomically and functionally connected to those structures (Bolzoni, Bączyk, & Jankowska, 2013; Polanía, Nitsche, Korman, Batsikadze, & Paulus, 2012; Takano et al., 2011; Weber, Messing, Rao, Detre, & Thompson-Schill, 2014). Even if subcortical modulations are achieved, they occur through targeting the cortex. Therefore, the cognitive function of interest needs to be subserved by a cortical region, and thus choosing the correct region becomes a critical concern, as discussed in the next point.

Choosing the Brain Region

Previous neuroimaging research (e.g., *f*MRI) and meta-analytic reviews of neuroimaging research often yield reliable targets for brain stimulation by indicating where activation occurs during particular types of tasks or experiences. In the case of numerical cognition, two brain regions that have repeatedly been implicated across a large number of studies are the parietal lobes (Cantlon, Brannon, Carter, & Pelphrey, 2006; Cohen Kadosh, Lammertyn, & Izard, 2008; Dehaene, Piazza, Pinel, & Cohen, 2003; Izard, Sann, Spelke, & Streri, 2009; Piazza, Izard, Pinel, Le Bihan, & Dehaene, 2004) and the dorsolateral prefrontal cortex (Arsalidou & Taylor, 2011; Zamarian, Ischebeck, & Delazer, 2009).

Furthermore, the participants' age and task-relevant experience may also play an important role in the relative contribution of a brain region to the cognitive function of interest. For example, arithmetic task processing in the child's brain activates frontal areas more than parietal ones, while in the adult brain, similar tasks activate parietal regions more heavily than frontal areas (Kaufmann, Wood, Rubinsten, & Henik, 2011; Rivera, Reiss, Eckert, & Menon, 2005). Similarly, imaging studies of cognitive disability, such as developmental dyscalculia (Berteletti, Prado, & Booth, 2014; Qin et al., 2014) either alone, or in interaction with development, may suggest other brain targets for tES intervention. These differences need to be assessed in the development of a stimulation paradigm.

Size, Number, and Placement of the Electrodes

Electrodes are typically 25 or 35 cm^2. These electrode sizes are quite large, and there is always some concern that additional brain regions are being inadvertently stimulated. Recently, the possibility of smaller electrodes has also emerged (e.g., round Pi electrodes of size 3.14 cm^2) to increase the focality of stimulation (Ruffini, Fox, Ripolles, Miranda, & Pascual-Leone, 2014). An important aspect of future research will be a consideration of how electrode size relates to the smaller head size of children. This is largely absent as a point of discussion, because of the relative absence of tES studies in children compared to adults. However, as the number of studies using children increases, the use of smaller electrodes seems more appropriate.

Because tES essentially occurs through the closing of a circuit, the minimum number of electrodes that must be used is two, and most published studies use two-electrode montages (see Figure 1 for a four-electrode tDCS setup, which also makes it simple to visualize the typical two-electrode setup). Thus, a crucial consideration is the placement of the second electrode, as many studies involving cognitive enhancement hypothesize a relationship between just one brain region and the cognitive function of interest. Where to place the second electrode is an important consideration. One possibility is to place both electrodes on the scalp. While this seems to yield stronger results (Moliadze, Antal, & Paulus, 2010), it also has the drawback of stimulating more than one brain

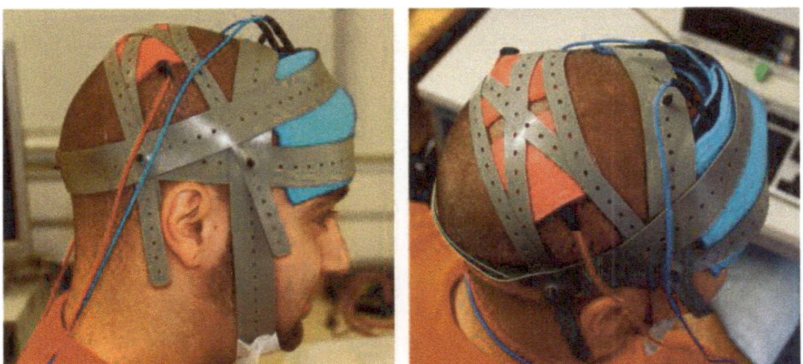

FIGURE 1 A four-electrode montage. The red electrodes are conventionally the anodes, the blue electrodes are conventionally the cathodes. The most common type of setup is to have just one anode and one cathode. The kind of setup shown in this photograph can be obtained by using two separate stimulators, or a single multichannel stimulator. *Photograph taken from Klein et al. (2013).*

region, and therefore precludes a straightforward assessment of the relationship between the targeted brain region and the behavioral effect. This scenario is also likely when researchers stimulate one region of interest and put the second electrode (which is termed a reference electrode) on a site that is supposedly inactive, such as the vertex or the supraorbita. This may be a faulty assumption, as placing the electrode on any region of the scalp is likely to have an effect on the brain regions underneath it. With both the supraorbita and the vertex, for example, the reference electrode does seem to affect the brain, defeating, or at least compromising, their utility as inactive sites (Truong, Magerowski, Blackburn, Bikson, & Alonso-Alonso, 2013). Another possibility is to place just one electrode on the scalp and place the other electrode on an extracephalic site, such as the arm, shoulder, or cheek. However, this may produce milder effects, or null results (Moliadze et al., 2010). As of yet, there is no clear way to account for where to place the second electrode.

Another issue is that tES has relatively poor spatial resolution. Its effects may not be localized. Not only does stimulating one brain region cause alterations in the activity of other functionally networked brain regions (Keeser et al., 2011), but the electrodes themselves are often so large that other brain regions may also be incidentally stimulated. This makes attributing performance or changes thereof to modulation of a given brain region difficult. For tDCS, the placement and number of the electrodes may provide one solution. In this case, referred to as high-density tDCS, a single anodal electrode is surrounded by multiple cathodal electrodes (usually four electrodes). This perforce limits the current distribution in a given brain region through inhibiting the surrounding brain regions (Borckardt et al., 2012). Compared to the more common two-electrode montage, high-density tDCS offers an important advantage in terms of spatial resolution. However, this is a more complex procedure and may require individual magnetic resonance imaging of each participant's brain to position the electrodes. This dramatically increases both the cost and time of the experiment. In this respect, one of the paradoxical advantages of more traditional tES electrode montages is *poor* spatial resolution. Even with a relatively crude localization of the electrodes, one is highly likely to target the brain area of interest. A common example is the list of measurements from the international EEG 10-20 system. The purpose of the 10-20 system is to provide an easily replicable method of locating scalp positions that correspond to major cranial structures (Jasper, 1958). The method is straightforward and easy to both learn and perform (often requiring as little as a measuring tape and a marker), but permits a high degree of spatial error.

While more precise forms of measurement and stimulation are being developed, methods of higher spatial resolution and precision, such as transcranial magnetic stimulation, can pose a different set of problems. For example, higher spatial resolution also leads to higher error variance, and therefore requires larger sample sizes to gain the necessary power to demonstrate effects (Sack et al., 2009).

Choosing the Type of Stimulation

Given the availability of different forms of stimulation, there is an inevitable choice between tDCS, tRNS, and tACS. As previously outlined, each method has relative advantages and disadvantages. tDCS may be useful when the hypotheses is framed around the simultaneous up- and downregulation of different brain regions, as in the case of left and right dorsolateral prefrontal cortex in emotional regulation (Sarkar et al., 2014), or the right and left parietal lobes in visuospatial attention (Sparing et al., 2009). Because tACS and tRNS do not have a conventional cathode, downregulation is not well understood, and therefore, experiments that require inhibition of particular brain regions are out typically of their reach. In contrast, when downregulation is unnecessary or problematic, both tACS and tRNS have important advantages. If the researcher can determine the specific frequencies associated with a particular cognitive function, then tACS may be applied at that frequency to entrain oscillations and enhance outcomes. In contrast, tRNS is particularly useful when there is limited a priori knowledge of the frequencies. It is also useful when the hypothesis requires the simultaneous excitation of multiple regions.

EVIDENCE OF tES-INDUCED ENHANCEMENT OF NUMERICAL COGNITION

Numerosity

An important construct within numerical cognition is sensitivity to quantity or numerosity, also known as number acuity or approximate number sense (ANS), an evolved mechanism that allows for a generalized representation of the number of objects in an array (either auditory or visual) that is present in human infants and adults, and in myriad animals (Barth, Kanwisher, & Spelke, 2003; Brannon & Terrace, 1998; Cordes, Gelman, Gallistel, & Whalen, 2001; Dehaene, Dehaene-Lambertz, & Cohen, 1998; Gordon, 2004; Meck & Church, 1983; Nieder & Miller, 2003; Pica, Lemer, Izard, & Dehaene, 2004; Xu & Spelke, 2000; see also Volume 1 of this series). The prevalence of sensitivity to numerosity in a large number of nonhuman animals is highly suggestive of its evolutionary significance and the necessity of at least a rudimentary number system that contributes to survival and mating prospects in many species (Geary, Berch, & Mann Koepke, 2015). It is an approximation of the actual quantities in these arrays, and can be explicitly assessed by presenting arrays of differing size and number and eliciting relative quantity judgments. Importantly, these judgments can be made without item-by-item counting, and appear to be based on a rapid perceptual assessment rather than a more effortful cognitive one. The ANS that underlies numerosity judgments also appears to be closely associated with the parietal lobes (Hubbard, Piazza, Pinel, & Dehaene, 2005; Piazza & Izard, 2009; Piazza, Pinel, Le Bihan, & Dehaene, 2007).

Previous studies have suggested that sensitivity to numerosity may be linked to mathematical achievement. Halberda, Mazzocco, and Feigenson (2008) demonstrated that sensitivity to numerosity is an individual difference variable, with considerable discrepancies between participants in terms of ANS acuity, as evidenced by their ability to judge the quantities of two sets of objects. Furthermore, they revealed a link between an individual's ANS acuity and explicit, symbolic mathematics of the kind taught in schools, showing that greater acuity predicts higher academic mathematical achievement (Halberda et al., 2008; Libertus, Feigenson, & Halberda, 2011; Libertus, Odic, & Halberda, 2012). The tantalizing question, therefore, is whether sensitivity to numerosity can be enhanced through brain stimulation, and whether this enhancement can transfer to more advanced mathematical performance.

Cappelletti et al. (2013) investigated this question in a study with four experimental conditions, one of which combined stimulation with quantity discrimination training, while the other three ruled out prominent alternative explanations. In particular, Cappelletti et al. divided 40 participants (healthy adults) into the following groups: quantity discrimination training with (1) stimulation to the parietal lobes, (2) stimulation to a control region (the motor cortex), (3) sham stimulation, and (4) a group that received stimulation to parietal lobes with no quantity discrimination training. Quantity discrimination training (for the first three groups only) was administered for 5 days, with pre- and posttesting before and after the first and last sessions, respectively. tRNS was applied (high frequency, 0-250 Hz) for 20 min (or 20 s, in the placebo condition) concurrently with the training sessions (see Figure 2 for the complete study design). Each participant in the training groups was trained on a quantity discrimination task every day for 5 days, where they were presented with intermixed sets of blue and yellow dots. The dots appeared for only 200 ms, and the participants had to judge which set of dots (blue or yellow) was more numerous. Easier trials were characterized by large differences in the numbers of blue and yellow dots. Trials became progressively more difficult as the distance between dots became smaller. This ability is indexed using the Weber fraction, or w. A smaller w represents greater finesse in quantity discrimination (this construct is discussed in greater detail in Halberda et al., 2008).

At the posttest following the 5 days of training, the authors found that, overall, training substantially enhanced quantity discrimination (approximately 18%). However, the greatest benefits were observed in the participants who received training with stimulation to the parietal lobes, who improved by approximately 33%. Participants receiving tRNS alone with no training did not show any significant improvements. Furthermore, when participants were tested 16 weeks later, only those who received quantity discrimination training combined with tRNS showed maintenance of learning effects (see Figure 3 for the results), suggesting that tRNS assisted the cortical processes that accompany and characterize learning, enhancing both the initial learning and the longevity of the effect.

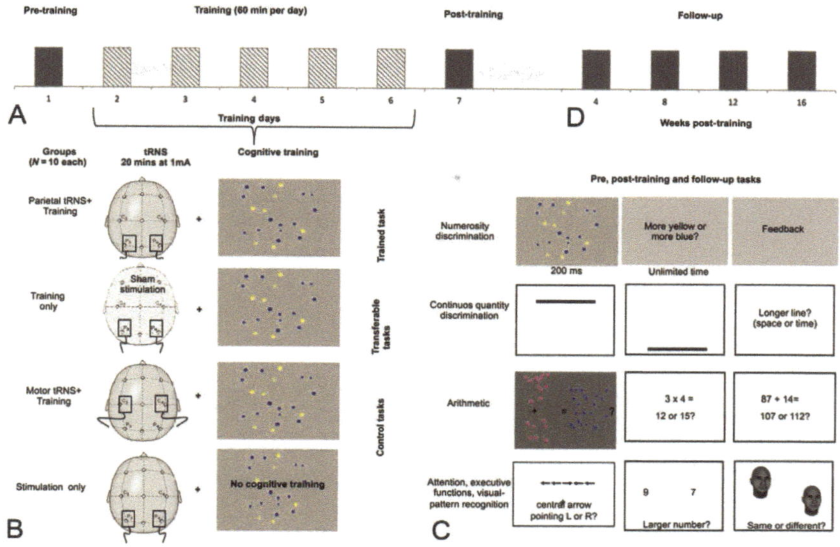

FIGURE 2 The training paradigm and the tasks used. (A) Participants in the parietal or motor tRNS + Training and in the Training Only groups were trained intensively (560 trials per day) on a numerosity discrimination task for 5 consecutive days, i.e., day 2-6 while (B) receiving real or sham stimulation to the parietal or motor areas. Participants on the Stimulation Only group received no training. Before the training (day 1, pretraining), all participants were tested with (C) the numerosity discrimination task in addition to other continuous quantity-based tasks (time and space discrimination), arithmetic and control tasks. The same cognitive tasks were repeated at the end of the training (day 7, posttraining) to test for any training-induced change, and (D) in the parietal tRNS + Training and Training Only groups again at week 4, 8, 12, and 16 posttraining to test for any long-term effect of the training. *Figure and legend taken from Cappelletti et al. (2013).*

The experiment also accounted for possible transfer of effects by assessing a variety of changes on several other cognitive competencies. The authors found evidence of transfer to two variables that, like numerosity, require processing of quantity: time and space discrimination (i.e., determining whether a line is presented for a shorter or longer duration of time, or is shorter or longer in length, compared to a reference line). Perceptions and evaluations of time and space, themselves continuous variables, are theorized to share neural architecture in the parietal lobes with the ANS (Bueti & Walsh, 2009; Cantlon, 2012; Cantlon, Platt, & Brannon, 2009; Cohen Kadosh et al., 2008; Walsh, 2003) and may therefore benefit from enhancements in quantity discrimination. The authors found evidence of enhanced time and space discrimination during end-of-training posttests in only the participants given quantity discrimination training combined with parietal tRNS. However, these transfers were short-lived and were not apparent at follow-up testing. At the same time, they demonstrate that even this transient transfer is tES induced, as the other groups that received training did not show any evidence of transfer to time and space discrimination.

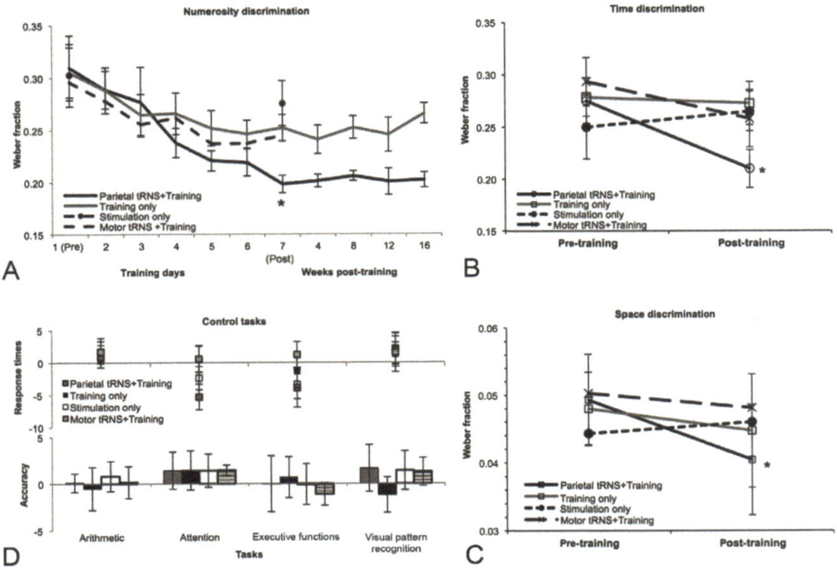

FIGURE 3 Results. Performance in the four groups expressed as Weber fraction in (A) the numerosity discrimination task at pretraining and at posttraining, during each of the training days, and at week 4, 8, 12, and 16 after training; in the transfer tasks, i.e., time (B) and space (C) discrimination at pretraining and posttraining; and (D) in tasks measuring arithmetic, attention, executive functions, and visual pattern recognition showing no training-induced changes in accuracy (correct answers) or RTs measured as percentage change from pretraining. Smaller *wf* indicate better performance. *Figure and legend taken from Cappelletti et al. (2013).*

This highlights the potential for tES to enable successful transfer. There was no evidence of transfer to other tested functions, including attention and visual perception. The transfer effect cannot be explained by a general regional effect of tRNS, as other cognitive abilities subserved by the same structure did not show any changes as a function of stimulation.

Of greater relevance to mathematical achievement was the *absence* of transfer to measures of arithmetic performance (e.g., determining whether 104 or 112 is closer to the solution for the sum 87 + 14, which is 101). Therefore, while these findings are of substantial scientific interest, their ecological validity remains untested. There are several possible explanations for a lack of transfer, both conceptual and methodological. From a conceptual standpoint, one explanation for lack of transfer to arithmetic performance is that such transfer may require concurrent stimulation of other regions where more complex operations are carried out that draw more extensively on higher-level cognition, such as working memory (e.g., the dlPFC). Another explanation is that quantity discrimination itself may not be essential for arithmetic performance (Butterworth, 2010), or that the connection between quantity discrimination and mathematical achievement may be closer/stronger in the developing brain compared to the adult brain (Price, Palmer, Battista, & Ansari, 2012). Alternatively, quantity discrimination might be based on sensory cues, rather than numerical processing (Gebuis, Gevers, & Cohen Kadosh, 2014; Gebuis & Reynvoet, 2012).

Other explanations for lack of transfer revolve around methodological issues, such as the training dose and the time between the end of training and follow-ups. In the case of Cappelletti et al. (2013), the training regime lasted 5 days. In a prominent study demonstrating the transfer of *n*-back working memory training to fluid reasoning (assessed through matrix completion tasks), Jaeggi et al. (2008) also found that the size of the transfer effect was dependent on the amount of working memory training. In another study, Holmes et al. (2009) demonstrated that adaptive working memory training in children with low working memory skills over 6 weeks produced substantial improvements in both verbal and visuospatial working memory. The defining finding of the Holmes et al. (2009) study, however, was the transfer to mathematical achievement on a standardized test 6 months after the completion of the training, but not at the end-of-training posttests. The authors predicted that transfer of working memory improvements onto mathematical performance would require time to work through the cognitive support structure and enhance performance. The methods are too different to compare Holmes et al. (2009) and Cappelletti et al. (2013) directly (e.g., the former used children with identifiably low working memory, while the latter used healthy adults with no evidence of compromises in cognitive functions). However, these findings do suggest that the dose and the time between the posttest and the follow-up may be relevant to the assessment of transfer. Indeed, it may be possible that even if there was successful transfer to arithmetic performance, the follow-up was too soon to detect changes. Or that the training, while transferring directly to other forms of continuous quantity processing, required a larger dose to transfer to more distally related arithmetic performance.

Yet another possibility is that the transfer task itself was not appropriate. For example, Park and Brannon (2013) also trained participants on quantity discrimination and assessed transfer to more advanced symbolic mathematics. In this case, the training involved an additional layer of nuance: participants viewed moving dots of two colors, then had to estimate their added or subtracted values and compare that estimate to a third set of dots and state whether it was greater or smaller. The transfer task was then to solve multidigit addition and subtraction problems, performance on which was greater for individuals trained on quantity judgment compared to those who did not receive the training. By this logic, perhaps a more appropriate transfer task for Cappelletti et al. (2013) would have been numerical comparisons (e.g., which is more, 47 or 81?) of varying difficulties, which more closely resembled the training.

Though Cappelletti et al. (2013) did not find transfer to arithmetic performance, the reader will appreciate the methodological rigor of this design, which overcomes several important issues that are frequently found in tES studies. For example, it controls for the placebo effect, establishes regional specificity and benefits of stimulation to a particular area, and establishes the importance of combining training with stimulation to show both that stimulation by itself cannot produce effects, and that the combination of training and tES is superior to training alone (Chick, 2014).

Given the poor spatial resolution of tES and possible expectancy effects, steps should be taken to bolster the internal validity of any set of findings. The most frequent measure is the inclusion of a sham condition, which can help to rule out the placebo effect. However, viable alternative explanations abound. For example, the possibility persists that stimulation effects are not region-specific, but are rather generalized effects of altering neural network activity *anywhere* on the cortex. If an effect is specific to a particular region, as is assumed, then stimulating another region not involved in that cognitive process should not produce the same effect. This makes stimulating a second, presumably unrelated, region an important measure. Cappelletti et al. (2013) investigated regional specificity, concluding that it was indeed only the stimulation to the parietal lobes in combination with training that enhanced quantity discrimination.

Symbolic and Magnitude Processing

Symbolic-magnitude mapping

An important feature of numerical cognition in humans is the association of symbols with magnitude (e.g., "1" and "9" symbolically represent different magnitudes). One method of assessing basic numerical competence is pitting these numbers against each other in a variation of the Stroop task. In the classic version, participants have to identify the color a word is printed in, say BLUE. The answer, of course, is "red." However, the automaticity of processing the meaning of the word takes priority over assessing its printed ink color, and the resultant interference, a lengthening of the time taken to name the ink color when it conflicts with the written color name, is the Stroop effect (MacLeod, 1991; Stroop, 1935). In one variation of the Stroop task, digits are presented side-by-side and the participant determines which one is *physically* larger. For a pair of numbers on a congruent trial, say 1 and 9, the 9 is physically larger than the 1, and there is no interference between semantic and perceptual processing. On an incongruent trial, the 1 is physically larger than the 9, and the physical and semantic features of the stimuli are in conflict. It is on these trials that basic numerical competence can be inferred. If a Stroop effect is apparent (longer reaction times on the incongruent trials), it is taken as evidence of a suitably well-learned and automatized system of symbolic magnitude whose semantic features take precedence over more basic perceptual features, and hence, is interpreted as a behavioral signature of proficient numerical processing (Girelli, Lucangeli, & Butterworth, 2000; Rubinsten, Henik, Berger, & Shahar-Shalev, 2002; Schwarz & Ischebeck, 2003). In contrast, deficits in this core capability are a behavioral signature of numerical difficulties (Cohen Kaodsh et al., 2007; Rubinsten & Henik, 2005), and of young children without well-developed numerical processing skills (Girelli et al., 2000; Rubinsten et al., 2002).

Cohen Kadosh et al. (2010) employed this paradigm to assess the effects of tES on the learning of an *artificial* system of symbolic magnitudes. In other words, the digits 1-9 were replaced with novel symbols taken from Gibson, Gibson, Pick, and Osser (1962) and used by Tzelgov, Yehene, Kotler, and Alon (2000) to constitute a new set of numbers, each with a new magnitude (see Figure 4). Cohen Kadosh et al. (2010) applied 20 min of tDCS at 1 mA to three groups of healthy adult participants (15 in total) using three different stimulation configurations. One group was given anodal tDCS to the right parietal lobe and cathodal stimulation to the left parietal lobe (RA-LC), the second was given the reverse montage (RC-LA), and the third group was a sham condition. Participants learned the association between the arbitrary symbols and the magnitudes they represented over 6 days of training. At posttest, participants in the group given anodal stimulation to the right parietal lobe and cathodal stimulation to the left parietal lobe showed the strongest Stroop effect when given the modified version of the Stroop task described earlier (further modified by

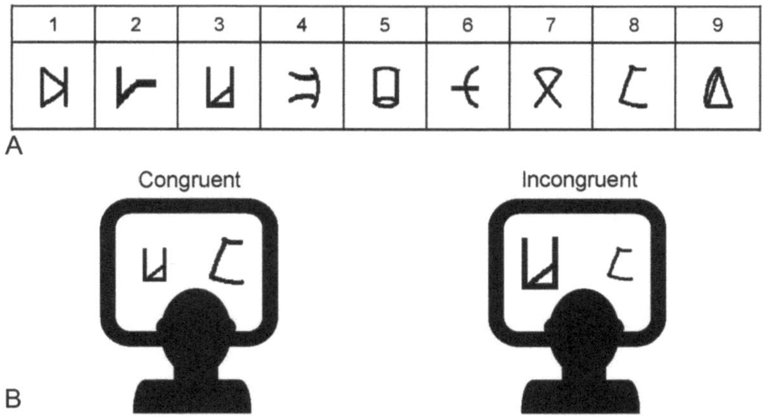

FIGURE 4 Example of stimuli and the numerical Stroop task. (A) Everyday digits and their corresponding artificial digits that were used to create the new numerical system. Each everyday digit appears above its corresponding artificial digit. Over 6 days of training, participants were instructed to refer to the artificial digits as representing various magnitudes and to decide on each learning trial which one of the two stimuli has a larger magnitude, while receiving visual feedback for their decision. Only adjacent pairs have been presented during this phase. (B) An example for congruent and incongruent trials with artificial digits from the numerical Stroop task. In the current example, the symbols corresponding to the Arabic digits 3 and 8 appear on the left and right sides of a congruent and an incongruent trial, respectively. Participants were asked to compare the stimuli according to their physical size while ignoring their numerical meaning. On congruent trials, the larger number appeared in larger font, whereas on incongruent trials the larger number appeared in smaller font. The numerical Stroop effect (incongruent vs. congruent) indicates the slowing in the decision time when numbers are irrelevant to the task and are therefore processed automatically (Henik & Tzelgov, 1982). Nonadjacent pairs have been presented in this task to examine transitive inference from the learned material (Tzelgov et al., 2000). *Figure and legend taken from Iuculano and Kadosh (2013).*

replacing Arabic digits with Gibson figures). This suggests that these participants had acquired a familiarity with these new representations that resembled, in terms of the Stroop effect, their familiarity with conventional Arabic digits. In contrast, the group that received anodal stimulation to the left parietal lobe and cathodal stimulation to the right parietal lobe underperformed on this task, revealing poor acquisition of this new system of numbers. The performance of the group that received sham stimulation fell in between (see Figure 5).

Cohen Kadosh et al. (2010) further investigated these effects using a second measure: number-to-space mapping. For a well-learned system of numbers, when individuals are asked to map or point out numbers on a horizontal number line, they do so in a linear fashion. For example, when indicating the positions of 1-9

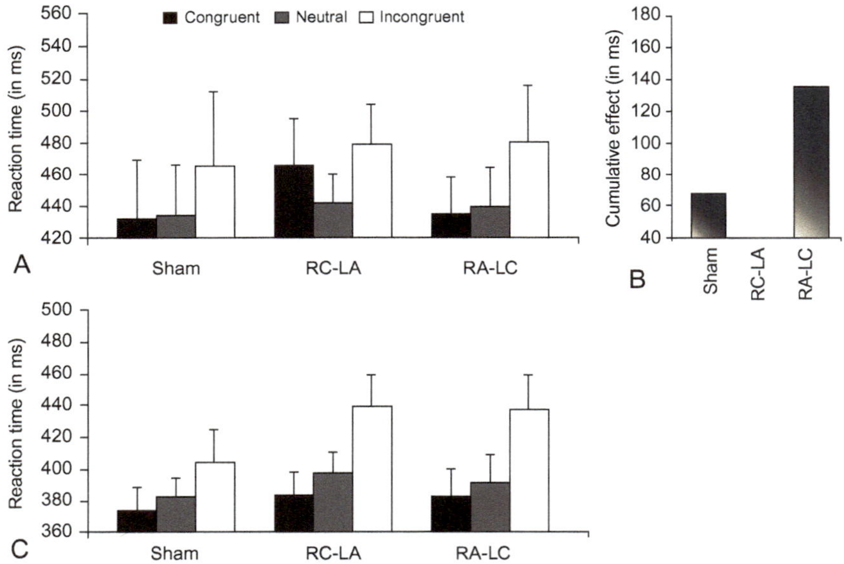

FIGURE 5 The Congruity Effect for the Artificial Digits, the Cumulative Congruity Effect over Training, and the Congruity Effect for Everyday Digits for the Sham, RC-LA, and RA-LC Groups in the Numerical Stroop Task. The data of the artificial digits for each group are averaged across the sessions that showed a significant congruity effect (three sessions for the RA-LC group, two sessions for the sham group, and five sessions for the RC-LA group; note that the latter group showed an abnormal congruity effect that was not changed as a function of learning). (A) Whereas the RA-LC group and the sham group showed a typical congruity effect, the RC-LA group showed an abnormal effect that mirrored the performance of children at the age of 6 years and might reflect perceptual rather than semantic interference (Girelli et al., 2000). (B) The cumulative congruity effect demonstrates the emergence of a consistent automatic numerical processing already from the fourth day for the RA-LC group ($p=0.005$, Table S1), whereas it occurred only later for the sham group ($p=0.049$, Table S1) and did not appear for the RC-LA group. (C) All groups showed a consistent and typical congruity effect for everyday digits ($p=0.00009$; group × congruity interaction, $p=0.46$), as reflected by slower RTs for the incongruent condition versus the congruent condition. Data are mean ± standard error (SE) of the mean. Note the different scaling in each panel. *Figure and legend taken from Cohen Kadosh et al. (2010) see Table S1 in Cohen Kadosh et al. (2010).*

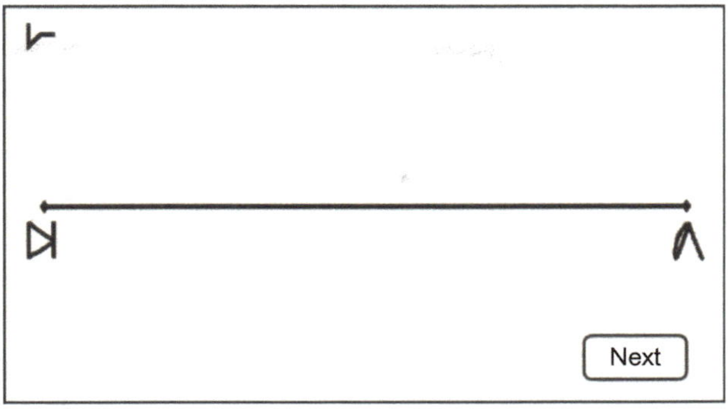

FIGURE 6 Artificial digits. Symbols used as stimuli during the learning phase and the numerical Stroop task and their equivalent as everyday digits—adapted from Tzelgov et al., 2000. *Figure from Cohen Kadosh et al. (2010), Current Biology, 20, 2016-2020, as cited in Iuculano and Cohen Kadosh (2014)*

on a number line, individuals place the successive digits at approximately equal distances from one another, suggesting an understanding of the magnitude relations between the numbers (see Figure 6). When these points are modeled, they are best described by a straight line. Like a more pronounced Stroop effect with an incongruent number-size task, linear number-to-space maps are a behavioral signature of intact and advanced numerical cognition (Booth & Siegler, 2008; Geary, Hoard, Nugent, & Byrd-Craven, 2008; Siegler & Opfer, 2003). In this respect, one of the indications for poorer numerical skills is a logarithmic number-to-space mapping, characteristic of young children, children with a mathematical learning disability, and persons with minimal, if any, formal mathematics education (Booth & Siegler, 2008; Dehaene, Izard, Spelke, & Pica, 2008; Geary et al., 2008). An example would be marking 8 farther away from 7 than 7 from 6. When these points are modeled, they are best described by a logarithmic function. In this experiment, all participants produced linear number-to-space maps when asked to work with Arabic digits. However, only the participants who received anodal stimulation to the right parietal lobe and cathodal stimulation to the left parietal lobe produced linear number-to-space maps for the arbitrary symbols. The other two groups produced the expected logarithmic maps (see Figure 7).

Overall, therefore, the results suggest that tDCS (with anodal stimulation to the right parietal lobe and cathodal stimulation applied to the left parietal lobe) enhanced the acquisition of a new system of numbers. Furthermore, these effects appear to persist long term, as they were apparent at a 6-month follow-up, at a similar magnitude to the posttest that immediately followed training. However, while it is tempting to conclude that anodal tDCS to the right parietal lobe led to the enhancement effect, the electrode montage in this experiment does not permit a firm conclusion. As discussed in the previous section, bilateral tDCS requires that another region, cephalic or extra-cephalic, be stimulated as well. In this case, cathodal stimulation was provided to the left parietal lobe, and

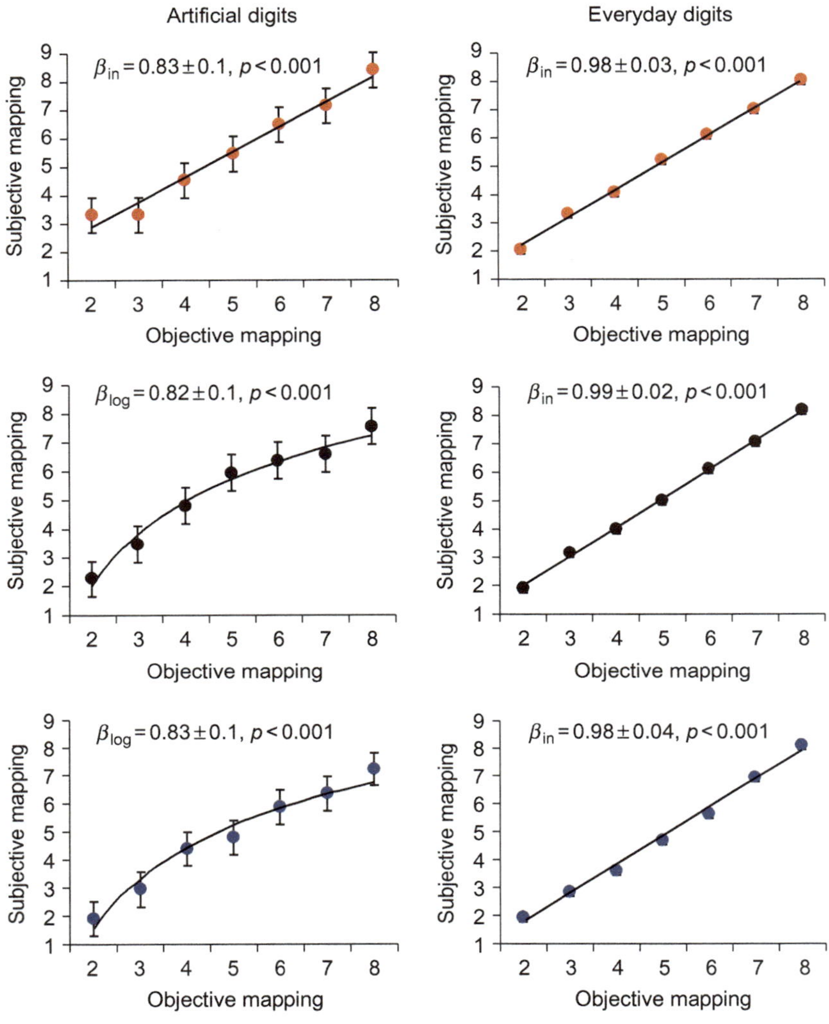

FIGURE 7 Average location of artificial digits on the horizontal segment, shown separately for artificial digits in the left column, everyday digits in the right column, and type of stimulation. β represents the selection of the best weight, whether it was logarithmic (β_{log}) or linear (β_{lin}), in stepwise regression analysis with linear and logarithmic predictors. Data are mean ± SE of the mean. The first row reflects the performance of the RA-LC group (red circles), the middle row reflects the performance of the sham group (black circles), and the bottom row presents the performance of the RC-LA group (blue circles). Whereas the performance with artificial digits was affected by the type of brain stimulation and showed a linear fit only for the RA-LC group, the performance with everyday digits was independent of the type of brain stimulation and showed a linear fit for all the groups. *Figure and legend taken from Cohen Kadosh et al. (2010).*

this bilateral pattern of stimulation means that the enhancement may have come from independent (e.g., excitation of activity in the right parietal lobe or inhibition of activity in the left parietal lobe) or interactive effects of such a montage. This issue becomes clearer in the next subsection.

Multiple electrodes

One method to have a greater number of stimulation sites for tDCS is the use of more than two electrodes, and some studies have begun employing four electrodes. This could be achieved through two anodes and two cathodes with two tDCS devices concurrently, or a multichannel stimulator (Ruffini et al., 2014). This kind of montage is useful in stimulating, for example, both regions of a laterlized brain structure such as the dlPFC with the same polarity. The experiments discussed here up to now have all implemented two-electrode setups. A bilateral tDCS montage means that the two regions receive opposite stimulation polarities. Four electrodes can overcome this limitation, as both the left and right areas can be given tDCS of the same polarity (see Figure 1). Here, the reference electrodes may be placed over sites that are apparently inactive, or at least not involved in that cognitive process (however, see Truong et al., 2013).

Hauser, Rotzer, Grabner, Mérillat, and Jäncke (2013) employed this type of four-electrode setup (with the reference electrode located over the supraorbital regions) in two experiments. In the first experiment, 21 healthy adult participants were given two types of stimulation: a sham session, and one of anodal tDCS to both left and right parietal lobes, anodal stimulation to only the left parietal lobe, or cathodal tDCS to both left and right parietal lobes. The study implemented a within-subjects arrangement, with every participant performing tasks in a sham condition and one of the three stimulation conditions in a counterbalanced format. In a second experiment, 16 healthy adult participants were given either anodal tDCS to the right parietal lobe or sham stimulation. In these experiments, participants completed two numerical processing tasks. In the magnitude processing task, they decided whether a number presented from 31 to 99 was larger in magnitude than 65. In the second task, participants performed two-digit subtractions (e.g., 63-29), and chose a solution from among three options, two of which were distractors (e.g., the correct response to 63-29 is 34, which appeared with two distractors, 44 and 32). Twenty-five minutes of tDCS at 1 mA was applied before these tasks were performed. Interestingly, only anodal tDCS to the left parietal lobe improved performance in the numerical tasks. For magnitude processing, only the left-anodal tDCS improved both accuracy and reaction time. For subtractions, left-anodal stimulation reduced reaction times without changes in accuracy.

Prima facie, a benefit from only stimulating the left parietal lobe seems inconsistent with the implications of previous research, especially the conclusions from meta-analytical reviews describing the neural underpinnings of numerical cognition (e.g., Arsalidou & Taylor, 2011), which show that both parietal lobes should be involved in these types of tasks. Therefore, bilateral anodal tDCS to the parietal lobes should also have produced performance benefits. The authors propose that the absence of benefits in bilateral stimulation relates to a specific pattern of communications between the right and left parietal lobes during numerical cognition: interhemispheric inhibition (for more detailed discussion on interhemispheric inhibition in the context of numerical cognition, see Cappelletti, Barth, Fregni, Spelke, & Pascual-Leone, 2007; Cohen Kadosh et al., 2010). In the framework of these results, interhemispheric inhibition would suggest that the left parietal lobe may be naturally inhibited by the right parietal lobe. Anodal stimulation showed that the left parietal lobe was involved in the tasks, while the right parietal lobe may be less involved. Inhibition from the right to the left parietal lobe would therefore contribute to performance deterioration. Anodal stimulation to the left parietal lobe should therefore help it overcome inhibitory signals from the right parietal lobe. However, bilateral anodal stimulation, or increasing activity in both left and right parietal lobes, which would by definition also increase activity in the right parietal lobe, would thus increase the right parietal lobe's ability to inhibit the left parietal lobe's activity. This would essentially cancel out any beneficial effects of tDCS. The results are in line with such a principle, but of course a priori tests of this hypothesis are necessary before firmer conclusions can be drawn.

These results also run counter to the Cohen Kadosh et al. (2010) findings described earlier. While Cohen Kadosh et al. (2010) detected evidence for improved performance when anodal tDCS was given to the right parietal lobe, Hauser et al. (2013) did not. At the same time, a direct comparison of results is not possible because of substantial methodological differences. For example, Cohen Kadosh et al. (2010) used a multiday training paradigm to study the effects of tDCS (with anodal stimulation to the right parietal lobe) on emergence of associations between symbols and magnitudes, and not the effect of tDCS on the performance of arithmetic tasks. In contrast, Hauser et al. (2013) applied anodal tDCS to the right parietal lobe only once in a two-session study to examine the effects on arithmetic performance. Clearly, both the overall goal of the experiments and the number of stimulation sessions differed between the studies. The tDCS protocols also differed. Cohen Kadosh et al. (2010) gave online tDCS (i.e., stimulation *during* the task), whereas Hauser et al. (2013) applied offline tDCS with participants performing tasks before and after, but not during, stimulation. This sort of difference in method precludes both a straightforward assessment of the studies and a comparison of their conclusions. However, they serve to illustrate two important points: (1) a general convergence of evidence for enhancement from two different research groups that employ different formats of task and tDCS, and (2) the diversity of tDCS effects on numerical cognition and performance that clearly points to the need for further research in this area.

Klein et al. (2013) also used a four-electrode tDCS montage to target both the left and right parietal cortices with the +, or sham tDCS at 1 mA. During stimulation, participants performed two tasks, an arithmetic task involving two-digit addition problems, and a color-word Stroop task. The addition problems were composed of two tasks: In one, the participant was presented with the addition of two numbers and had to choose the correct response, which appeared alongside an incorrect distractor (e.g., 25+31=56 or 51). In the other task, participants were again presented with an addition problem and two options. But this time, they had to select the option that was *closer* to the correct answer. For example, 25+31=54 or 51. While 56 is the actual solution, 54 is closer to 56 than 51; therefore, in this type of task, the correct response is 54. The distance between 54 and 51 was the distractor distance. Clearly, smaller distractor distances are more difficult to work with than larger ones—the distractor distance effect.

tDCS did not affect performance on the Stroop task and also had no effect on the detection of correct solutions. However, the authors did find that bilateral anodal stimulation lowered the distractor distance effect. This was interpreted as an improvement in processing the magnitude of numbers. In contrast, bilateral cathodal stimulation heightened the distractor distance effect, suggesting impaired number magnitude processing. This design has several interesting and important features. The use of the Stroop task shows that stimulating a brain region may not have equal effects on different tasks that reflect different cognitive processes. This point is further supported by the results from the two subtasks within the addition task. In combining two types of tasks—distractor distances and choosing the correct response—and finding that tDCS affected only the latter, the results provide evidence for the specificity of tDCS effects on a given cognitive operation. These results also show that a brain region associated with a broad cognitive function (in this case, numerical processing), may not be involved to an equal degree for different tasks (in this case, estimating distances and choosing correct solutions).

Finally, the benefit accrued from bilateral anodal stimulation conflicts with a prediction made from the standpoint of interhemispheric interference, which would suggest an abolishing of effects as seen in Hauser et al. (2013). This is an important demonstration of contradictory results in the tES research field. At this early stage, the technique is generating at least as many questions as it is answering. At the same time, it adds to a more general picture that, despite the large amount of noise, research is converging on possibilities for enhancement.

Arithmetic Operations

Recent research is moving into the use of more challenging numerical tasks. For example, Snowball et al. (2013) went beyond numerosity discrimination and number-space mappings to a more advanced aspect of numerical performance: arithmetic operations. They combined this with near-infrared spectroscopy (NIRS), which was used to examine concurrent physiological changes in the stimulated and associated brain regions.

Before discussing the experiment itself, it will be useful to describe brain imaging with NIRS. The technique employs light from the near-infrared region of the electromagnetic spectrum, which is absorbed by tissue and provides information about the changes in hemoglobin concentrations (Villringer & Chance, 1997; Villringer, Planck, Hock, Schleinkofer, & Dirnagl, 1993). Active brain regions consume more oxygen, which in turn require a supply of blood, causing changes in local blood volume levels. NIRS is able to record changes in blood hemoglobin levels, making inferences about regional activity possible. NIRS is noninvasive in the sense that measurements can be made through the skull, and provides similar information to *f*MRI. However, it is much more convenient to use and administer, and is also more portable. Furthermore, NIRS data closely overlaps and correlates with simultaneously acquired *f*MRI data (Cui, Bray, Bryant, Glover, & Reiss, 2011), and depending on various constraints such as scalp-brain distance, may be used as at least a partial substitute for *f*MRI (see Cui et al., 2011; Strangman, Boas, & Sutton, 2002).

In the Snowball et al. (2013) experiment, 25 healthy adult participants completed 5 days of training while performing arithmetic operations that were learned via drills or calculations. Drilling results in the practiced ability to retrieve arithmetic facts from long-term memory (e.g., $7 \times 6 = 42$), and is based on rote learning, without engagement with the numerical relationships that produce the solution. Calculation learning, in contrast, requires the participant to actively manipulate the numbers according to specific algorithms or rules in order to produce a specific result, or to apply several mathematical operations to produce the result (e.g., $32 - 17 + 5 = 20$) (see Figure 8). While participants performed these tasks, tRNS at 1 mA was applied to the bilateral dlPFC, which has been implicated in arithmetic learning (Delazer et al., 2005; Rivera et al., 2005; Zamarian et al., 2009).

Snowball et al. (2013) found that participants receiving real tRNS improved on both drill and calculation performance. NIRS data were collected on the first and last days of training, and also at a six month follow-up. They revealed specific, stimulation-induced changes in the left, but not right, dlPFC with respect to amplitude and latency of oxygenated and deoxygenated hemoglobin. These results suggest that, as compared to receiving sham stimulation, the genuinely stimulated left-dlPFC required reduced quantities of oxygen, changes that were maintained long term. The coupling of enhanced behavioral performance with a decreased neurophysiological requirement for oxygen implies more efficient brain functioning following stimulation (see Figures 9–11). This is also consistent with the general finding that dlPFC engagement decreases as individuals gain arithmetic competence (Rivera et al., 2005).

The neurophysiological and behavioral effects were still apparent at a 6-month follow-up. These long-term benefits were specific to calculations, and were not observed for drill learning, indicating that the nature of learning influences the longevity of the effects. The authors suggested that this might be due to deeper-level learning in the case of calculation compared to shallower

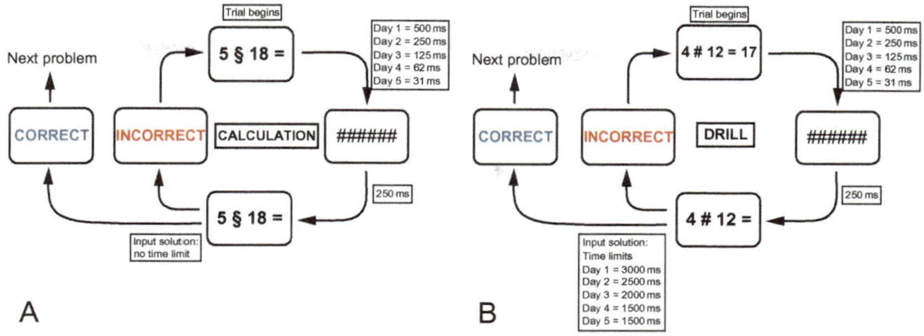

FIGURE 8 Schematic representation of the arithmetic learning regimes. (A) The calculation task. Answers to each calculation problem were obtained by manipulation of two numerical operands according to a particular algorithm (either [(right number-left number) + 1] + right number or [(right number + left number) − 10] + right number). Participants were instructed to enter their two-digit solution on the number pad of a standard QWERTY keyboard. Positive and negative feedback was provided for each answer (500 ms duration), and participants were only allowed to progress to subsequent problems once they had obtained the correct solution. (B) The drill task. Each trial began with the presentation of two numerical operands accompanied by the problem's answer. After the initial presentation, the problem would disappear from the screen and reappear without the answer, at which point subjects were required to enter their two-digit solution. Positive and negative feedback was provided (500 ms duration), and if participants answered incorrectly, the whole presentation cycle would repeat. Calculation and drill problems were presented in alternating groups of 18, referred to as "blocks." In line with previous studies (Delazer et al., 2005), the total number of blocks varied according to the day of training: 10 blocks on the first day, 12 on the second, 14 on the third, 16 on the fourth, and 14 on the fifth. The ratio of calculation to drill blocks was the same on each day, at 1:1. For both tasks, the round-edged boxes represent the individual presentation screens, and the square-edged boxes the time delays between each presentation screen. *Figure and legend taken from Snowball et al. (2013).*

learning in drilling (Craik & Lockhart, 1972). The calculation problems entailed implementation of the cardinal arithmetic operations, while for drilling, the association between the digits and the solution was arbitrary.

Furthermore, improved performance at follow-up for the participants given tRNS transferred to problems that were not presented during the training. As mentioned earlier, generalizations beyond the tested stimuli are relatively uncommon in cognitive training studies (Green & Bavelier, 2003; Owen et al., 2010), and their appearance in these results signals the potential of tES to facilitate transfer. Note that only about 50% of the participants (of the original 25) were available at the 6-month follow-up. Nevertheless, in addition to providing evidence for enhanced numerical cognition, these findings also provide important biological insight by demonstrating that brain stimulation may increase neurophysiological efficiency. It is also important to note that participants, being healthy young adults and university students in the main, possessed a certain level of competence before beginning the experiment. It is very possible that the results may have been different for adults with compromised numerical

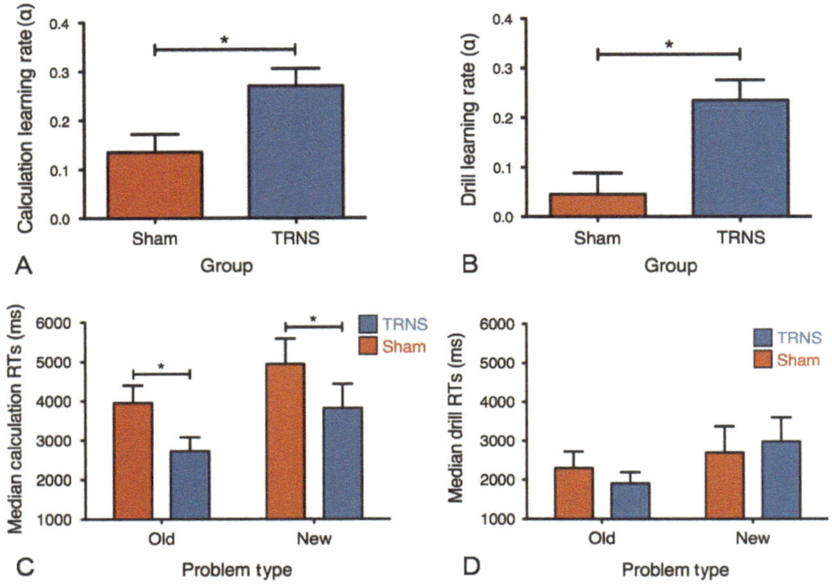

FIGURE 9 The effect of TRNS on arithmetic performance. (A) Calculation learning rates during training were significantly higher in the TRNS group relative to sham controls. (B) Drill learning rates during training were significantly higher in the TRNS group relative to sham controls. (C) Calculation RTs during testing were significantly faster in the TRNS group relative to sham controls for both old and new problems. (D) Drill RTs during testing did not differ between TRNS and sham groups for either old or new problems. Error bars indicate one SEM. Significant differences are marked with asterisks. *Figure and legend taken from Snowball et al. (2013).*

abilities, or with children still learning basic arithmetic facts. In such cases, perhaps the drill learning would have benefited instead of the calculations.

Another study (Looi et al., submitted) innovatively combined video game training with brain stimulation. The video game required participants to work with fractions. Competence in handling fractions has been shown to predict mathematical achievement (Bailey, Hoard, Nugent, & Geary, 2012; Siegler, Fazio, Bailey, & Zhou, 2013). Participants had to map fractions onto the number line as they played the game. The video game was adaptive, becoming more difficult as the participants gained proficiency. The game also required participants to respond with increasing precision as they improved on the task. The game drew on principles of embodied learning of numerical concepts presented in Link, Moeller, Huber, Fischer, and Nuerk (2013), and employed a motion detection device that required participants to physically move from side to side to solve the problems.

Thirty healthy adult participants received a total of 60 min of tDCS at 1 mA (broken into two 30-min sessions spread over 2 days), with anodal stimulation given to the right dlPFC and cathodal stimulation given to the left dlPFC.

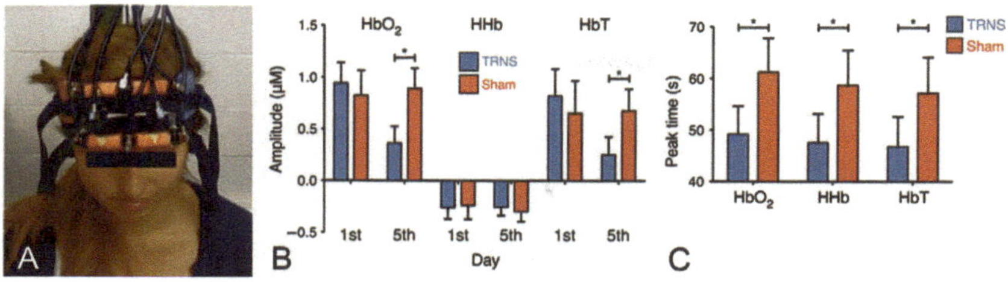

FIGURE 10 NIRS: The effect of TRNS on hemodynamic response amplitudes and latencies within the left LPFC during training. (A) Combined TRNS-NIRS setup. The NIRS plate (orange) is embedded with two transmitting and six receiving optodes and secured to the forehead: transmitting optodes are capped with blue labels, and receiving optodes are unmarked. Recordings were taken during the training phase on the first day and the fifth (last) day, as well as during the testing phase 6 months later. The fiber-optic cables connecting the optodes to the NIRS device can be seen emanating from the top of the image. TRNS electrodes are positioned over the bilateral DLPFC and encased in blue and red saline-soaked sponges, as shown. This innovative TES-NIRS combination allowed us to quantify the hemodynamic response to functional activation and to assess how it varied as a function of brain stimulation and arithmetic training. (B) TRNS reduced the amplitude of HbO_2 and HbT responses by the end of training. A significant three-way interaction between hemodynamic measure, day, and group in the left LPFC indicates a significant decrease in peak amplitude for HbO_2 and HbT at the end of the training (fifth day) in the TRNS group relative to sham controls. (C) Reduced peak latencies emerged in the TRNS group compared to sham controls, for HbO_2, HHb, and HbT responses, independent of day. Both these effects were restricted to the left LPFC. Error bars indicate one SEM. Significant differences are marked with asterisks. *Figure and legend taken from Snowball et al. (2013).*

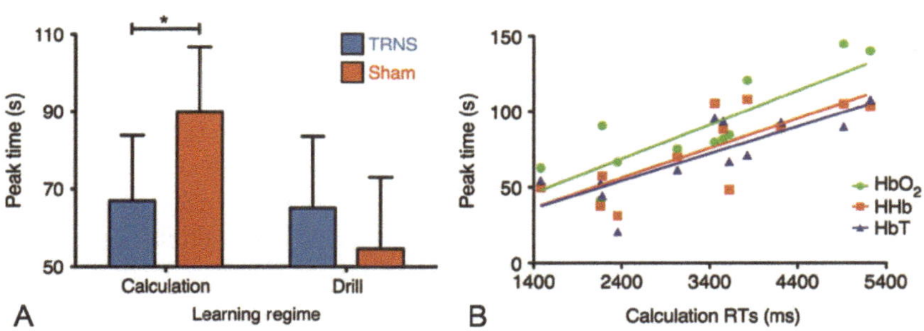

FIGURE 11 The effect of TRNS on hemodynamic response latencies during testing: relationship with behavioral performance. (A) A significant two-way interaction existed between learning regime and group for peak latency in the left LPFC, 6 months after the end of training. While the TRNS and sham groups did not differ for drill problems, the TRNS group showed a significant decrease in peak latency compared to sham controls for calculation problems. Error bars indicate one SEM. Significant differences are marked with asterisks. (B) Significant correlations existed between calculation RTs and the peak time of changes in HbO_2, HHb, and HbT concentrations 6 months after the completion of training. *Figure and legend taken from Snowball et al. (2013).*

Participants were divided into three groups (real tDCS + training, sham tDCS + training, active control). The group receiving real stimulation outperformed the sham stimulation group in terms of reaction time, especially in the higher difficulty levels, where the sham group displayed a typical reduction in speed but the real stimulation group did not. The results also suggested that this enhanced speed was not the product of a speed-accuracy trade-off. One interpretation of this outcome is that stimulation facilitated a greater or more efficient deployment of cognitive resources toward the task, while training by itself (sham stimulation) did not.

After only 2 days of training, the participants given real stimulation were approximately 13% quicker in placing fractions on number lines than participants who received sham stimulation during the training. Another striking result was an improvement on an untrained task, verbal working memory span, for only the participants who received real stimulation, corroborating what Cappelletti et al. (2013) and Snowball et al. (2013) have suggested—that tES may facilitate transfer to untrained tasks. In this case, the fractions training, which demanded working memory resources, may have resulted in an expansion of working memory capacity through tES. This implies a bidirectional relationship between an aspect of numerical cognition and working memory. Even more striking, the benefits on fraction placement and working memory capacity were apparent at a 2-month follow-up, providing clear evidence that the effects are long lasting. The transfer to working memory is particularly intriguing. Similar capacity improvements in behavioral training studies have taken up to 100 h of training to induce (Schmiedek, Lövdén, & Lindenberger, 2010). Because this improvement was restricted to the group that received real tDCS during training, with no such effects observed in the sham group or the active control group, the increase in working memory capacity may be a product of a synergistic interaction between tDCS and fraction training.

All or Null? The Case of Nonsignificant Results

The results discussed until here all show relatively strong behavioral effects, at least at a statistical and laboratory level. But as with all research, null results are frequently unpublished. Clemens, Jung, Zvyagintsev, Domahs, and Willmes (2013) applied tDCS to the angular gyrus in the posterior parietal cortex (PPC) for 10 healthy participants in a single-session study to improve the recollection of numerical facts. There were no significant differences in task performance between real and sham stimulation. Interestingly, however, the study entailed pre- and poststimulation *f*MRI. Neurophysiological changes were noted for participants receiving real tDCS, consistent with findings from Snowball et al. (2013). Studies such as Clemens et al. (2013) reveal that neuronal changes may emerge after even a single session of stimulation, and may not be accompanied at once by a behavioral outcome. Indeed, it is quite possible that such effects may have emerged had the authors provided stimulation in further sessions.

This is an example of an important principle in brain stimulation research, and for the use of convergent measures in general: the absence of behavioral results, such as changes in reaction times and accuracy, does not indicate the absence of underlying neurophysiological changes. Such changes can only be detected by neuroimaging methods, such as *f*MRI, EEG, or NIRS. The cost of such techniques may prove prohibitive in many cases, but null behavioral results should not discourage researchers from developing new tES designs to ask and answer questions.

EVIDENCE OF tES-INDUCED ENHANCEMENTS IN DYSFUNCTIONAL NUMERICAL COGNITION

The experiments described earlier focus on participants without difficulties in numerical processing. However, there is a substantial population that faces difficulties, sometimes extraordinary difficulties, in numerical cognition and performance. Such deficits impose significant constraints on several aspects of daily life. Perhaps as many as 15-20% of the adult population faces generalized numerical difficulties (Bynner & Parsons, 1997; Parsons & Bynner, 2005), and some estimates go as high as 20-25%. Nowadays, with a growing dependence on technology, increasing educational emphasis on the importance of quantitative aptitude, and a heightened occupational need for numerical skills, difficulties in perceiving and using numbers may seriously limit individual achievement, with costs in educational and job opportunities, salaries, and perhaps even physical and mental health (Beddington et al., 2008; Duncan et al., 2007; Parsons & Bynner, 2005). For example, poor mathematical achievement at as early as 7 years is associated with lower educational and socioeconomic attainment at 42 years (Ritchie & Bates, 2013). Individuals with numerical deficits are over 2.5 times more likely to face unemployment as individuals with competent numeracy (Parsons & Bynner, 2005), likely a secondary effect due to difficulties in living associated with poor academic competencies. Numerical abilities and deficits therefore have important implications for both individual and societal achievement. Analogous to the tES-induced enhancements in numerical cognition in numerically competent adults, there is also emerging evidence that tES may enhance performance in individuals with numerical deficits and difficulties.

Mathematics Anxiety

Mathematics anxiety refers to the persistent negative arousal, particularly apprehension or fear, elicited by engaging in numerical tasks (Ashcraft & Kirk, 2001; Richardson & Suinn, 1972). A numerical task may be a conventional final exam where mathematical performance is explicitly assessed, or may be implicit in everyday situations, such as calculating tips at restaurants, checking bills and receipts, thinking about distances, or even reading a timetable at a bus stop. The research in this area has identified a variety of cognitive, emotional, and behavioral consequences of mathematics anxiety in the performance of mathematical tasks. These include feelings of tension (Richardson & Suinn, 1972) that range from slight discomfort to complete avoidance (Hembree, 1990). Mathematics anxiety also manifests in a compromised ability to perform basic counting operations (Maloney, Risko, Ansari, & Fugelsang, 2010), as well as in the precision of mental magnitude representations (Maloney, Ansari, & Fugelsang, 2011).

A recent study tested the possibility that tDCS would ameliorate the negative emotional reaction that characterizes mathematics anxiety, and thereby improve numerical performance (Sarkar et al., 2014). Twenty-five participants with high mathematics anxiety were given tDCS at 1 mA for 30 min (or sham stimulation) to their left and right dorsolateral prefrontal cortices to enhance cognitive control over the negative emotional response elicited by mathematical stimuli. A low mathematics anxiety group of 20 participants was included for comparative purposes. All participants received both real and sham stimulation in counterbalanced order. Previous research has established the role of the dlPFC in affective control (Lorenz, Minoshima, & Casey, 2003). Very interestingly, participants with high self-reported mathematics anxiety who showed frontoparietal network activity, including the dlPFC, during mathematical performance actually perform relatively well on those tasks, especially when compared to high mathematics anxiety individuals who did not show such activation (Lyons & Beilock, 2012). This suggests that some high mathematics anxiety individuals are able to exert greater control over their negative response to mathematical tasks, likely through frontoparietal activation, and thereby enhance their performance.

To examine more closely how tDCS modulates stress, Sarkar et al. (2014) also investigated changes in salivary cortisol, a biomarker of psychological stress and anxiety (Hellhammer, Wüst, & Kudielka, 2009). Anodal stimulation was given to the left dlPFC, while cathodal stimulation was delivered to the right dlPFC. This montage was chosen based on previous evidence of lateralized emotional processing properties of the dlPFC. In particular, the left dlPFC has been linked to positive emotional processing and emotional control (e.g., Herrington et al., 2005; Wolkenstein, Zeiller, Kanske, & Plewnia, 2014), and the right dlPFC has been associated with negative emotional processing, and variance in activity in this region is correlated with threat-evoked anxiety and avoidance-oriented, withdrawal behaviors (e.g., Dalton, Kalin, Grist, & Davidson, 2005; Shackman, McMenamin, Maxwell, Greischar, & Davidson, 2009). It was expected that joint anodal stimulation to the left dlPFC and cathodal stimulation to the right dlPFC would result in the largest reduction in negative emotional reactions. Indeed, Brunoni et al. (2013) confirmed this by comparing all three cases (anodal left dlPFC, cathodal right dlPFC; anodal right dlPFC, cathodal left dlPFC; sham), showing that the first of these three montages produced the greatest decrease in salivary cortisol concentrations.

tDCS reduced reaction times (RT) in the arithmetic decision task for high mathematics anxiety participants, compared to sham stimulation. The researchers further examined the effect of the identical stimulation and task on those with low mathematics anxiety, finding the *opposite* pattern: In comparison to sham stimulation, the reactions times during real stimulation were slower. These findings were paralleled by changes in salivary cortisol. In particular, for the high mathematics anxiety group, salivary cortisol concentrations decreased from pretest to posttest during real tDCS, but not during sham stimulation. For the low mathematics anxiety group, salivary cortisol concentrations decreased from pretest to posttest during sham tDCS, but not during real stimulation (see Figure 12). While the neural underpinnings of this double-dissociation were beyond the scope of the study, the authors speculated that differences in traits (in this case, mathematics anxiety) produce neural state differences that interact with stimulation to influence the outcome. In the case of mathematics anxiety, it is quite reasonable that high and low mathematics anxiety participants would experience the task with different levels and types of neuronal activation, as already shown in previous work (e.g., Lyons & Beilock, 2012).

These results have two important implications: (1) Taking mathematics anxiety as a trait, tDCS effects may be trait-dependent; and (2) tDCS may be able to alleviate the stress associated with mathematics anxiety, thereby improving mathematical performance. Its potential in enhancing performance for individuals with high mathematics anxiety should be investigated with the possibility of developing interventions in mind. The arithmetic decision task Sarkar et al. (2014) employed involved stating whether very basic arithmetic equations were

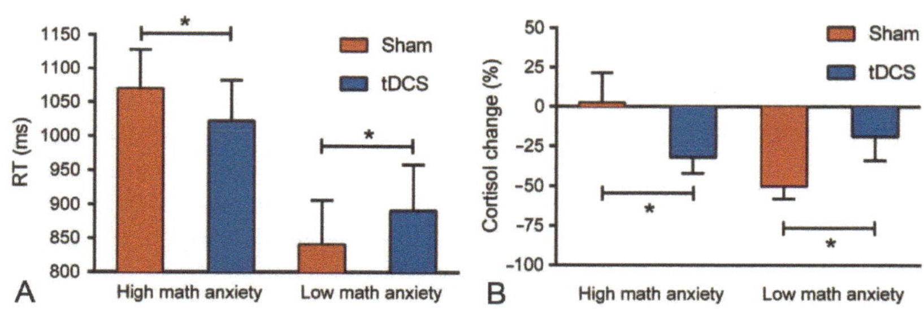

FIGURE 12 Mathematics anxiety-dependent effects of tDCS. (A) Behavioral double-dissociation between high and low mathematics anxiety groups. For the high mathematics-anxiety group ($n=25$), tDCS produced faster arithmetic decisions compared to sham stimulation. For the low mathematics-anxiety group ($n=20$), tDCS impaired arithmetic decision RTs compared to sham (both $ps<.001$). (B) Double-dissociation in salivary cortisol concentration between high and low mathematics anxiety. Negative cortisol change indicates lower cortisol concentrations at posttest compared to pretest. High mathematics-anxiety participants showed greater cortisol reductions after real stimulation, while low mathematics-anxiety participants showed greater cortisol reductions after sham stimulation (both $ps<.05$). Error bars represent 1 SEM. *Figure and legend taken from Sarkar et al. (2014).*

correct or incorrect (e.g., 8×2=16, true or false). For most adults, this type of task does not involve making true calculations as much as it involves verifying it against numerical information that is already known. Despite the apparent simplicity of the task, there was a clear difference between high and low mathematics anxiety groups: While both groups performed the task equally accurately in both real and sham stimulation, the low mathematics anxiety group performed substantially faster than the high mathematics anxiety group. This is evidence that mathematics anxiety affects performance even on basic tasks that do not require true computation, and that tES may be used to enhance performance in participants with high mathematics anxiety. Once again, the issue of ecological validity is important, especially as the improvement in RT in the high mathematics anxiety group was only 50 ms. While these findings represent an important first step, extension of this method to more difficult mathematical problems is of particular importance in establishing tDCS as an effective intervention for mathematics anxiety.

Dyscalculia

Developmental dyscalculia is characterized by severe difficulties in learning about numbers and arithmetic (Butterworth, Varma, & Laurillard, 2011), and affects approximately 3-7% of the population (Reigosa-Crespo et al., 2012; Shalev, 2007). Although it exerts many of the same consequences as mathematics anxiety (e.g., poor performance in school mathematics, difficulties in dealing with numerical concepts, difficulties in everyday numerical situations), it appears to be a specific developmental disorder, the numerical analogue of dyslexia, which mathematics anxiety is not. The causes of developmental dyscalculia are unknown, though some research has linked neuroanatomical and neurophysiological deficiencies in the PPC (Butterworth et al., 2011).

In a recent study, Iuculano and Cohen Kadosh (2014) attempted to mitigate some of these deficits in individuals with developmental dyscalculia by applying tDCS to the PPC. The right PPC has been more closely linked with the development of numerical cognition in infancy and early childhood (Ansari, Garcia, Lucas, Hamon, & Dhital, 2005; Cantlon et al., 2006; Hyde, Boas, Blair, & Carey, 2010). A right-to-left shift in PPC activity over the developmental course has also been described in recent work, which has been associated with the mastery of more sophisticated numerical concepts through enculturation, including magnitude comparisons using numerical symbols (Ansari, 2008; Ansari & Dhital, 2006; Ansari et al., 2005).

Two adult participants with developmental dyscalculia, DD1 and DD2, were given 20 min of bilateral tDCS to the PPC during training at 1 mA. The participants were randomly assigned to two stimulation conditions, where one participant was given anodal stimulation to the left PPC and cathodal stimulation to the right PPC, and the other received anodal and cathodal stimulation to the right PPC and left PPC, respectively. The study comprised six sessions and resembled Cohen Kadosh et al. (2010), described earlier. Participants were trained on the Gibson number task used by Cohen Kadosh et al. (2010). The data revealed that, for the participant who received anodal stimulation to the left PPC and cathodal stimulation to the right PPC, there was clear emergence of the interference effect that characterizes a basic aspect of numerical competence. When modeling performance on the number-space mapping task, the data were also best described by a linear function, which, as discussed earlier, has been identified as another behavioral signature of intact numerical competence (as seen in Figures 6 and 7). For the other participant, who was given the reverse stimulation montage, there was no statistically significant difference in performance between congruent and incongruent trials on the modified Stroop task, and performance on the number-space task was more or less random (see Figures 13–15). Therefore, for this participant, the montage appears to have had no effect, which also highlights the need to select lateralized subregions of the cortex with care. Taken together, this pattern of results is precisely the opposite of what was found with healthy adults in Cohen Kadosh et al. (2010). Such results hint at potential differences in cortical organization that might characterize typical and atypical numerical ability, leading to opposing results. They may also hint at interference by the right-PPC with engagement of the left-PPC during the mapping of number symbols to representations of associated quantities.

With just two participants, the findings can only be regarded as preliminary. However, the results are also promising. This research appears to be the first attempt to remedy an important cognitive characteristic of developmental dyscalculia with a noninvasive, physiologically relevant intervention. The majority of studies in this area have focused on behavioral interventions for the developing brain (Dowker, 2009). Though further investigations are necessary, the work

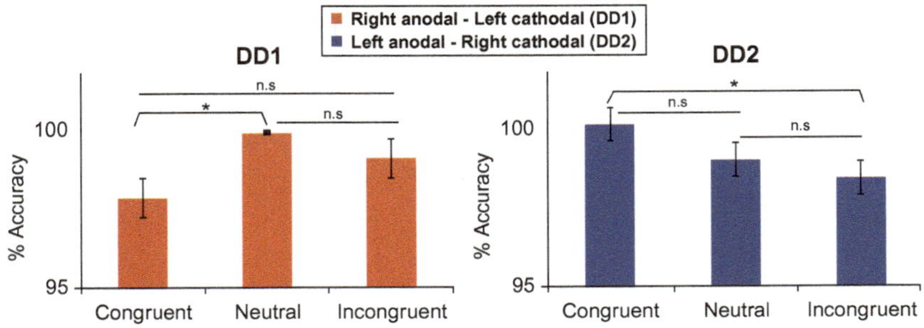

FIGURE 13 Numerical Stroop task. Congruency effect (measured in terms of accuracy) for the two DD individuals. DD1 did not exhibit the canonical Congruency effect (Congruent > Neutral > Incongruent), while DD2 showed a clear Congruency pattern in the predicted direction. For DD1: Neutral > Congruent ($p < 0.05$); Congruent vs. Incongruent ($p = 0.13$); Incongruent vs. Neutral ($p = 0.08$). For DD2: Congruent > Incongruent ($p < 0.05$); Congruent vs. Neutral ($p = 0.07$); Incongruent vs. Neutral ($p = 0.16$). Data are mean ± standard error (SE) of the mean. $^*p < 0.05$. *Figure and legend taken from Iuculano and Cohen Kadosh (2014).*

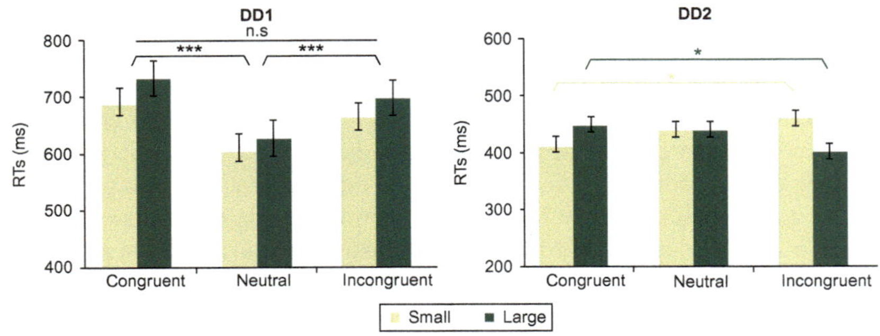

FIGURE 14 Numerical distance by congruency effects. Effects measured in terms of RTs for each DD individual. DD1 did not exhibit the canonical Congruency effect (Incongruent > Neutral > Congruent), while DD2 showed a Congruency pattern related to the numerical distance between stimuli. For DD1 both Congruent as well as Incongruent trials were slower than Neutral trials ($p<0.001$) and no effect of numerical distance was evident ($F<1$). In DD2, the canonical pattern typical of the Congruency effect (Incongruent > Congruent) was only present for small distances (e.g., 2-4) ($p<0.05$); while the reverse pattern (Congruent > Incongruent) characterized DD2's performance with large numerical distances (e.g., 2-7) ($p<0.05$). Main effects are shown in black (DD1's profile). Interaction is shown in shades of green (DD2's profile). Data are mean ± standard error (SE) of the mean. *$p<0.05$; ***$p<0.001$. *Figure and legend taken from Iuculano and Cohen Kadosh (2014).*

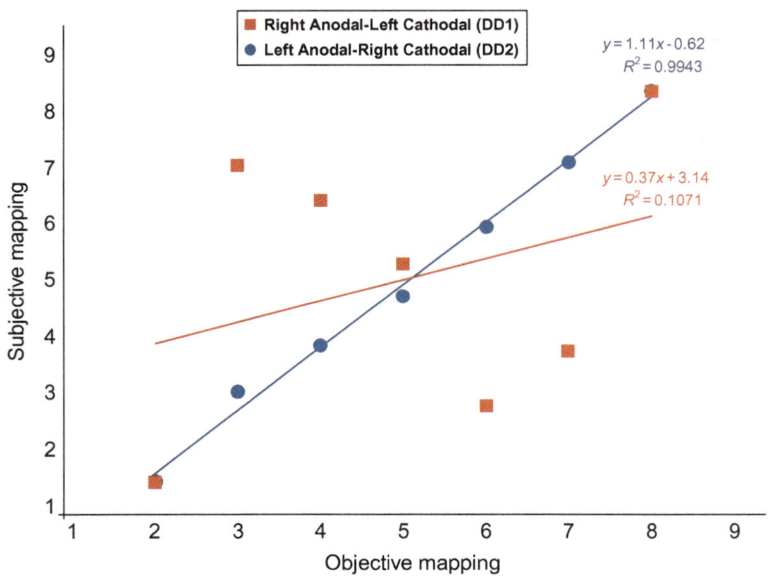

FIGURE 15 Average location of subjective responses on the number line task plotted for each type of stimulation. Linear regression lines and equations are indicated for each type of stimulation (red line—right anodal-left cathodal stimulation received by DD1; blue line—left anodal-right cathodal stimulation received by DD2). *Figure and legend taken from Iuculano and Cohen Kadosh (2014).*

described in Iuculano and Cohen Kadosh (2014) represents a promising step toward establishing a new intervention for adults.

THE TO-DO LIST

tES has made large strides in the direction of cognitive enhancement, especially in the last decade. However, it must progress even further if it is to live up to the potential ascribed to it. Whether it will be able to fulfill the promise it has shown will depend on the accumulation of research findings in the coming years. There are several important directions that research efforts should focus on in order to validate tES as a means for cognitive enhancement, four of which are discussed below.

Cognitive Cost

Economists state there are no free lunches. That is, there are always trade-offs between costs and benefits. The search for cognitive enhancement has, to a large extent, ignored the possibility that enhancement in one cognitive domain may compromise performance in another. Given what is known about tES and changes in regional activation, metabolic activity and resource consumption are likely to change during stimulation, especially since the brain receives a relatively fixed supply of nutrients and oxygen. Increasing or decreasing excitation in one region may, therefore, affect other brain regions and associated cognitive functions (Brem, Fried, Horvath, Robertson, & Pascual-Leone, 2014; Krause et al., 2013; Pascual-Leone, Horvath, & Robertson, 2012). Essentially, this cognitive cost offsets the enhancement in another area. Presently, the evidence for such a view is sparse, owing to the lack of research on this topic. However, a few recent findings have provided some initial support for this perspective (Iuculano & Cohen Kadosh, 2013; Sarkar et al., 2014).

Iuculano and Cohen Kadosh (2013) targeted the dorsolateral prefrontal cortex with tDCS, using a left anodal-right cathodal montage, and the same montage to the parietal lobe using the Gibson number-learning training paradigm described earlier in Cohen Kadosh et al. (2010) and Iuculano and Cohen Kadosh (2014) (see Figure 4). Participants determined which of the symbols represented a larger magnitude, and also performed the modified Stroop task of making relative quantity judgments between two symbols while ignoring the magnitudes they represented. Performance on the first task reflects the actual learning and recognition of the symbol. Performance on the second task represents the extent to which this system of numbers has become automatically accessible, and therefore produces Stroop-like interference. Compared to the participants who were given sham stimulation, those who received left anodal-right cathodal stimulation to the parietal cortex displayed impaired automaticity on the modified Gibson number Stroop task, but at the same time, displayed an increased learning rate compared to sham. In contrast, participants given real stimulation to the dorsolateral prefrontal cortex displayed a higher level of automaticity, but a lower learning rate compared to the sham group (Iuculano & Cohen Kadosh, 2013). These findings reveal a double-dissociation between two key aspects of learning: the memorization and recognition of stimuli, and the automaticity with which knowledge of those stimuli is accessed (see Figures 16 and 17). Therefore, there were costs in both cases: enhanced learning and recognition came at the cost of impaired automaticity, and enhanced automaticity came at the cost of impaired learning and recognition.

Although the Sarkar et al. (2014) study was not designed explicitly to test the question of cognitive costs, they did examine this possibility. After completing the arithmetic decision task, participants completed the attentional networks task (ANT) (Fan, McCandliss, Sommer, Raz, & Posner, 2002). This permitted an investigation of the effects of tDCS on other cognitive functions that are subserved by the dlPFC. The ANT examines three aspects of attention: alerting (reaching and sustaining a state of alertness), orienting (choosing stimuli from an array of perceptual information), and executive control (deciding between conflicting responses). For both high and low mathematics anxiety participants, scores on executive control, which recruits the dlPFC, were lower after real compared to sham stimulation. For the low mathematics anxiety participants, whose performance on the arithmetic decision task was better in sham compared to real stimulation, lowered executive control in real compared to sham stimulation suggests that there were no gains in *any* of the variables examined. This finding is particularly important, as it suggests that there are traits and situations that may interact with stimulation to produce only costs with no benefits. One theory, the zero-sum effect of brain stimulation (Brem et al., 2014), predicts that enhancements and costs will offset one another. Given fixed neuronal processing capacity, the enhancement in one domain is ultimately offset by impairment in another, leading to no overall change in the system. As of yet it is difficult to predict which abilities and domains may be impaired by brain stimulation (as costs predicted by a zero-sum effect), and which ones may be enhanced, as by transfer (e.g., Cappelletti et al., 2013). This will become clearer as greater understanding is developed about the independence and interdependence of various cognitive abilities.

Transfer Effects

Transfer is a rarely glimpsed and much sought-after bird in the cognitive science forest. At present, experimental explorations of tES-induced transfer effects have yielded only mixed findings. Part of the reason, as discussed earlier, may be the inherent difficulty in defining what would count as transfer, and in selecting the appropriate tasks. A related difficulty is that of identifying which brain regions would facilitate transfer effects through training, and the determination of which cognitive functions activate which brain regions. This is a general problem in transfer studies (Cohen Kadosh, 2013). Clearly transfer is possible, as demonstrated by some of studies discussed in this chapter. Indeed, it seems

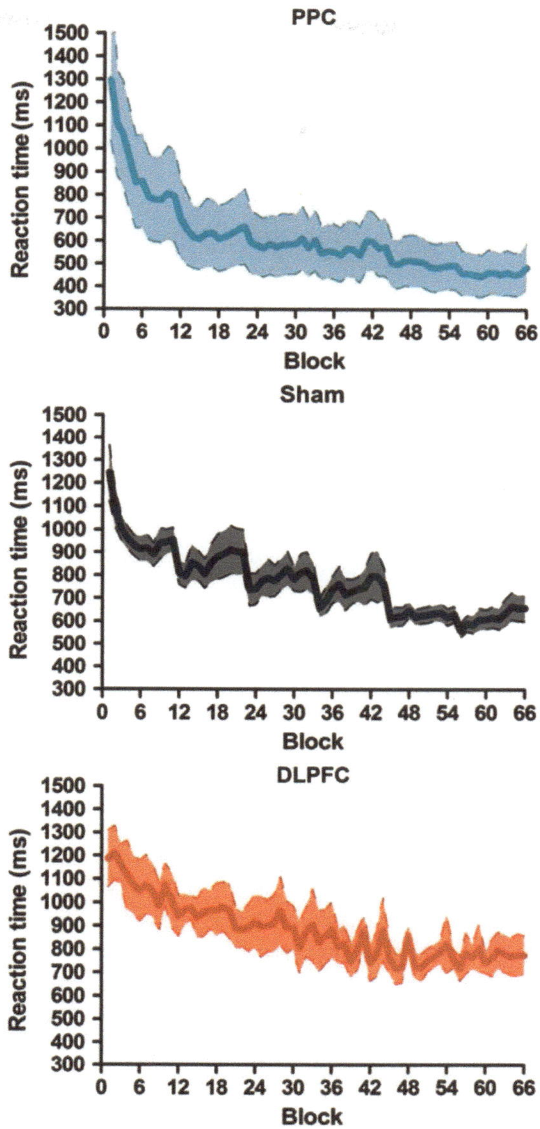

FIGURE 16 Learning functions for the three groups. Top, PPC group. Middle, Sham group. Bottom, DLPFC group. The slope for the PPC group (top) is steeper than the other two (middle, bottom), indicating faster learning. Analysis of the learning rate gave statistical support to this result by showing a significant difference between the learning rate for each group that was best explained by a linear trend: PPC > sham > DLPFC ($p < 0.006$); 92% of the variance explained. *Figure and legend taken from Iuculano and Cohen Kadosh (2013).*

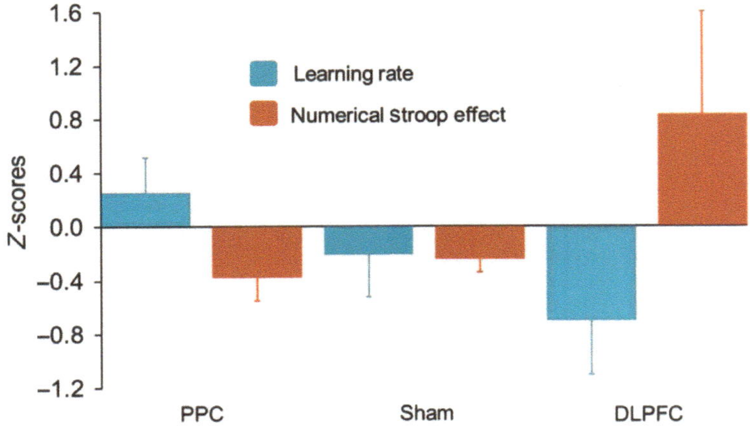

FIGURE 17 Double dissociation between learning rate and automaticity as a function of brain stimulation. Although the speed of learning has been the fastest for PPC stimulation and slowest for the DLPFC stimulation, the automatic processing of the learned material showed exactly the opposite pattern. A linear trend analysis (group [PPC < sham < DLPFC], and task [learning rate ≠ numerical Stroop effect, $RT_{incongruent} - RT_{congruent}$]) explained 99% of the variance ($p < 0.006$). Data are mean ± SEM. *Figure and legend taken from Iuculano and Cohen Kadosh (2013).*

that tES combined with training increases the likelihood of successful transfer beyond that of training by itself (Cappelletti et al., 2013; Looi et al., submitted; Snowball et al., 2013). Numerical cognition in particular is a valuable model to study transfer, given its very close relations to other executive functions and other forms of quantity processing.

Individual Differences

A prominent and unresolved question that is only just gaining currency, and which has methodological and translational implications for tES, is that of individual differences. Namely, it remains unclear how tES itself is modulated by individual psychological and physiological differences. The vast majority of studies in this field use a one-size-fits-all approach that is characteristic of research at early stages. However, participants differ along a number of variables that bear clear implications on the outcomes of brain stimulation. For example, differences in gray matter, neurochemicals such as GABA or glutamate, optimality of brain function and neural oscillations, and developmental period.

Consider Sarkar et al. (2014), where benefits of stimulation were restricted to the high mathematics-anxiety participants, while low mathematics-anxiety participants did not show any gains from stimulation, performing slower on the arithmetic decision task. For Tseng et al. (2012), participants with lower visuospatial short-term memory improved, while those with higher capacity did not. Similar results were also observed by Furuya et al. (2014) in skilled and unskilled musicians, with unskilled musicians improving substantially, but skilled musicians not demonstrating any benefits. These studies point to an intuitively appealing ceiling effect, with individuals of lower ability having greater room to improve. However, this would be too simplistic an explanation, and is at best only a partial explanation for the individual difference-driven variance in the data. For example, Berryhill and Jones (2012) found that elderly participants with *greater* educational attainment benefited more from tDCS in terms of working memory performance compared to those with lower levels of education. Therefore, at least in some cases, the conventional ceiling is either absent, or of a different form. What is abundantly clear is that individual differences play a substantial role in how, and if, brain stimulation affects performance.

Other individual differences abound. For example, Krause and Cohen Kadosh (2014) also predict that differences in sex-hormone concentrations in female participants (progesterone and estradiol) will affect cortical excitability, thereby producing differential responses to brain stimulation. Yet another source of influence is genetic variance. For example, polymorphisms in the catechol-*O*-methyltransferase (COMT) gene influence how executive functions respond to tDCS, with some COMT variants associated with greater individual sensitivity to stimulation effects (Plewnia et al., 2013). A less obvious but equally important difference is the participant's cultural background. For example, Tang et al. (2006) showed that the brain regions active during arithmetic processing in native English and Chinese speakers share some key neural circuitry, but more strikingly, reveal important neural differences as well. Therefore, the isolation of cortical regions "responsible" for particular functions, already a vexed question, is further vexed by the influence of culture. But taking this into account would be extremely important in studies of tES and numerical cognition.

Aside from broader generalizability, individual differences play an important role within the context of tES research in the laboratory as well. Discrepancies between published studies and replicability problems may be due to issues of sample size, as frequently mentioned in discussion sections of published research. However, another source of this variance may be individual differences. Different research groups recruit participants, often university students, as is highly prevalent across psychological science. Just a few differences these participants represent are varying socioeconomic and cultural backgrounds, not to mention different ability levels on the processes the experiment is designed to affect. Any or all of these differences could be influencing stimulation outcomes. One individual difference many studies do take into account is handedness. However, this often amounts to excluding left-handed participants because of hypothesized differences in brain organization, rather than exploring how handedness and the associated differences in the brain may play a role in the hypothesis.

Consider another example: All of the studies described in detail here are studies of tES and numerical cognition. However, aside from Sarkar et al. (2014), who explicitly tested mathematics anxiety as an individual difference, none of the other reports screened for this construct. Considering the high prevalence of mathematics anxiety in the general population, differing levels of mathematics anxiety among participants across studies may be playing an invisible but powerful role in the results that may make replication in the future difficult.

The point is not, of course, to suggest that individual differences are a Pandora's box that must not be opened. Rather, these constitute a crucial source of variance that we should meaningfully explore. The ultimate aim of tES-induced cognitive enhancement is to design sensitive enhancers and interventions. This sensitivity is logically tied to a consideration of the individual differences that modulate the effects of brain stimulation. The point is also not to suggest that all individual differences can or should be deeply explored. However, in framing hypotheses and designing experiments, there are clearly some individual differences that will play an important role (e.g., mathematics anxiety in studies of numerical cognition). These individual differences, once appropriately accounted for, become part of the signal rather than the noise.

Ecological Validity

The emphasis in many psychological research methods textbooks, psychological research, and the research enterprise in general, is on the internal validity of experiments. However, the entire case of cognitive enhancement is couched in the assumption that it will eventually have utility in everyday life. Therefore, ecological validity is very important. Even with the requisite internal validity that allows for a causal inference, without real-world relevance, the notion of tES-induced cognitive enhancement remains a presumption in need of verification. This is not to say that internal validity can be sacrificed. Indeed, without strong internal validity, the basis for asserting causal statements and ruling out alternative explanations is compromised. Many of the critiques of tES studies revolve around internal validity issues, and several of the studies described in this chapter have taken care to assess and rule out competing explanations. Without sound internal validity, explorations of ecological validity would most definitely be putting the cart before the horse.

However, ensuring high internal validity should not be where the discussion ends. Instead, establishing internal validity must be the first and essential step, and must then be followed by assessing the ecological validity of the findings. The importance of testing ecological validity is frequently raised in research reports, either directly by stating its necessity, or indirectly, by describing the implications and applications of the findings. However, researchers often do not test the findings in the real world. As Pascual-Leone et al. (2012) point out, most studies involving tES report improvements in reaction times in the range of tens of milliseconds, changes that are large and interesting in the context of a laboratory environment, but which are difficult to justify as true enhancement in terms of how an individual has actually benefited from the stimulation. For example, the improvement of real over sham stimulation on arithmetic decision performance for the high mathematics anxiety participants noted by Sarkar et al. (2014) was about 50 ms. Furthermore, there were no changes in accuracy on the tasks between real and sham stimulation (though this may be attributable to a ceiling effect, with the arithmetic task being too easy). Though the relative decrease in cortisol concentrations in real compared to sham stimulation does suggest that participants were less anxious, a behavioral improvement of 50 ms yields little, if any, discernible significance in everyday life. Therefore, the next logical next step is to assess the effect of tDCS on more difficult mathematics problems for participants with high mathematics anxiety, and whether potential improvements are linked to decreases in markers of stress or anxiety (e.g., salivary cortisol). Such steps will meaningfully advance the goal of establishing greater ecological validity for tDCS as a tool to address the emotional and performance consequences of mathematics anxiety.

In general, it seems the internal validity of tES experiments will continue to increase. It is more difficult to assess when ecological validity will come to the fore. Part of the reason for this, of course, is the immense difficulty in establishing *what* would constitute a suitable test of ecological validity. For example, the very innovative and interesting work of Dockery et al. (2009) demonstrates that a short training program on a Tower of London task (a well-validated measure of planning ability and a marker for successful skill acquisition; Peretti, Danion, Gierski, & Grangé, 2002; Shallice, 1982) during the application of tDCS to the dlPFC enhanced performance on both reaction time and accuracy in the task, even 12 months after the original stimulation sessions. These results, once replicated and extended, have immense potential to improve the lives of both patient and healthy populations, given the importance of planning in daily life. However, it is an open and challenging question of how to assess enhanced planning outside the laboratory. In the host of relevant planning activities (e.g., schedules, goals, structuring schedules to fit goals, changing schedules and goals, and so on), how can some of these be operationalized appropriately into response variables? An intriguing possibility is the use of virtual reality, which was combined with *f*MRI to study the neural bases of planning behavior in an ecologically valid context (Campbell et al., 2009). Perhaps the addition of tES to such a paradigm would enable studies of more complex cognitive functions? Questions such as these must form an integral part of the research agenda when assessing the potential of tES for a particular type of cognitive enhancement.

CONCLUSIONS

In spite of substantial variability in the research findings, there is clear evidence that tES is able to enhance cognition in laboratory settings, including some aspects of numerical cognition. This is particularly well established for numerical cognition in adults with intact numerical abilities, and there is emerging evidence that it may prove effective for individuals with numerical deficits as well. At the same time, there are several challenges facing the brain stimulation community, which center on the internal validity and ecological validity of the findings. Internal validity measures can be implemented with relative ease compared to those for ecological validity, of which there are still no examples in the tES-induced cognitive enhancement research. With advances in technology, and with combinations of different technologies (e.g., smaller and more focal electrodes, combining tES with *f*MRI, EEG, and so on), as well as stricter criteria for assessing findings (e.g., accounting for the placebo effect, ensuring regional and task specificity, proper blinding), internal validity across the tES experimental literature is sure to continue increasing.

This is less clear for ecological validity. For example, in the case of numerical cognition and mathematical achievement, there is now sufficient evidence to conclude that tES can enhance numerical cognition in healthy individuals. This has been demonstrated with several processes, including numerosity, learning of artificial numbers, perceiving magnitudes, and in more advanced types of arithmetic operations. Furthermore, some of these effects are apparent up to even 6 months after the conclusion of training. For individuals with deficits in numerical cognition, such as mathematics anxiety or developmental dyscalculia, tES may in the future offer potential means to improve existing interventions, as well as the development of new types of interventions. Taken together, these results suggest that tES has a promising role in this area. However, translating that promise is the challenge. There is still no evidence to suggest that it can enhance mathematical achievement and performance in the classroom or the workplace, and *this*, therefore, must become the next step in bridging research with the world outside the laboratory.

REFERENCES

Ambrus, G. G., Paulus, W., & Antal, A. (2010). Cutaneous perception thresholds of electrical stimulation methods: Comparison of tDCS and tRNS. *Clinical Neurophysiology*, *121*, 1908–1914.

Anguera, J. A., Boccanfuso, J., Rintoul, J. L., Al-Hashimi, O., Faraji, F., Janowich, J., et al. (2013). Video game training enhances cognitive control in older adults. *Nature*, *501*, 97–101.

Ansari, D. (2008). Effects of development and enculturation on number representation in the brain. *Nature Reviews Neuroscience*, *9*, 278–291.

Ansari, D., & Dhital, B. (2006). Age-related changes in the activation of the intraparietal sulcus during nonsymbolic magnitude processing: An event-related functional magnetic resonance imaging study. *Journal of Cognitive Neuroscience*, *18*, 1820–1828.

Ansari, D., Garcia, N., Lucas, E., Hamon, K., & Dhital, B. (2005). Neural correlates of symbolic number processing in children and adults. *Neuroreport*, *16*, 1769–1773.

Antal, A., Boros, K., Poreisz, C., Chaieb, L., Terney, D., & Paulus, W. (2008). Comparatively weak after-effects of transcranial alternating current stimulation (tACS) on cortical excitability in humans. *Brain Stimulation*, *1*, 97–105.

Antal, A., Kovács, G., Chaieb, L., Cziraki, C., Paulus, W., & Greenlee, M. W. (2012). Cathodal stimulation of human MT+ leads to elevated fMRI signal: A tDCS-fMRI study. *Restorative Neurology and Neuroscience*, *30*(3), 255.

Antal, A., Nitsche, M. A., Kruse, W., Kincses, T. Z., Hoffmann, K.-P., & Paulus, W. (2004). Direct current stimulation over V5 enhances visuomotor coordination by improving motion perception in humans. *Journal of Cognitive Neuroscience*, *16*, 521–527.

Antal, A., & Paulus, W. (2013). Transcranial alternating current stimulation (tACS). *Frontiers in Human Neuroscience*, *7*, 317. http://dx.doi.org/10.3389/fnhum.2013.00317.

Arsalidou, M., & Taylor, M. J. (2011). Is 2 + 2 = 4? Meta-analyses of brain areas needed for numbers and calculations. *NeuroImage*, *54*, 2382–2393.

Ashcraft, M. H., & Kirk, E. P. (2001). The relationships among working memory, math anxiety, and performance. *Journal of Experimental Psychology: General*, *130*, 224–237.

Bailey, D. H., Hoard, M. K., Nugent, L., & Geary, D. C. (2012). Competence with fractions predicts gains in mathematics achievement. *Journal of Experimental Child Psychology*, *113*, 447–455.

Ball, K., Berch, D. B., Helmers, K. F., Jobe, J. B., Leveck, M. D., Marsiske, M., et al. (2002). Effects of cognitive training interventions with older adults: A randomized controlled trial. *Journal of the American Medical Association*, *288*, 2271–2281.

Barth, H., Kanwisher, N., & Spelke, E. (2003). The construction of large number representations in adults. *Cognition*, *86*, 201–221.

Battleday, R. M., Muller, T., Clayton, M. S., & Cohen Kadosh, R. (2014). Mapping the mechanisms of transcranial alternating current stimulation: A pathway from network effects to cognition. *Frontiers in Psychiatry*, *5*, 162. http://dx.doi.org/10.3389/fpsyt.2014.00162.

Beddington, J., Cooper, C. L., Field, J., Goswami, U., Huppert, F. A., Jenkins, R., et al. (2008). The mental wealth of nations. *Nature*, *455*, 1057–1060.

Berryhill, M. E., & Jones, K. T. (2012). tDCS selectively improves working memory in older adults with more education. *Neuroscience Letters*, *521*, 148–151.

Berteletti, I., Prado, J., & Booth, J. R. (2014). Children with mathematical learning disability fail in recruiting verbal and numerical brain regions when solving simple multiplication problems. *Cortex*, *57*, 143–155.

Bindman, J. L., Lippold, O. C. J., & Redfearm, J. W. T. (1964). The action of brief polarizing currents on the cerebral cortex of the rat (1) during current flow and (2) in the production of long-lasting after-effects. *Journal of Physiology*, *172*, 369–382.

Bolwig, T. (2011). How does electroconvulsive therapy work? Theories on its mechanism. *The Canadian Journal of Psychiatry*, *51*, 13–18.

Bolzoni, F., Bączyk, M., & Jankowska, E. (2013). Subcortical effects of transcranial direct current stimulation in the rat. *The Journal of Physiology*, *591*, 4027–4042.

Booth, J. L., & Siegler, R. S. (2008). Numerical magnitude representations influence arithmetic learning. *Child Development*, *79*, 1016–1031.

Borckardt, J. J., Bikson, M., Frohman, H., Reeves, S. T., Datta, A., Bansal, V., et al. (2012). A pilot study of the tolerability and effects of high-definition transcranial direct current stimulation (HD-tDCS) on pain perception. *The Journal of Pain*, *13*, 112–120.

Boyke, J., Driemeyer, J., Gaser, C., Büchel, C., & May, A. (2008). Training-induced brain structure changes in the elderly. *The Journal of Neuroscience*, *28*, 7031–7035.

Brannon, E. M., & Terrace, H. S. (1998). Ordering of the numerosities 1 to 9 by monkeys. *Science*, *282*, 746–749.

Brem, A. K., Fried, P. J., Horvath, J. C., Robertson, E. M., & Pascual-Leone, A. (2014). Is neuroenhancement by noninvasive brain stimulation a net zero-sum proposition? *NeuroImage*, *85*, 1058–1068.

Brunoni, A. R., Vanderhasselt, M. A., Boggio, P. S., Fregni, F., Dantas, E. M., Mill, J. G., et al. (2013). Polarity-and valence-dependent effects of prefrontal transcranial direct current stimulation on heart rate variability and salivary cortisol. *Psychoneuroendocrinology*, *38*, 58–66.

Bueti, D., & Walsh, V. (2009). The parietal cortex and the representation of time, space, number and other magnitudes. *Philosophical Transactions of the Royal Society, B: Biological Sciences*, *364*, 2369–2380.

Butterworth, B. (2010). Foundational numerical capacities and the origins of dyscalculia. *Trends in Cognitive Sciences, 14*, 534–541.

Butterworth, B., Varma, S., & Laurillard, D. (2011). Dyscalculia: From brain to education. *Science, 332*, 1049–1053.

Bynner, J., & Parsons, S. (1997). *Does numeracy matter?* London: The Basic Skills Agency.

Campbell, Z., Zakzanis, K. K., Jovanovski, D., Joordens, S., Mraz, R., & Graham, S. J. (2009). Utilizing virtual reality to improve the ecological validity of clinical neuropsychology: An FMRI case study elucidating the neural basis of planning by comparing the Tower of London with a three-dimensional navigation task. *Applied Neuropsychology, 16*, 295–306.

Cantlon, J. F. (2012). Math, monkeys, and the developing brain. *Proceedings of the National Academy of Sciences of the United States of America, 109*, 10725–10732.

Cantlon, J. F., Brannon, E. M., Carter, E. J., & Pelphrey, K. A. (2006). Functional imaging of numerical processing in adults and 4-y-old children. *PLoS Biology, 4*, e125. http://dx.doi.org/10.1371/journal.pbio.0040125.

Cantlon, J. F., Platt, M. L., & Brannon, E. M. (2009). Beyond the number domain. *Trends in Cognitive Sciences, 13*, 83–91.

Cappelletti, M., Barth, H., Fregni, F., Spelke, E. S., & Pascual-Leone, A. (2007). rTMS over the intraparietal sulcus disrupts numerosity processing. *Experimental Brain Research, 179*, 631–642.

Cappelletti, M., Gessaroli, E., Hithersay, R., Mitolo, M., Didino, D., Kanai, R., et al. (2013). Transfer of cognitive training across magnitude dimensions achieved with concurrent brain stimulation of the parietal lobe. *The Journal of Neuroscience, 33*, 14899–14907.

Chaieb, L., Kovacs, G., Cziraki, C., Greenlee, M., Paulus, W., & Antal, A. (2009). Short-duration transcranial random noise stimulation induces blood oxygenation level dependent response attenuation in the human motor cortex. *Experimental Brain Research, 198*, 439–444.

Chaieb, L., Paulus, W., & Antal, A. (2011). Evaluating aftereffects of short-duration transcranial random noise stimulation on cortical excitability. *Neural Plasticity, 2011*. http://dx.doi.org/10.1155/2011/105927.

Chick, C. F. (2014). Basic mechanisms of numerical processing: Cross-modal number comparisons and symbolic versus nonsymbolic numerosity in the intraparietal sulcus. *The Journal of Neuroscience, 34*, 1567–1569.

Clark, V. P., Coffman, B. A., Trumbo, M. C., & Gasparovic, C. (2011). Transcranial direct current stimulation (tDCS) produces localized and specific alterations in neurochemistry: A 1H magnetic resonance spectroscopy study. *Neuroscience Letters, 500*, 67–71.

Clemens, B., Jung, S., Zvyagintsev, M., Domahs, F., & Willmes, K. (2013). Modulating arithmetic fact retrieval: A single-blind, sham-controlled tDCS study with repeated fMRI measurements. *Neuropsychologia, 51*, 1279–1286.

Cohen Kadosh, R. (2013). Using transcranial electrical stimulation to enhance cognitive functions in the typical and atypical brain. *Translational Neuroscience, 4*, 20–33.

Cohen Kadosh, R., Cohen Kadosh, K., Schuhmann, T., Kaas, A., Goebel, R., Henik, A., et al. (2007). Virtual dyscalculia induced by parietal-lobe TMS impairs automatic magnitude processing. *Current Biology, 17*(8), 689–693.

Cohen Kadosh, R., Lammertyn, J., & Izard, V. (2008). Are numbers special? An overview of chronometric, neuroimaging, developmental and comparative studies of magnitude representation. *Progress in Neurobiology, 84*, 132–147.

Cohen Kadosh, R., Levy, N., O'Shea, J., Shea, N., & Savulescu, J. (2012). The neuroethics of non-invasive brain stimulation. *Current Biology, 22*, R108–R111.

Cohen Kadosh, R., Soskic, S., Iuculano, T., Kanai, R., & Walsh, V. (2010). Modulating neuronal activity produces specific and long lasting changes in numerical competence. *Current Biology, 20*, 2016–2020.

Cordes, S., Gelman, R., Gallistel, C. R., & Whalen, J. (2001). Variability signatures distinguish verbal from nonverbal counting for both large and small numbers. *Psychonomic Bulletin & Review, 8*, 698–707.

Craik, F. I., & Lockhart, R. S. (1972). Levels of processing: A framework for memory research. *Journal of Verbal Learning and Verbal Behavior, 11*, 671–684.

Cui, X., Bray, S., Bryant, D. M., Glover, G. H., & Reiss, A. L. (2011). A quantitative comparison of NIRS and fMRI across multiple cognitive tasks. *NeuroImage, 54*, 2808–2821.

Dalton, K. M., Kalin, N. H., Grist, T. M., & Davidson, R. J. (2005). Neural-cardiac coupling in threat-evoked anxiety. *Journal of Cognitive Neuroscience, 17*(6), 969–980.

Dehaene, S., Dehaene-Lambertz, G., & Cohen, L. (1998). Abstract representations of numbers in the animal and human brain. *Trends in Neurosciences, 21*, 355–361.

Dehaene, S., Izard, V., Spelke, E., & Pica, P. (2008). Log or linear? Distinct intuitions of the number scale in Western and Amazonian indigene cultures. *Science, 320*, 1217–1220.

Dehaene, S., Piazza, M., Pinel, P., & Cohen, L. (2003). Three parietal circuits for number processing. *Cognitive Neuropsychology, 20*, 487–506.

Delazer, M., Ischebeck, A., Domahs, F., Zamarian, L., Koppelstaetter, F., Siedentopf, C. M., et al. (2005). Learning by strategies and learning by drill—Evidence from an fMRI study. *NeuroImage, 25*, 838–849.

Ditye, T., Jacobson, L., Walsh, V., & Lavidor, M. (2012). Modulating behavioral inhibition by tDCS combined with cognitive training. *Experimental Brain Research, 219*, 363–368.

Dockery, C. A., Hueckel-Weng, R., Birbaumer, N., & Plewnia, C. (2009). Enhancement of planning ability by transcranial direct current stimulation. *The Journal of Neuroscience, 29*, 7271–7277.

Dowker, A. (2009). What works for children with mathematical difficulties? *The effectiveness of intervention schemes*. http://dera.ioe.ac.uk/2505/1/ma_difficulties_0008609.pdf.

Draganski, B., Gaser, C., Busch, V., Schuierer, G., Bogdahn, U., & May, A. (2004). Neuroplasticity: Changes in grey matter induced by training. *Nature, 427*, 311–312.

Dubljević, V., Saigle, V., & Racine, E. (2014). The rising tide of tDCS in the media and academic literature. *Neuron, 82*, 731–736.

Duncan, G. J., Dowsett, C. J., Claessens, A., Magnuson, K., Huston, A. C., Klebanov, P., et al. (2007). School readiness and later achievement. *Developmental Psychology, 43*, 1428–1446.

Elliott, P. (2014). Electricity and the brain: An historical evaluation. In R. Cohen Kadosh (Ed.), *The stimulated brain: Cognitive enhancement using non-invasive brain stimulation*. Amsterdam: Academic Press.

Fan, J., McCandliss, B. D., Sommer, T., Raz, A., & Posner, M. I. (2002). Testing the efficiency and independence of attentional networks. *Journal of Cognitive Neuroscience, 14*, 340–347.

Fertonani, A., Pirulli, C., & Miniussi, C. (2011). Random noise stimulation improves neuroplasticity in perceptual learning. *The Journal of Neuroscience, 31*, 15416–15423.

Fink, M. (1984). The origins of convulsive therapy. *American Journal of Psychiatry, 141*, 1034–1041.

Flöel, A., Rösser, N., Michka, O., Knecht, S., & Breitenstein, C. (2008). Noninvasive brain stimulation improves language learning. *Journal of Cognitive Neuroscience, 20*, 1415–1422.

Floyer-Lea, A., Wylezinska, M., Kincses, T., & Matthews, P. M. (2006). Rapid modulation of GABA concentration in human sensorimotor cortex during motor learning. *Journal of Neurophysiology, 95*, 1639–1644.

Fritsch, B., Reis, J., Martinowich, K., Schambra, H. M., Ji, Y., Cohen, L. G., & Lu, B. (2010). Direct current stimulation promotes BDNF-dependent synaptic plasticity: Potential implications for motor learning. *Neuron, 66*(2), 198–204.

Furuya, S., Klaus, M., Nitsche, M. A., Paulus, W., & Altenmüller, E. (2014). Ceiling effects prevent further improvement of transcranial stimulation in skilled musicians. *The Journal of Neuroscience, 34*, 13834–13839.

Gandiga, P., Hummel, F., & Cohen, L. (2006). Transcranial DC stimulation (tDCS): A tool for double-blind sham-controlled clinical studies in brain stimulation. *Clinical Neurophysiology, 117*, 845–850.

Geary, D. C., Berch, D. B., & Mann Koepke, K. (2015). The evolution of number systems. In D. C. Geary, D. B. Berch, & K. Mann Koepke (Eds.), *Evolutionary origins and early development of number processing* (pp. 337–355). San Diego, CA: Academic Press/Elsevier.

Geary, D. C., Hoard, M. K., Nugent, L., & Byrd-Craven, J. (2008). Development of number line representations in children with mathematical learning disability. *Developmental Neuropsychology, 33*, 277–299.

Gebuis, T., Gevers, W., & Cohen Kadosh, R. (2014). Topographic representation of high-level cognition: Numerosity or sensory processing? *Trends in Cognitive Sciences, 18*, 1–3.

Gebuis, T., & Reynvoet, B. (2012). The interplay between nonsymbolic number and its continuous visual properties. *Journal of Experimental Psychology: General, 141*, 642–648.

Gibson, E. P., Gibson, J. J., Pick, A. D., & Osser, H. (1962). A developmental study of the discrimination of letter-like forms. *Journal of Comparative and Physiological Psychology, 55*, 897–906.

Gilula, M. F., & Kirsch, D. L. (2005). Cranial electrotherapy stimulation review: A safer alternative to psychopharmaceuticals in the treatment of depression. *Journal of Neurotherapy, 9*, 7–26.

Girelli, L., Lucangeli, D., & Butterworth, B. (2000). The development of automaticity in accessing number magnitude. *Journal of Experimental Child Psychology, 76*, 104–122.

Gordon, P. (2004). Numerical cognition without words: Evidence from Amazonia. *Science, 306*, 496–499.

Green, C. S., & Bavelier, D. (2003). Action video game modifies visual selective attention. *Nature, 423*, 534–537.

Halberda, J., Mazzocco, M. M., & Feigenson, L. (2008). Individual differences in non-verbal number acuity correlate with maths achievement. *Nature, 455*, 665–668.

Hattori, Y., Moriwaki, A., & Hori, Y. (1990). Biphasic effects of polarizing current on adenosine-sensitive generation of cyclic AMP in rat cerebral cortex. *Neuroscience Letters, 116*, 320–324.

Hauser, T. U., Rotzer, S., Grabner, R. H., Mérillat, S., & Jäncke, L. (2013). Enhancing performance in numerical magnitude processing and mental arithmetic using transcranial direct current stimulation (tDCS). *Frontiers in Human Neuroscience, 7*, 244. http://dx.doi.org/10.3389/fnhum.2013.00244.

Helfrich, R. F., Schneider, T. R., Rach, S., Trautmann-Lengsfeld, S. A., Engel, A. K., & Herrmann, C. S. (2014). Entrainment of brain oscillations by transcranial alternating current stimulation. *Current Biology, 24*, 333–339.

Hellhammer, D. H., Wüst, S., & Kudielka, B. M. (2009). Salivary cortisol as a biomarker in stress research. *Psychoneuroendocrinology, 34*, 163–171.

Hembree, R. (1990). The nature, effects, and relief of mathematics anxiety. *Journal for Research in Mathematics Education, 21*, 33–46.

Henik, A., & Tzelgov, J. (1982). Is three greater than five: The relation between physical and semantic size in comparison tasks. *Memory & Cognition, 10*(4), 389–395.

Herrington, J. D., Mohanty, A., Koven, N. S., Fisher, J. E., Stewart, J. L., Banich, M. T., et al. (2005). Emotion-modulated performance and activity in left dorsolateral prefrontal cortex. *Emotion, 5*(2), 200.

Herrmann, C. S., Rach, S., Neuling, T., & Strüber, D. (2013). Transcranial alternating current stimulation: A review of the underlying mechanisms and modulation of cognitive processes. *Frontiers in Human Neuroscience, 7*, 279. http://dx.doi.org/10.3389/fnhum.2013.00279.

Holland, R., Leff, A. P., Josephs, O., Galea, J. M., Desikan, M., Price, C. J., et al. (2011). Speech facilitation by left inferior frontal cortex stimulation. *Current Biology*, *21*, 1403–1407.

Holmes, J., Gathercole, S. E., & Dunning, D. L. (2009). Adaptive training leads to sustained enhancement of poor working memory in children. *Developmental Science*, *12*, F9–F15.

Hubbard, E. M., Piazza, M., Pinel, P., & Dehaene, S. (2005). Interactions between number and space in parietal cortex. *Nature Reviews Neuroscience*, *6*, 435–448.

Hyde, D. C., Boas, D. A., Blair, C., & Carey, S. (2010). Near-infrared spectroscopy shows right parietal specialization for number in pre-verbal infants. *NeuroImage*, *53*, 647–652.

Iuculano, T., & Kadosh, R. C. (2013). The mental cost of cognitive enhancement. *The Journal of Neuroscience*, *33*, 4482–4486.

Iuculano, T., & Kadosh, R. C. (2014). Preliminary evidence for performance enhancement following parietal lobe stimulation in developmental dyscalculia. *Frontiers in Human Neuroscience*, *8*, 38. http://dx.doi.org/10.3389/fnhum.2014.00038.

Izard, V., Sann, C., Spelke, E. S., & Streri, A. (2009). Newborn infants perceive abstract numbers. *Proceedings of the National Academy of Sciences of the United States of America*, *106*, 10382–10385.

Jacobson, L., Koslowsky, M., & Lavidor, M. (2012). tDCS polarity effects in motor and cognitive domains: A meta-analytical review. *Experimental Brain Research*, *216*, 1–10.

Jaeggi, S. M., Buschkuehl, M., Jonides, J., & Perrig, W. J. (2008). Improving fluid intelligence with training on working memory. *Proceedings of the National Academy of Sciences of the United States of America*, *105*, 6829–6833.

Jaeggi, S. M., Buschkuehl, M., Jonides, J., & Shah, P. (2011). Short- and long-term benefits of cognitive training. *Proceedings of the National Academy of Sciences of the United States of America*, *108*, 10081–10086.

Jasper, H. H. (1958). The ten twenty electrode system of the international federation. *Electroencephalography and Clinical Neurophysiology*, *10*, 371–375.

Kanai, R., Chaieb, L., Antal, A., Walsh, V., & Paulus, W. (2008). Frequency-dependent electrical stimulation of the visual cortex. *Current Biology*, *18*, 1839–1843.

Kaufmann, L., Wood, G., Rubinsten, O., & Henik, A. (2011). Meta-analyses of developmental fMRI studies investigating typical and atypical trajectories of number processing and calculation. *Developmental Neuropsychology*, *36*, 763–787.

Keeser, D., Meindl, T., Bor, J., Palm, U., Pogarell, O., Mulert, C., et al. (2011). Prefrontal transcranial direct current stimulation changes connectivity of resting-state networks during fMRI. *The Journal of Neuroscience*, *31*, 15284–15293.

Klein, E., Mann, A., Huber, S., Bloechle, J., Willmes, K., Karim, A. A., et al. (2013). Bilateral bi-cephalic tdcs with two active electrodes of the same polarity modulates bilateral cognitive processes differentially. *PLoS One*, *8*, e71607. http://dx.doi.org/10.1371/journal.pone.0071607.

Klingberg, T. (2010). Training and plasticity of working memory. *Trends in Cognitive Sciences*, *14*, 317–324.

Krause, B., & Cohen Kadosh, R. (2013). Can transcranial electrical stimulation improve learning difficulties in atypical brain development? A future possibility for cognitive training. *Developmental, Cognitive Neuroscience*, *6*, 176–194. http://dx.doi.org/10.1016/j.dcn.2013.04.001.

Krause, B., & Cohen Kadosh, R. (2014). Not all brains are created equal: The relevance of individual differences in responsiveness to transcranial electrical stimulation. *Frontiers in Systems Neuroscience*, *8*, 25. http://dx.doi.org/10.3389/fnsys.2014.00025.

Krause, B., Márquez-Ruiz, J., & Cohen Kadosh, R. (2013). The effect of transcranial direct current stimulation: A role for cortical excitation/inhibition balance? *Frontiers in Human Neuroscience*, *7*, 602. http://dx.doi.org/10.3389/fnhum.2013.00602.

Kuo, M. F., & Nitsche, M. A. (2012). Effects of transcranial electrical stimulation on cognition. *Clinical EEG and Neuroscience*, *43*, 192–199.

Libertus, M. E., Feigenson, L., & Halberda, J. (2011). Preschool acuity of the approximate number system correlates with school math ability. *Developmental Science*, *14*, 1292–1300.

Libertus, M. E., Odic, D., & Halberda, J. (2012). Intuitive sense of number correlates with math scores on college-entrance examination. *Acta Psychologica*, *141*, 373–379.

Link, T., Moeller, K., Huber, S., Fischer, U., & Nuerk, H. C. (2013). Walk the number line–An embodied training of numerical concepts. *Trends in Neuroscience and Education*, *2*, 74–84.

Looi, C. Y., Duta, M., Brem, A.-K., Huber, S., Nuerk, H.-C., & Cohen Kadosh, R. (submitted). Combining brain stimulation and video gaming to promote long-term transfer and cognitive enhancement.

Lorenz, J., Minoshima, S., & Casey, K. L. (2003). Keeping pain out of mind: The role of the dorsolateral prefrontal cortex in pain modulation. *Brain*, *126*, 1079–1091.

Lyons, I. M., & Beilock, S. L. (2012). Mathematics anxiety: Separating the math from the anxiety. *Cerebral Cortex*, *22*, 2102–2110.

MacLeod, C. M. (1991). Half a century of research on the Stroop effect: An integrative review. *Psychological Bulletin*, *109*, 163–203.

Maloney, E. A., Ansari, D., & Fugelsang, J. A. (2011). The effect of mathematics anxiety on the processing of numerical magnitude. *The Quarterly Journal of Experimental Psychology*, *64*, 10–16.

Maloney, E. A., Risko, E. F., Ansari, D., & Fugelsang, J. (2010). Mathematics anxiety affects counting but not subitizing during visual enumeration. *Cognition*, *114*, 293–297.

Márquez-Ruiz, J., Leal-Campanario, R., Sánchez-Campusano, R., Molaee-Ardekani, B., Wendling, F., Miranda, P. C., et al. (2012). Transcranial direct-current stimulation modulates synaptic mechanisms involved in associative learning in behaving rabbits. *Proceedings of the National Academy of Sciences of the United States of America*, *109*, 6710–6715.

Meck, W. H., & Church, R. M. (1983). A mode control model of counting and timing processes. *Journal of Experimental Psychology: Animal Behavior Processes*, *9*, 320–334.

Meinzer, M., Antonenko, D., Lindenberg, R., Hetzer, S., Ulm, L., Avirame, K., et al. (2012). Electrical brain stimulation improves cognitive performance by modulating functional connectivity and task-specific activation. *The Journal of Neuroscience*, *32*, 1859–1866.

Meinzer, M., Jähnigen, S., Copland, D. A., Darkow, R., Grittner, U., Avirame, K., et al. (2014). Transcranial direct current stimulation over multiple days improves learning and maintenance of a novel vocabulary. *Cortex*, *50*, 137–147.

Meinzer, M., Lindenberg, R., Antonenko, D., Flaisch, T., & Flöel, A. (2013). Anodal transcranial direct current stimulation temporarily reverses age-associated cognitive decline and functional brain activity changes. *The Journal of Neuroscience*, *33*, 12470–12478.

Moliadze, V., Antal, A., & Paulus, W. (2010). Electrode-distance dependent after-effects of transcranial direct and random noise stimulation with extracephalic reference electrodes. *Clinical Neurophysiology*, *121*, 2165–2171.

Moss, F., Ward, L. M., & Sannita, W. G. (2004). Stochastic resonance and sensory information processing: A tutorial and review of application. *Clinical Neurophysiology*, *115*, 267–281.

Neves, G., Cooke, S. F., & Bliss, T. V. P. (2008). Synaptic plasticity, memory and the hippocampus: a neural network approach to causality. *Nature Reviews Neuroscience*, *9*, 65–75.

Nieder, A., & Miller, E. K. (2003). Coding of cognitive magnitude: Compressed scaling of numerical information in the primate prefrontal cortex. *Neuron*, *37*, 149–157.

Nissen, M. J., & Bullemer, P. (1987). Attentional requirements of learning: Evidence from performance measures. *Cognitive Psychology*, *19*, 1–32.

Nitsche, M. A., Cohen, L. G., Wassermann, E. M., Priori, A., Lang, N., Antal, A., et al. (2008). Transcranial direct current stimulation: State of the art 2008. *Brain Stimulation*, *1*, 206–223.

Nitsche, M. A., & Paulus, W. (2000). Excitability changes induced in the human motor cortex by weak transcranial direct current stimulation. *Journal of Physiology*, *527*, 633–639.

Nitsche, M., & Paulus, W. (2001). Sustained excitability elevations induced by transcranial DC motor cortex stimulation in humans. *Neurology*, *57*, 1899–1901.

Owen, A. M., Hampshire, A., Grahn, J. A., Stenton, R., Dajani, S., Burns, A. S., et al. (2010). Putting brain training to the test. *Nature*, *465*, 775–778.

Park, J., & Brannon, E. M. (2013). Training the approximate number system improves math proficiency. *Psychological Science*, *24*, 2013–2019.

Parsons, S., & Bynner, J. (2005). *Does numeracy matter more?* London: National Research and Development Centre for Adult Literacy and Numeracy.

Pascual-Leone, A., Horvath, J. C., & Robertson, E. M. (2012). Enhancement of normal cognitive abilities through noninvasive brain stimulation. In R. Chen, & J. C. Rothwell (Eds.), *Cortical connectivity* (pp. 207–249). Berlin: Springer-Verlag.

Peretti, C. S., Danion, J. M., Gierski, F., & Grangé, D. (2002). Cognitive skill learning and aging: A component process analysis. *Archives of Clinical Neuropsychology*, *17*, 445–459.

Piazza, M., & Izard, V. (2009). How humans count: Numerosity and the parietal cortex. *The Neuroscientist*, *15*, 261–273.

Piazza, M., Izard, V., Pinel, P., Le Bihan, D., & Dehaene, S. (2004). Tuning curves for approximate numerosity in the human intraparietal sulcus. *Neuron*, *44*, 547–555.

Piazza, M., Pinel, P., Le Bihan, D., & Dehaene, S. (2007). A magnitude code common to numerosities and number symbols in human intraparietal cortex. *Neuron*, *53*, 293–305.

Pica, P., Lemer, C., Izard, V., & Dehaene, S. (2004). Exact and approximate arithmetic in an Amazonian indigene group. *Science*, *306*, 499–503.

Plewnia, C., Zwissler, B., Längst, I., Maurer, B., Giel, K., & Krüger, R. (2013). Effects of transcranial direct current stimulation (tDCS) on executive functions: Influence of COMT Val/Met polymorphism. *Cortex*, *49*, 1801–1807.

Polanía, R., Nitsche, M. A., Korman, C., Batsikadze, G., & Paulus, W. (2012). The importance of timing in segregated theta phase-coupling for cognitive performance. *Current Biology*, *22*, 1314–1318.

Price, G. R., Palmer, D., Battista, C., & Ansari, D. (2012). Nonsymbolic numerical magnitude comparison: Reliability and validity of different task variants and outcome measures, and their relationship to arithmetic achievement in adults. *Acta Psychologica*, *140*(1), 50–57.

Priori, A. (2003). Brain polarization in humans: A reappraisal of an old tool for prolonged non-invasive modulation of brain excitability. *Clinical Neurophysiology*, *114*, 589–595.

Qin, S., Cho, S., Chen, T., Rosenberg-Lee, M., Geary, D. C., & Menon, V. (2014). Hippocampal-neocortical functional reorganization underlies children's cognitive development. *Nature Neuroscience*, *17*, 1263–1269.

Reigosa-Crespo, V., Valdés-Sosa, M., Butterworth, B., Estévez, N., Rodríguez, M., Santos, E., et al. (2012). Basic numerical capacities and prevalence of developmental dyscalculia: The Havana Survey. *Developmental Psychology*, *48*, 123–135.

Reis, J., Prichard, G., & Fritsch, B. (2014). Improving functions in the typical brain: Motor system. In R. Cohen Kadosh (Ed.), *The stimulated brain: Cognitive enhancement using non-invasive brain stimulation*. Amsterdam: Academic Press.

Reis, J., Schambra, H. M., Cohen, L. G., Buch, E. R., Fritsch, B., Zarahn, E., et al. (2009). Noninvasive cortical stimulation enhances motor skill acquisition over multiple days through an effect on consolidation. *Proceedings of the National Academy of Sciences of the United States of America*, *106*, 1590–1595.

Richardson, F. C., & Suinn, R. M. (1972). The mathematics anxiety rating scale: Psychometric data. *Journal of Counseling Psychology*, *19*, 551–554.

Ritchie, S. J., & Bates, T. C. (2013). Enduring links from childhood mathematics and reading achievement to adult socioeconomic status. *Psychological Science*, *24*, 1301–1308.

Rivera, S. M., Reiss, A. L., Eckert, M. A., & Menon, V. (2005). Developmental changes in mental arithmetic: Evidence for increased functional specialization in the left inferior parietal cortex. *Cerebral Cortex*, *25*, 1779–1790.

Robertson, E. M. (2007). The serial reaction time task: Implicit motor skill learning? *The Journal of Neuroscience*, *27*, 10073–10075.

Rubinsten, O., & Henik, A. (2005). Automatic activation of internal magnitudes: A study of developmental dyscalculia. *Neuropsychology*, *19*, 641–648.

Rubinsten, O., Henik, A., Berger, A., & Shahar-Shalev, S. (2002). The development of internal representations of magnitude and their association with Arabic numerals. *Journal of Experimental Child Psychology*, *81*, 74–92.

Ruff, C. C., Ugazio, G., & Fehr, E. (2013). Changing social norm compliance with noninvasive brain stimulation. *Science*, *342*, 482–484.

Ruffini, G., Fox, M. D., Ripolles, O., Miranda, P. C., & Pascual-Leone, A. (2014). Optimization of multifocal transcranial current stimulation for weighted cortical pattern targeting from realistic modeling of electric fields. *NeuroImage*, *89*, 216–225.

Sack, A. T., Kadosh, R. C., Schuhmann, T., Moerel, M., Walsh, V., & Goebel, R. (2009). Optimizing functional accuracy of TMS in cognitive studies: A comparison of methods. *Journal of Cognitive Neuroscience*, *21*, 207–221.

Santarnecchi, E., Polizzotto, N. R., Godone, M., Giovannelli, F., Feurra, M., Matzen, L., et al. (2013). Frequency-dependent enhancement of fluid intelligence induced by transcranial oscillatory potentials. *Current Biology*, *23*, 1449–1453.

Sarkar, A., Dowker, A., & Cohen Kadosh, R. (2014). Cognitive enhancement or cognitive cost: Trait-specific outcomes of brain stimulation in the case of mathematics anxiety. *The Journal of Neuroscience*, *34*, 16605–16610.

Schmiedek, F., Lövdén, M., & Lindenberger, U. (2010). Hundred days of cognitive training enhance broad cognitive abilities in adulthood: Findings from the COGITO study. *Frontiers in Aging Neuroscience*, *2*, 27. http://dx.doi.org/10.3389/fnagi.2010.00027.

Schwarz, W., & Ischebeck, A. (2003). On the relative speed account of number-size interference in comparative judgments of numerals. *Journal of Experimental Psychology: Human Perception and Performance*, *29*, 507–522.

Shackman, A. J., McMenamin, B. W., Maxwell, J. S., Greischar, L. L., & Davidson, R. J. (2009). Right dorsolateral prefrontal cortical activity and behavioral inhibition. *Psychological Science*, *20*, 1500–1506.

Shalev, R. S. (2007). Prevalence of developmental dyscalculia. In D. B. Berch & M. M. M. Mazzocco (Eds.), *Why is math so hard for some children? The nature and origins of mathematical learning difficulties and disabilities* (pp. 49–60). Baltimore, MD: Paul H. Brookes Publishing Co.

Shallice, T. (1982). Specific impairments of planning. *Philosophical Transactions of the Royal Society of London B, Biological Sciences*, *298*, 199–209.

Siegler, R. S., Fazio, L. K., Bailey, D. H., & Zhou, X. (2013). Fractions: The new frontier for theories of numerical development. *Trends in Cognitive Sciences*, *17*, 13–19.

Siegler, R. S., & Opfer, J. E. (2003). The development of numerical estimation: Evidence for multiple representations of numerical quantity. *Psychological Science*, *14*, 237–243.

Slagter, H. A., Lutz, A., Greischar, L. L., Francis, A. D., Nieuwenhuis, S., Davis, J. M., et al. (2007). Mental training affects distribution of limited brain resources. *PLoS Biology*, *5*, e138. http://dx.doi.org/10.1371/journal.pbio.0050138.

Snowball, A., Tachtsidis, I., Popescu, T., Thompson, J., Delazer, M., Zamarian, L., et al. (2013). Long-term enhancement of brain function and cognition using cognitive training and brain stimulation. *Current Biology*, *23*, 987–992.

Sparing, R., Thimm, M., Hesse, M. D., Küst, J., Karbe, H., & Fink, G. R. (2009). Bidirectional alterations of interhemispheric parietal balance by non-invasive cortical stimulation. *Brain*, *132*, 3011–3020.

Stagg, C. J., Bachtiar, V., & Johansen-Berg, H. (2011). The role of GABA in human motor learning. *Current Biology*, *21*, 480–484.

Stagg, C. J., Best, J. G., Stephenson, M. C., O'Shea, J., Wylezinska, M., Kincses, Z. T., et al. (2009). Polarity-sensitive modulation of cortical neurotransmitters by transcranial stimulation. *Journal of Neuroscience*, *29*, 5202–5209.

Strangman, G., Boas, D. A., & Sutton, J. P. (2002). Non-invasive neuroimaging using near-infrared light. *Biological Psychiatry*, *52*, 679–693.

Stroop, J. R. (1935). Studies of interference in serial verbal reactions. *Journal of Experimental Psychology*, *18*, 643–662.

Takano, Y., Yokawa, T., Masuda, A., Niimi, J., Tanaka, S., & Hironaka, N. (2011). A rat model for measuring the effectiveness of transcranial direct current stimulation using fMRI. *Neuroscience Letters*, *491*, 40–43.

Tang, Y., Zhang, W., Chen, K., Feng, S., Ji, Y., Shen, J., et al. (2006). Arithmetic processing in the brain shaped by cultures. *Proceedings of the National Academy of Sciences of the United States of America*, *103*, 10775–10780.

Terhune, D. B., Tai, S., Cowey, A., Popescu, T., & Cohen Kadosh, R. (2011). Enhanced cortical excitability in grapheme-color synesthesia and its modulation. *Current Biology*, *21*, 2006–2009.

Terney, D., Chaieb, L., Moliadze, V., Antal, A., & Paulus, W. (2008). Increasing human brain excitability by transcranial high-frequency random noise stimulation. *The Journal of Neuroscience*, *28*, 14147–14155.

Truong, D. Q., Magerowski, G., Blackburn, G. L., Bikson, M., & Alonso-Alonso, M. (2013). Computational modeling of transcranial direct current stimulation (tDCS) in obesity: Impact of head fat and dose guidelines. *NeuroImage*, *2*, 759–766.

Tseng, P., Hsu, T.-Y., Chang, C. F., Tzeng, O. J., Hung, D. L., Muggleton, N. G., et al. (2012). Unleashing potential: Transcranial direct current stimulation over the right posterior parietal cortex improves change detection in low-performing individuals. *The Journal of Neuroscience*, *32*, 10554–10561.

Tzelgov, J., Yehene, V., Kotler, L., & Alon, A. (2000). Automatic comparisons of artificial digits never compared: Learning linear ordering relations. *Journal of Experimental Psychology: Learning, Memory, and Cognition*, *26*, 103–120.

Villringer, A., & Chance, B. (1997). Non-invasive optical spectroscopy and imaging of human brain function. *Trends in Neurosciences*, *20*, 435–442.

Villringer, A., Planck, J., Hock, C., Schleinkofer, L., & Dirnagl, U. (1993). Near infrared spectroscopy (NIRS): A new tool to study hemodynamic changes during activation of brain function in human adults. *Neuroscience Letters*, *154*, 101–104.

Walsh, V. (2003). A theory of magnitude: Common cortical metrics of time, space and quantity. *Trends in Cognitive Sciences*, *7*, 483–488.

Waters-Metenier, S., Husain, M., Wiestler, T., & Diedrichsen, J. (2014). Bihemispheric transcranial direct current stimulation enhances effector-independent representations of motor synergy and sequence learning. *The Journal of Neuroscience*, *34*, 1037–1050.

Weber, M. J., Messing, S. B., Rao, H., Detre, J. A., & Thompson-Schill, S. L. (2014). Prefrontal transcranial direct current stimulation alters activation and connectivity in cortical and subcortical reward systems: A tDCS-fMRI study. *Human Brain Mapping*, *35*, 3673–3686.

Wirth, M., Rahman, R. A., Kuenecke, J., Koenig, T., Horn, H., Sommer, W., et al. (2011). Effects of transcranial direct current stimulation (tDCS) on behaviour and electrophysiology of language production. *Neuropsychologia*, *49*, 3989–3998.

Wolkenstein, L., Zeiller, M., Kanske, P., & Plewnia, C. (2014). Induction of a depression-like negativity bias by cathodal transcranial direct current stimulation. *Cortex*, *59*, 103–112.

Xu, F., & Spelke, E. S. (2000). Large number discrimination in 6-month-old infants. *Cognition*, *74*, B1–B11.

Zamarian, L., Ischebeck, A., & Delazer, M. (2009). Neuroscience of learning arithmetic—Evidence from brain imaging studies. *Neuroscience & Biobehavioral Reviews*, *33*, 909–925.

Part II

Genetic Influences

Chapter 11

Individual Differences in Mathematics Ability: A Behavioral Genetic Approach

Stephen A. Petrill[1], **Yulia Kovas**[2,3,4]

[1]Department of Psychology, The Ohio State University, Columbus, Ohio, USA, [2]Tomsk State University, Tomsk, Russia, [3]Department of Psychology, Goldsmiths, University of London, London, UK, [4]Social, Genetic and Developmental Psychiatry Centre, Institute of Psychiatry, King's College London, London, UK

INTRODUCTION

Competence in mathematics is important in modern society and is associated with higher job quality, positive social relationships, better health care decisions, and enhanced mental health (Geary, 2011). Unfortunately, however, there is a significant gap between the importance of mathematics and performance in mathematics. For example, the most recent National Assessment of Educational Progress (2013) suggested that 36% of Grade 12 students performed below basic levels of mathematical performance. Moreover, a study by the U.S. Department of Education (2014) found that only 16% of high school seniors have the required math proficiency and the desire to pursue a career in science, technology, engineering, or mathematics. Results are similar in the UK. For example, in the summer of 2014, 42% of candidates for the General Certificate of Secondary Education (GCSE) Math exam did not meet the minimum mathematical competencies required for a whole range of educational and employment opportunities (Joint Council for Qualifications, 2013).

In response, a large and growing literature has examined the cognitive skills that influence mathematical learning and achievement. Behavioral studies have identified two groups of cognitive processes that contribute to mathematics performance. The first covers a range of *Domain-General* processes—such as phonological processing, working memory, inhibitory control, and visuospatial abilities—that are associated with multiple areas of academic performance and are not specific to mathematical learning (e.g., Geary, Bailey, & Hoard, 2009; Swanson & Jerman, 2007). A second *Domain-Specific* set of processes is

Development of Mathematical Cognition. http://dx.doi.org/10.1016/B978-0-12-801871-2.00011-3

specific to mathematics, including numerical quantity approximation and discrimination (Fuchs, Geary, Compton, Fuchs, & Hamlett, 2014; Houde, Rossi, Lubin, & Joliot, 2010; Mazzocco, Feigenson, & Halberda, 2011).

In this chapter, we examine mathematics and associated measures of cognitive and neurobiological performance from the perspective of quantitative and molecular genetics. We first describe the logic of quantitative genetics, including twin/adoption studies as well as Genome-wide Complex Trait Analysis (GCTA). Second, we discuss the genetic and environmental etiology of mathematics performance and mathematical cognition. We then examine how genes and environments influence the longitudinal development of mathematics achievement as well as the relationship between mathematics, mathematical cognition, and other academic skill—in particular, reading, motivation, and emotional characteristics. Finally, we end with a discussion of molecular genetic findings and implications for future directions integrating neurobiological studies of mathematics within genetically sensitive designs.

INTRODUCTION TO QUANTITATIVE GENETICS

Quantitative genetic research uses multiple analytic and statistical methodologies to examine the genetic and environmental contributions to mathematical development. Quantitative genetic methods include a variety of family designs, such as twin studies, adoption studies, and more recently, GCTA. GCTA estimates genetic influences on complex traits using genome-wide genotypes in large samples of unrelated individuals (Plomin & Deary, 2014). Comparing GCTA results to the results of family studies provides important insights into the genetic architecture of complex traits.

Studies using the classical adoption method have already provided important insights into the etiology of variation in ability and achievement (see Rhea, Bricker, Wadsworth, & Corley, 2013). The method is based on comparing biological and environmental (adoptive) relatives, which disentangles postnatal environmental and genetic influences on development. The association between biological relatives who do not live together (e.g., parents and their adopted away children) is attributed to genetic factors, whereas the correlation between adoptive relatives is attributed to environmental factors in the absence of selective placement (children placed into families nonrandomly, so that their adoptive family resembles their biological family for reasons including genetic similarity; see Plomin, DeFries, Knopik, & Neiderhiser, 2013). When data from both biological (e.g., biological parents) and adoptive (e.g., adoptive parents and their adopted child) relatives are available, the adoption method provides a particularly powerful tool for disentangling genetic and environmental effects, as well as for studying the processes of gene-environment correlations. For example, an evocative gene-environment correlation is suggested when adoptive children exhibit traits that are also present in their biological parents and adoptive parents respond to these traits by providing specific suitable or matching environments (e.g., Plomin et al., 2013).

The twin method remains the most commonly used quantitative genetic design because it allows for large representative samples and high precision of estimates. As with any other approach, the twin method has its limitations, such as potential differences between twins and nontwins, and differences in the degree to which environment is shared by identical and nonidentical twins (e.g., Boomsma, Busjahn, & Peltonen, 2002; Martin, Boomsma, & Machin, 1997; Plomin et al., 2013). Despite these limitations, the twin method has been shown to provide consistent results which have been supported through multiple replications in different samples. Moreover, the results from twin studies have recently gained further support through convergence with results from other methods, such as adoption studies, the GCTA method (described more fully below) and molecular genetics (Bates, 2008; Visscher et al., 2006; Yang, Lee, Goddard, & Visscher, 2011, 2013).

Comparing the resemblance of identical (monozygotic, MZ) and nonidentical (dizygotic, DZ) twin pairs in the classical twin design addresses the origins of individual differences by estimating the proportion of variance that can be attributed to genetic, shared environment, and nonshared environment factors (Plomin et al., 2013). In the case of complex traits that are likely to be influenced by multiple factors, the genetic component of variance refers to the influence of alleles at all gene loci that affect the trait. The similarity between twins for any particular trait can be due wholly or in part to these genetic effects. Twin similarity may also be due wholly or in part to shared environment, which refers to environmental influences that vary between families, but are experienced similarly by twins within families. For example, pairs of twins experience similar conditions during gestation, have the same socioeconomic status if they live in the same home, live with the same family, and go to the same school. These factors could reasonably be expected to increase similarity between co-twins, above and beyond their genetic similarity. Nonshared environment refers to any aspect of environmental influence that is experienced differently by the two twins and contributes to measureable differences between them, including measurement error. Such influences involve aspects of experience that are specific to an individual, such as traumas and diseases, idiosyncratic experiences, different peers, differential treatment by the parents and teachers, and, importantly, different perceptions of such experiences, even if the events appear to be ostensibly the same for the two children.

Genetic influence can be estimated by comparing intraclass correlations for identical (MZ) twins and fraternal (DZ) twins (see Figure 1). MZ twins result from the division of the same fertilized egg (zygote) and therefore are 100% genetically similar in most cases. DZ twins develop at the same time from two, separately fertilized, zygotes and therefore, like other siblings, they share about 50% of the segregating (variable) DNA on average. As shown in Figure 1, the phenotypic variance of a trait can be attributed to genetic variance to the extent that the MZ twin correlation exceeds the DZ twin correlation. Specifically, heritability (or h^2), which is the proportion of phenotypic variance attributed to genetic variance, can be estimated as twice the difference between the MZ and DZ twin correlations. The relatedness for shared (common, or c^2) environmental influences is assumed to be 1.0 for both MZ and DZ twin pairs who grow up in the same family because they experience equally similar prenatal and postnatal environments.

Shared environmental influences are evident to the extent that the DZ twins' correlation is more than half of the MZ correlation (see Figure 1). Nonshared environment (or e^2) refers to those environmental influences that are not shared by family members. In the case of the twin design, nonshared environment (including measurement error) is estimated by subtracting the MZ correlation from 1.0. For example, if the correlation in mathematical achievement for MZ twins is 0.68 (68%) and for DZ twins is 0.41 (41%) (see Luo, Kovas, Haworth, & Plomin, 2011), heritability of mathematical achievement is estimated as $2\times(0.68-0.41)=0.54$. In other words, 54% of the phenotypic variance in mathematical achievement, as measured in this study, is explained by genetic factors. The effect of shared environment is $0.68-0.54=0.14$ (14% of the phenotypic variance); and of nonshared environment is $1-0.68=0.32$ (32% of the phenotypic variance). The variance components can also be estimated using structural equation model fitting, which estimates parameters and their confidence intervals by fitting the models to variance/covariance matrices using the model-fitting program Mx (Neale, Boker, Xie, & Maes, 2006).

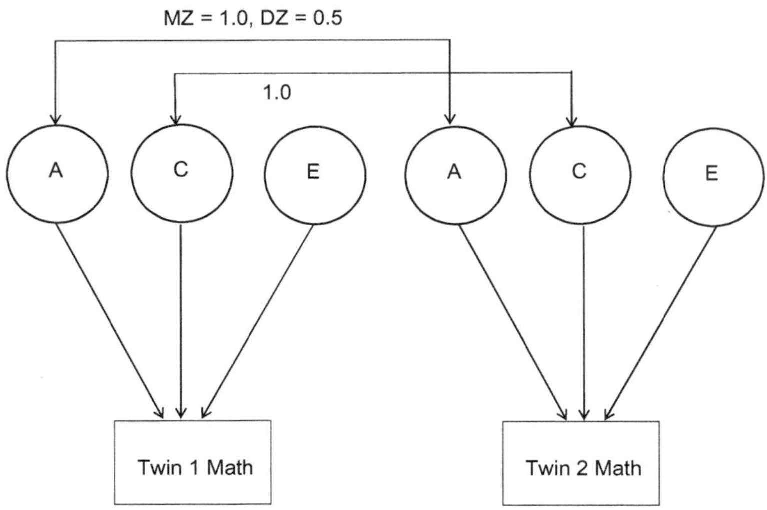

FIGURE 1 Univariate ACE model. A = additive genetics, C = shared environment, E = nonshared environment. Identical (MZ) twins share 100% additive genetic variance whereas fraternal (DZ) twins share 50% additive genetic variance. Both MZ and DZ twins share 100% shared environment.

ETIOLOGY OF INDIVIDUAL DIFFERENCES IN MATHEMATICS

In contrast to the long history of behavioral genetic studies in domains such as language and reading, until the past decade, the behavioral genetic literature on mathematics consisted of a relatively small number of published papers with varying results (e.g., Alarcón, DeFries, Light, & Pennington, 1997; Knopik, Alarcón, & DeFries, 1997; Loehlin & Nichols, 1976; Thompson, Detterman, & Plomin, 1991). For example, Thompson et al. (1991) examined the heritability of performance on standardized achievement tests for 278 pairs of twins ranging in age from 6 to 12 years. Results suggested that heritability was modest (h^2=0.20) with shared and nonshared environment explaining most of the variation. In contrast, in a study of more than 1000 twin pairs aged 8-20 years, Alarcón, Knopik, and DeFries (2000) found that the heritability of mathematics was very high (h^2=0.91) with no evidence for shared environment. The remaining studies fell between this very wide range of heritability estimates.

There were several factors in this first generation of studies that contributed to these diverse results. First, age range varied both across and within samples. This was problematic because quantitative genetic studies in other domains such as intelligence and reading suggest that heritability increases with age at the expense of the shared environment (e.g., Davis, Haworth, & Plomin, 2009; Haworth et al., 2010). Moreover, samples varied with respect to degree of selection for mathematical impairment as well as the type of mathematics measures that were employed.

In response, a second generation of quantitative genetic studies has examined mathematics: (1) using narrower age ranges; (2) systematically examining differences in heritability in unselected versus selected samples; and (3) examining a wide range of mathematical achievement as well as cognition (e.g., Hart, Petrill, Thompson, & Plomin, 2009; Haworth, Kovas, Petrill, & Plomin, 2007; Kovas, Haworth, Dale, Petrill, & Plomin, 2007; Oliver et al., 2004). For example, in our work we have examined the genetic and environmental etiology of mathematics achievement using two related samples. The U.S.-based Western Reserve Reading and Math Project (WRRMP) is based on 468 pairs of twins assessed across 10 measurement occasions, beginning when children entered school. Each twin was tested in their home via a three-hour in-home assessment at each time point by separate examiners. Mathematics has been assessed on the final six measurement occasions, beginning at 8½ years. The UK-based Twins Early Development Study (TEDS) is based on 5000 pairs of twins who were tested on mathematics at ages 7, 9, 10, 12, 14, and 16 years via teacher ratings and online assessments.

Results suggest that heritability estimates for the U.S. sample are somewhat lower than the UK sample. For example, Hart, Petrill, and Thompson (2010) found univariate heritability estimates from h^2=0.34 to h^2=0.49, whereas the UK sample estimates ranged from h^2=0.63 to h^2=0.73 (Oliver & Plomin,

2007). The most recent TEDS study has found heritability of over 60% for 16-year-old students' GCSE exam grades (Rimfeld, Kovas, Dale, & Plomin, in press). Conversely, estimates of shared environment have been higher in the U.S. sample (0.32-0.46) than the UK sample (0.07-0.23), whereas nonshared influences were lower in the U.S. sample (0.19-0.25) than in the UK (0.42-0.48).

This discrepancy is interesting because it may reflect different curricula or other school differences between the two countries. The UK educational system maintains a national curriculum, whereas individual states determine the curriculum in the U.S., and school districts within each state select their own textbook. This could result in greater environmental variability in the U.S. context, leading to higher estimates of shared environment. Another possibility is that these differences are driven by differences in demographics, age, and/or measures across U.S. and UK samples. Cross-cultural research using identical measures of mathematical ability and performance in large samples of the same age are needed to make meaningful comparisons and to establish the true sources of differences in estimates from different studies.

In addition to examining the relative contribution of genetic and environmental influences on mathematics, these studies have examined whether mathematical disability is categorically different from the rest of the distribution of mathematics or instead represents the lower tail of a normal distribution of mathematical skills. At issue is whether specific genetic risks may be present at the low end of math performance, above and beyond the overall heritability of math performance. In contrast, it is also possible that the same set of genetic and environmental influences operate as risk and protective factors across the entire distribution. Consider Williams Syndrome, a rare neurodevelopmental disorder associated with a set of deletions on Chromosome 7 (Martens, Wilson, & Reutens, 2008). The math cognition literature has suggested that the deficits in mathematical performance in Williams Syndrome may be linked to deficits in numerical cognition (e.g., Libertus, Feigenson, Halberda, & Landau, 2014). These findings are important in that they elucidate mechanisms of poor math performance within a clinical population with a known genetic condition (see Chapter 12, for an extensive review of findings using this approach).

However, these findings do not necessarily extend to individuals with poor math performance who do not have Williams Syndrome. A recent study has examined whether variation in the DNA regions affected in Williams Syndrome is involved in normal variation in mathematical ability (Rodic et al., in press). None of the over one million genotyped single nucleotide polymorphisms (SNPs, which refers to a difference at a single nucleotide on the genome) were significantly associated with variation in mathematical ability and achievement in an unselected population, assessed using different measures at different school ages. These results suggest that genetic mutations involved in Williams Syndrome that affect their mathematical ability are unrelated to normal variation in mathematics. Alternatively, the results may suggest that the effect size of each genetic variant is so small that even greater samples are needed to detect them.

Beyond the insights of clinically derived samples such as Williams Syndrome, quantitative genetic methods have examined the issue of etiological continuity across the entire spectrum of mathematical variation. The purpose of this approach is to examine whether the genetic and environmental contributions to mathematical skills are consistent across the range of ability, or whether the genetic (or environmental) influences are higher or lower in individuals with mathematical disability. This general approach has employed multiple methods (see Plomin & Kovas, 2005), but most often using DeFries-Fulker (DF) analysis (DeFries & Fulker, 1985, 1988). DF analysis examines twin resemblance in disability by estimating *group heritability* (or h_g^2), which compares the degree of relationship between MZ and DZ twins when at least one pair meets the threshold for disability. In this case, mean ability scores are estimated for the selected twins, also called probands. These mean scores are then compared to the mean ability scores of their twin siblings, who are free to vary across the entire range of the distribution. Group heritability (h_g^2) is significant to the extent that MZ co-twin means are more similar to the proband mean than DZ co-twins and significantly lower (more extreme) than the mean of the unselected population. This indicates that genetic influences affect familial resemblance even when both members do not meet a clinical cutoff for disability. In this way, h_g^2 estimates the degree of genetic uniformity across the distribution. In the case of mathematics achievement, findings suggest that h_g^2 is not only significant but similar in magnitude to the h^2 of the entire population (see Haworth et al., 2009; Plomin & Kovas, 2005). Thus, beyond rare genetic effects such as those found in Williams Syndrome, there appears to be a large set of genetic effects that impact mathematical disability via their overall effect on mathematics performance across the distribution.

Further support for a normally distributed continuum underlying variation in mathematical ability comes from a recent study of 12-year-old children that used both the twin method and GCTA (Davis et al., 2014). The twin analysis showed that heritability of mathematics at this age in the UK children was 0.51. The GCTA analysis was then conducted on the data from the same sample (one twin from each pair), focusing on the DNA data (1,588,650 SNPs genotyped in this sample). GCTA uses genotypes available for a particular sample—participants in TEDS have been genotyped for hundreds of thousands of SNPs, using so called "gene chips," microarrays that enable genotyping for thousands of DNA markers simultaneously. GCTA then estimates genomic relationships between pairs of conventionally unrelated individuals (i.e., not relatives) using these genome-wide SNP data to estimate variance in a particular trait explained by all genotyped SNPs simultaneously (Yang et al., 2013).

The analysis in Davis et al. (2014) addressed the extent to which any observed similarity in mathematical achievement among unrelated individuals could be explained by similarity in their DNA—the number of the polymorphisms (out of the ones genotyped) for which the individuals had an identical variant. The heritability accounted for by the available SNPs was 0.52. The convergence between the two heritability estimates is remarkable and is a strong indication that multiple DNA polymorphisms of small effects are involved in mathematical variation and that low mathematical ability lies on the same etiological continuum as the rest of the variation.

ETIOLOGY OF THE LINKS BETWEEN MATHEMATICS ABILITY AND OTHER TRAITS

Multivariate Genetic Designs

Contemporary quantitative genetic studies often go beyond simply estimating the relative contribution of genetic and environmental factors to variation in traits. Multivariate genetic designs extend the principles of the twin method to address theoretically meaningful questions about the relationships between measures of educational interest. For example, multivariate analyses are used to estimate the extent to which covariation/comorbidity among traits, such as reading and math abilities/disabilities, result from common genetic and environmental factors (see Figure 2). These analyses can also be used to evaluate the extent to which the same (vs. different) genetic and environmental factors are involved in individual variation in traits throughout development (e.g., early and later school mathematics).

Multivariate models are used to decompose the covariance between two or more traits into genetic and environmental sources and to estimate the amount

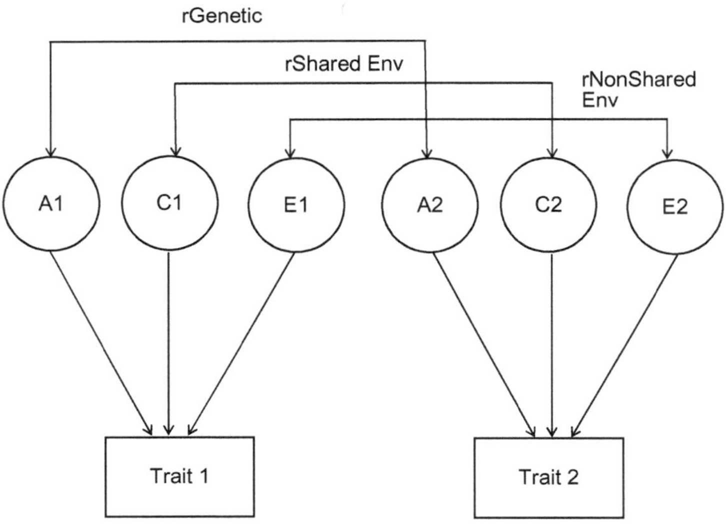

FIGURE 2 Bivariate ACE model. A = additive genetics, C = shared environment, E = nonshared environment. The genetic variance between two traits can be correlated (rGenetic) from 0 to 1.0. The same is true for shared environmental (rShared Env) and nonshared environmental (rNonShared Env) effects.

of overlapping genetic and environmental influences on the traits (genetic and environmental correlations). As shown in Figure 2, the absolute value of a genetic correlation can vary from 0 (no overlapping genetic influences) to 1 (genetic influences contributing to the twin similarity in Trait 1 are the same as those genetic influences contributing to the twin similarity in Trait 2). Similarly, the shared environmental correlation indexes environments common for Trait 1 and Trait 2, contributing to twin similarity in both traits; and the nonshared environmental correlation indexes environmental effects that make twins less similar in both traits. Multivariate genetic analysis is also used to estimate the relative contribution of genetic and environmental factors to the phenotypic correlation between two traits: bivariate heritability is the proportion of the phenotypic correlation between Trait 1 and Trait 2 that is explained by common genetic factors; bivariate shared and nonshared environmental estimates are proportions of the phenotypic correlation explained by shared and nonshared environmental factors. These bivariate heritability and environmental factors sum to 1, as they explain all the covariance between Trait 1 and Trait 2.

Multivariate genetic analyses rely on studies that assess multiple phenotypes in the same twins, such as measures of different abilities and measures of the same ability at different ages. Several ongoing longitudinal multivariate twin studies have already provided insights into the etiology of the associations between numerous traits, such as reading and mathematical abilities and disabilities (Hart, Petrill, & Kamp Dush, 2010; Hart et al., 2009; Kovas, Haworth, Dale, Petrill, et al., 2007; Kovas, Haworth, Dale, & Plomin, 2007; Kovas, Haworth, Harlaar, et al., 2007; Kovas, Petrill, & Plomin, 2007); or mathematical, reading, and attentional skills (Hart, Petrill, & Kamp Dush, 2010; Hart, Petrill, & Thompson, 2010). As described below, significant genetic correlations have been found across all of these traits, meaning that many genetic effects are general rather than specific to any one trait. Substantial genetic overlap has also been found for learning disabilities: genetic correlations are high between reading, mathematics, and language disabilities; and moderate between learning disabilities and other developmental difficulties, such as attention deficit hyperactivity disorder (ADHD) (Butterworth & Kovas, 2013; Greven, Kovas, Willcutt, Petrill, & Plomin, 2014; Plomin & Kovas, 2005).

Etiology of Relationships between Mathematics and Reading as well as Language-Related Skills

One of the first studies that investigated the relationships between mathematics and English, and mathematics and vocabulary, using the U.S. National Merit Scholarship Qualifying Test data from a large sample of 2000 twin pairs, reported genetic correlations of 0.52 and 0.39, respectively (Martin, Jardine, & Eaves, 1984). The results suggest that when genes associated with mathematical abilities are identified, many of the same genes will be associated with English and, to a lesser extent, with vocabulary. Another study (Thompson et al., 1991) reported strong genetic correlations between mathematics and reading (0.98), and mathematics and language (0.98), as measured by standardized tests, suggesting that largely the same genetic factors contributed to variation in all of the examined traits. Shared environments were also largely the same for the three traits (the shared environmental correlation was 0.93 on average). Nonshared environments were largely specific to each trait (correlations ranged from 0.28 to 0.54), explaining most of the differences among the traits.

Hart et al. (2009) examined several components of mathematics, reading, and general intelligence from 10-year-old twin pairs of the WRRMP study. They found no significant genetic overlap between "Reading Fluency" (how fast the child could read a series of words and nonwords), general intelligence, and the mathematical subcomponent of "Calculation" (an untimed test involving mathematical calculation). Conversely, the overlap in shared environmental influences on "Reading Fluency" with those on "Calculation" and general intelligence was statistically significant, suggesting that the within-families environments that are important for "Reading Fluency" and general intelligence also influence the learning of "Calculation." "Mathematical Fluency" (a timed test involving mathematical calculation) shared genetic influences with "Reading Fluency" (0.43) and *g* (0.20). However, some genetic influences on "Mathematical Fluency" (0.59) were independent from reading and general intelligence. Further, "Mathematical Fluency" showed no significant environmental overlap with "Reading Fluency" and general intelligence, indicating a degree of independence from these two abilities.

Moreover, the assessment in TEDS at 7 years found a genetic correlation between mathematics and *g* of 0.67; and between mathematics and reading of 0.74 (Kovas, Harlaar, Petrill, & Plomin, 2005). In the same sample at 10 years, the genetic correlation between mathematics and *g* was 0.68, while the genetic correlation between mathematics and reading was 0.73 (Davis et al., 2008). Overall, the results of several TEDS studies suggest that, to a large extent, the same genetic factors and the same shared environmental factors contribute to mathematics and aspects of reading and general intelligence—explaining most of the observed correlations among these traits. In contrast, nonshared environmental overlap was very small across the measures, indicating the contribution of nonshared environments to differences among the measures (Davis et al., 2008; Haworth, Kovas, Dale, & Plomin, 2008; Kovas et al., 2005).

Origin of Relationships between Mathematics and Spatial Ability

The etiology of the relationship between mathematics and spatial abilities (e.g., Wai, Lubinski, & Benbow, 2009; Webb, Lubinski, & Benbow, 2007) has been examined using data collected in TEDS when the twins were 12 years old. The study investigated the relationship between three components of mathematics (Understanding Numbers, Non-Numerical Processes and Computation and Knowledge) and spatial abilities (Tosto et al., 2014). Overall, spatial abilities showed modest (0.26) heritability, with similar heritability for boys and girls, and no additional genetic factors related to gender. The phenotypic correlation between spatial abilities and the three mathematical components was on average 0.43. Common genetic factors explained ~60% of the correlation between the three components of mathematics and spatial ability; shared and nonshared environmental factors explained 26% and 14% of the phenotypic correlation, respectively. The computational components of mathematics showed less genetic influence in common with spatial abilities (genetic correlation 0.66) compared to processes such as symmetry or mental rotation, indexed by the "Non-Numerical Processes" component (genetic correlation 0.91). These results suggest that mostly the same genetic factors contribute to spatial and mathematical skills (genetic correlation between the composite of the three mathematical components and spatial ability was 0.75), but shared and nonshared environmental factors also contribute to the covariation between mathematical and spatial abilities.

Etiology of Relationships among Mathematical Subskills

The etiology of the links among different aspects of mathematics has also been tested using the TEDS sample. At age 10, phenotypic correlations among five different aspects of mathematics ("Mathematical Application," "Understanding Number," "Computation and Knowledge," "Mathematical Interpretation," "Non-Numerical Processes") ranged from 0.45 ("Computation and Knowledge" and "Non-Numerical Processes") to 0.68 ("Mathematical Application" and "Understanding Number") (Kovas, Petrill, et al., 2007). On average, the genetic correlation among the five subtests was 0.91, indicating that the same genetic factors contribute to individual differences in these different aspects of mathematics. For example, the genetic correlation between "Understanding Number" and "Mathematical Application" was 0.94, meaning that the genetic influences involved in these two mathematical components were almost the same. The examination of the bivariate heritability and environmental factors suggested that the observed covariation among different aspects of mathematics is largely explained by genetic factors. For example, the bivariate heritability of "Understanding Number" and "Mathematical Application" was 0.49 indicating that half of the phenotypic correlation of 0.68 is genetically mediated. The bivariate shared environment explained 29% of the phenotypic correlation, while 22% was explained by nonshared environment. Overall, these results suggest that observed correlations among different components of the mathematical domain are largely explained by the same genetic factors affecting them. The dissociation in performance across the different components largely stems from component-specific influences of nonshared environments.

Overlap in Genetic Influences on Academic Subjects

Two recent multivariate twin studies examined the genetic architecture of academic achievement at age 16 and found several results important for education.

Heritability of achievement (measured by the state exams performance) was moderate for all school subjects, including mathematics, language, science, art, and business (Rimfeld et al., in press). The moderate to high observed correlations in performance in different academic subjects were largely explained by overlapping genetic influences. This large overlap in genetic influences on different academic subjects was not reduced (or reduced by little) once intelligence was controlled for. In other words, to a large extent the same genetic factors affected exam performance in different subjects (e.g., mathematics and art), even after accounting for any shared effects with intelligence. Another recent study examined the relationship between core school subjects, including mathematics, and a whole range of other traits measured in the same sample (Krapohl et al., 2014). Genetic influences on achievement in core school subjects were found to be partly overlapping with genetic factors that affect intelligence as well as other characteristics, including self-efficacy, psychopathology, behavioral problems, health, well-being, and even perceptions of home and school environment.

The mechanisms underlying the observed genetic overlap remain poorly understood. Genetic correlations may result from pleiotropy—whereby one gene is expressed in multiple organ systems, and/or in multiple brain areas, contributing to covariation between different traits at a very general level. In contrast, causal paths from one trait to another may also be involved. For example, genetic factors may contribute to low reading ability which in turn affects mathematical learning, resulting in the observed phenotypic association between reading and math.

One recent study (Davis et al., 2014), mentioned earlier, applied both twin and GCTA analyses to the question of covariation between mathematics and reading. The multivariate twin analyses replicated previous results, showing that half of the observed correlation in these traits in 12-year-old children is due to shared genetic effects. Moreover, these results were supported by the GCTA analysis that showed a genetic correlation of 0.74, indicating that much of the same polymorphisms related to individual differences in both mathematics and reading achievement.

Genetic Effects Specific to Mathematics

Despite the large amount of genetic overlap, some genetic effects seem to be specific to mathematics (Hart et al., 2009; Kovas et al., 2005). It is possible that this genetic specificity is related to some specific aspect of mathematically relevant cognition. Numerical estimation is a particularly salient cognitive skill that is correlated with mathematics achievement (e.g., Halberda, Mazzocco, & Feigenson, 2008; Siegler & Opfer, 2003). Estimation abilities (at least some forms of estimation) are evolutionarily conserved, as they are present in many animal species and in human infants (e.g., Agrillo, Piffer, & Bisazza, 2010; Feigenson, Carey, & Spelke, 2002; Geary, Berch, & Mann Koepke, 2015; Xu & Arriaga, 2007). In other words, something in the genetic code of humans and other animals enables the appreciation of approximate numerical information. It is possible that genetic factors involved in variation in estimation are uniquely associated with mathematical variation, explaining the portion of genetic effects in mathematics that is not shared with other abilities. A recent study (Tosto et al., 2014) examined the etiology of number sense, assessed through performance on a number line and comparison of large numerosities, and found that most of the individual differences in estimation skills (~70%) are influenced by nonshared environmental factors. Genetic influences on number sense were modest in this study, but these measures may not fully explain human number sense. However, the multivariate genetic analysis (Tosto et al., in preparation) showed that the phenotypic correlation between mathematics at 16 years and two Number Sense measures (0.40 for number line and 0.29 for estimation of large numerosities) was largely explained by genetic factors (average bivariate heritability 0.73). This suggested that the modest genetic influences on number sense abilities were to a large extent the same as those on mathematical ability at this age. The results also showed that, contrary to expectation, there were no genetic influences shared between number sense and mathematics independent from intelligence.

Etiology of Relationships between Mathematical Ability and Related Affective Factors

Research is emerging to address the etiology of the observed relationships between mathematical ability and emotional processes linked to mathematics. For example, it is possible that some of the variance in mathematical skills, not shared with reading and *g*, may be explained by genetic and environmental factors that influence mathematical anxiety. One recent study examined the etiology of the relationship between mathematical performance, reading performance, math anxiety, and general anxiety (Wang et al., 2014). Genetic factors accounted for roughly 40% of the variation in mathematical anxiety, with a portion of these genetic factors also involved in general anxiety, as well as a second set of genetic factors that was associated with math-based problem solving.

Motivational factors, such as self-perceived ability and enjoyment of mathematics, may also make unique contributions to mathematical development. Reciprocal relationships between academic achievement and self-evaluation have been previously reported (Marsh, Byrne, & Yeung, 1999; Marsh & Yeung, 1997), but the etiology of these relationships can only be examined in genetically sensitive studies. Several studies reported that motivational factors across school years are moderately heritable, with approximately 40% of the variance explained by genetic factors and 60% by nonshared environmental factors (Kovas et al., 2015; Luo et al., 2011; Spinath, Spinath, & Plomin, 2008). The relationship between liking mathematics and achievement was investigated in the TEDS sample at age 9 (Spinath et al., 2008). Liking mathematics was moderately influenced by genetic factors (0.40) and the rest of its variation was associated with nonshared environmental factors (0.60). Moreover, the relationship between liking mathematics and mathematical achievement was very small. In a regression analysis, liking mathematics explained 0.5% of the total variance in mathematical scores, and only for girls. The finding that self-perception of abilities and enjoyment of mathematics depend to a large extent on genetic and nonshared environmental factors, rather than shared (family-wide or school-wide) influences requires new theoretical models to explain individual differences in motivation and its links to achievement.

Despite the encouraging findings described earlier concerning the many inter-relationships between mathematical ability and other traits over the course of development, additional research is still needed to provide an in-depth account and more comprehensive understanding of the sources of such relationships. A better understanding of the etiology should inform the way we teach mathematics in the context of other cognitive and noncognitive characteristics (Haworth & Plomin, 2011; Kovas & Plomin, 2012).

MATHEMATICAL DEVELOPMENT

Using the multivariate genetic approaches discussed earlier, studies have also examined the genetic and environmental contributions to mathematical development. The logic of the method is the same, in that multiple outcome measures are examined. However, instead of analyzing a given set of measures at one point in time, measures are examined across two or more time points (e.g., mathematics at time 1 and mathematics at time 2). Using these methods, studies consistently suggest that genetic influences are largely responsible for the longitudinal stability of individual differences in mathematics performance over time, and that this stability increases from childhood into adolescence (see Haworth et al., 2007; Kovas, Haworth, Dale, Petrill, et al., 2007; Kovas, Haworth, Dale, & Plomin, 2007; Kovas, Haworth, Harlaar, et al., 2007; Kovas, Petrill, et al., 2007). These results are similar to the other domains such as reading (Byrne et al., 2005; Harlaar, Dale, & Plomin, 2007; Petrill et al., 2007; Wadsworth, Corley, Hewitt, Plomin, & DeFries, 2002), as well as general and specific cognitive abilities (see Tucker-Drob & Briley, 2014 for an excellent meta-analysis).

Taken together, the literature suggests that although mathematics is a learned set of skills and clearly influenced by environmental factors, the stability of mathematics performance is due largely to genetic influences. This is important because one cannot explain individual differences in the development of mathematical skills solely from differences in learning or other aspects of the environment.

Longitudinal multivariate studies can also address the relationship between different measures over time. For example, one TEDS investigation, using a cross-lagged design, addressed the etiology of the relationship between self-evaluation of mathematical abilities and mathematics performance over time (between 9 and 12 years; Luo et al., 2011). The results showed that self-evaluation at 9 predicted mathematical achievement at 12, and that achievement at 9 predicted self-evaluation at 12, although the effect size of these relationships was very small ($r = \sim 0.10$). These reciprocal relationships were found to be largely genetically mediated (Luo et al., 2011). Genetic influences on self-evaluation at 9 years were associated with the genetic influences on later mathematics at 12 years, independent of the genetic influences on mathematics at 9 years. These results suggest that new genetic influences on mathematics that emerge at age 12 may be partially driven by the same genetic factors influencing self-evaluation in previous years.

However, longitudinal research remains relatively sparse. Moreover, the findings are limited in that they have not yet identified the mechanisms through which genetic and environmental influences impact mathematics achievement, the relationship between mathematics and other domains, and how mathematical skills develop. The remainder of the chapter addresses two ways in which these mechanisms may be more fully elucidated: molecular genetics and neuroimaging.

MOLECULAR GENETIC STUDIES OF MATHEMATICS

The rapidly advancing field of molecular genetics aims to identify the actual genes that are involved in trait variation. Today's molecular genetics benefits from new and continuously improving technologies, such as microarrays that allow genotyping of hundreds of thousands of DNA markers simultaneously, or new generation whole-genome sequencing. Recent molecular genetic research has provided much evidence in support of the hypothesis that genetic factors contribute to individual differences in all complex traits, such as learning abilities, disabilities, motivation, and achievement (e.g., Davies et al., 2011; Docherty, Kovas, & Plomin, 2011; Plomin, Haworth, & Davis, 2009). Moreover, this genetic research provides firm support for the polygenic conceptualization of heritability of complex traits: many DNA markers contribute to variation in any complex trait and the effect of any individual marker on the outcome is very small. These genetic markers are often referred to as quantitative trait loci (QTLs) (Plomin et al., 2013; Plomin, Owen, & McGuffin, 1994). The QTL theory implies that if many DNA markers of very small effect contribute to any complex trait, each person's educational profile may be a product of unique combinations of many genetic markers, environments, and interactions among them. This supposition has implications for gene-hunting approaches, requiring shifts from candidate gene to genome-wide association studies (GWAS). In addition, QTL perspective views disabilities and disorders as multifactorial, polygenic complex traits that develop under probabilistic rather than deterministic genetic influences. This has implications for diagnosis, treatment, and prevention of learning disabilities and low motivation.

GWAS rely on microarray technology that enables simultaneous genotyping of millions of DNA markers from an individual. Then, genomes of hundreds or thousands of people are compared to establish any points of differences in the genomes associated with observed differences in any particular trait (e.g., The Psychiatric GWAS Consortium, 2009). GWAS have begun to identify DNA polymorphisms (variants) associated with educationally relevant traits, such as reading (e.g., Meaburn, Harlaar, Craig, Schalkwyk, & Plomin, 2008), mathematics (Docherty et al., 2010), and general cognitive ability (e.g., Butcher, Davis, Craig, & Plomin, 2008; Davies et al., 2011). These studies provide information on the genetic variants involved in variation in cognition and achievement, as well as support for the conclusions drawn from the twin method: a large portion of the observed individual differences can be explained by the effects of multiple genes of small effect.

More GWAS will be conducted using increasingly better technologies to identify more genetic variants involved in educationally relevant processes. There is also a possibility that some of the currently unexplained heritability may not be common in the population, but instead be individual- and family-specific (*de novo* mutations). In a few years' time, affordable whole-genome sequencing will be available, so that each person's genome can be fully sequenced. These studies will be a major asset for finding both common and rare variants. In parallel to the ongoing search for DNA markers associated with educationally relevant traits, the research into understanding the mechanisms by which each gene affects a trait is also taking place. However, the process of tracing a path from each genetic variant to behavior may take a long time as the exact mechanisms by which each gene affects a trait are complex and involves multiple processes, such as gene expression, specific protein functions, structure and function of the brain.

It is therefore likely that the information on our complete genomic profiles will be routinely available long before we can truly use this knowledge. This sequence will be stored in computer databases, where other pieces of information are already held, such as information on family history of diseases, birth complications, height, weight, and other health and well-being profiles. Perhaps, information on individual's unique brain organization profiles will also be held in these databases. The questions of ethics, access to such information, and its uses are of utmost importance and require much consideration. Although it might be difficult to foresee specific applications of molecular genetic research to education now, it is even more difficult to imagine that new genetic knowledge will not affect our understanding of instructional processes and our approaches to education.

NEUROBIOLOGICAL MECHANISMS

As the chapters in this volume acknowledge, neuroimaging techniques have built upon and influenced the mathematical cognition literature by elucidating plausible structural and functional mechanisms associated with mathematical thinking and learning. For example, research has suggested that the parietal area, in particular the bilateral intraparietal sulcus (IPS), is important for representing and processing the approximate quantity of collections of objects (e.g., Dehaene, Piazza, Pinel, & Cohen, 2003). Specifically, tasks that require approximation of quantity have been shown to activate an area around the IPS, whereas exact calculation does not—at least in well-educated adults (Cantlon, Brannon, Cater, & Pelphrey, 2006; Price, Holloway, Räsänen, Vesterinen, & Ansari, 2007; Stanescu-Cosson et al., 2000). To date, only a handful of studies has examined the neurobiology of the IPS within genetically sensitive designs. In a small adult twin study of genetic and environmental contributions to brain activation during mental calculation ($N=31$ pairs), Pinel and Dehaene (2013) found that genetic factors were significant for activation in the bilateral posterior superior parietal lobe, the right intraparietal sulcus, and the left superior frontal region. Although these findings are very preliminary, they highlight the power of examining functional measures of mathematical cognition, particularly if these measures are collected in the context of functional measures in other domains (e.g., left occipital temporal regions for reading), and multivariate behavioral measures of mathematics and reading achievement and cognition, both within and across time.

We are currently conducting such a study using $N=300$ pairs of 14- to 16-year-old twins, selected from the Western Reserve Reading and Math Project described previously, as well as from the International Longitudinal Twin Study sample from the University of Colorado (Christopher et al., 2015). Twins are being assessed on a comprehensive battery of reading and mathematics measures, structural imaging, and functional imaging that assesses numerical approximation, reading comprehension, and working memory. Thus, we will be able to systematically examine how structural and functional measures map onto at least some of the aspects of the genetic and environmental etiology of mathematics and reading comorbidity and independence. Furthermore, these samples of twins were recruited when children were entering school and have been assessed continuously since recruitment. Thus, it will be possible to examine the genetic and environmental etiology of development in the context of behavioral and neuroimaging measures.

Moreover, there is a growing literature examining brain structure and function using molecular genetic methods (see Kere, 2014). These studies are particularly exciting because they have begun to examine the potential role of genetics in neurobiological networks, which may enhance our understanding of gene-environment interplay in the development of individual differences in complex human cognition and behavior. In the case of mathematics, this work is only beginning; there is a handful of studies with mixed results (e.g., Ludwig et al., 2013; Pettigrew et al., 2015).

Similar to work in mathematics, there is a rich and extensive literature using functional neuroimaging in both skilled and impaired readers. Studies have explored the neural correlates of domain-general processing that have long been known to be critical for reading comprehension, such as sentence comprehension (Meyler et al., 2007; Rimrodt et al., 2009) and working memory

(Beneventi, Tønnessen, & Ersland, 2009; Wolf et al., 2010). These studies also show abnormalities between typically developing readers and those with reading disability (RD), although the areas of anomalous activation extend beyond those of traditional reading tasks, including the prefrontal cortex. In terms of domain-specific skills, much work has been devoted to better understanding the neural correlates of word recognition/decoding. In brief, neurobiological correlates of word recognition in RD show perisylvian region dysfunction, including underactivation of left temporo-parietal and occipito-temporal regions and greater right temporo-parietal and occipito-temporal activation (e.g., Pugh et al., 2000; Richlan, Kronbichler, & Wimmer, 2009) as compared to typically developing readers. Typically developing readers demonstrate consistent patterns of left-hemisphere perisylvian activation in temporo-parietal and inferior frontal regions, as well as occipito-temporal activation that are especially associated with fast and efficient reading (Fiez & Petersen, 1998; Georgiewa et al., 2002; Simos et al., 2002; Turkeltaub, Gareau, Flowers, Zeffiro, & Eden, 2003).

This brief review of the neuroimaging literature illustrates plausible domain-general and domain-specific mechanisms for the genetic overlap and discrepancy between reading disability and poor mathematics achievement found in the behavioral genetic literature. Consider a "domain-general" skill such as working memory that shows: (1) activation across widely distributed brain regions, (2) is highly heritable, and (3) is correlated with many complex behavioral outcomes (Koten et al., 2009). Within a twin design, one would be able to systematically examine whether structural and functional differences associated with working memory account for comorbidity between reading and mathematical skills. Conversely, it is also possible to examine functional differences in measures related to mathematics but not reading (e.g., numerical estimation) that demonstrate activation differences in more narrowly distributed regions (e.g., intraparietal sulcus) specific to math outcomes. When assessed in the same individuals, these relationships could be further examined by comparing patterns of genetic correlations within measures of brain function arising from same/different tasks, occurring in the same/different neurobiological regions. Moreover, beyond genetics, results indicate that shared environmental factors for mathematical ability are statistically significant (and substantial) for mathematics, but not for reading. Better understanding the structural and functional correlates of the shared environment may theoretically yield knowledge about the boundaries of genetic and environmental vulnerability at the level of the brain, in addition to the level of behavior.

One can also examine the question of domain-general and domain-specific factors in the context of selected groups. The extant literature suggests that a domain-general + domain-specific model appears most tenable (Cirino, Morris, & Morris, 2007; Moll, Fussenegger, Willburger, & Landerl, 2009; Rosselli, Matute, Pinto, & Ardila, 2006; van der Sluis, de Jong, & van der Leij, 2004), but what is unclear are the conditions under which the data are best explained by a model where: (1) different neurobiological factors account for seemingly similar behavioral mathematical and reading impairments, or (2) whether there is some overlap in neurobiological structure and function across mathematics and reading, in addition to unique neurobiological mechanisms for reading and mathematical difficulties.

CONCLUSIONS

Taken together, the quantitative genetic literature strongly suggests that genetic influences are important for the development of mathematical ability. These genetic influences contribute to the relationship among mathematics-related constructs; the relationship between mathematics and other outcomes such as reading, general cognitive ability, and language; and the longitudinal stability of mathematical ability over time. However, molecular genetic studies to date have not yielded specific replicable markers accounting for variation in mathematical ability. We argue that inclusion of neuroimaging and more careful attention to the cognitive and affective processes underlying mathematics may help with establishing a more comprehensive understanding of the etiology of individual differences in mathematical and related traits. This improved understanding of the etiology will help with finding new approaches to personalization of education that will take into account all levels of individual differences, including genetic, environmental, as well as neural and psychological (cognitive, motivational, and emotional) risk and protective factors.

ACKNOWLEDGMENT

YK's work is supported by the grant from the Government of the Russian Federation (Grant 11.G34.31.0043).

REFERENCES

Agrillo, C., Piffer, L., & Bisazza, A. (2010). Number versus continuous quantity in numerosity judgments by fish. *Cognition, 119*, 281–287.

Alarcón, M., DeFries, J. C., Light, J. G., & Pennington, B. F. (1997). A twin study of mathematics disability. *Journal of Learning Disabilities, 30*, 617–623.

Alarcón, M., Knopik, V. S., & DeFries, J. C. (2000). Covariation of mathematics achievement and general cognitive ability. *Journal of School Psychology, 38*, 63–77.

Bates, T. C. (2008). Current genetic discoveries and education: Strengths, opportunities, and limitations. *Mind, Brain, and Education*, *2*(2), 74–79.

Beneventi, H., Tønnessen, F. E., & Ersland, L. (2009). Dyslexic children show short-term memory deficits in phonological storage and serial rehearsal: An fMRI study. *International Journal of Neuroscience*, *119*(11), 2017–2043.

Boomsma, D., Busjahn, A., & Peltonen, L. (2002). Classical twin studies and beyond. *Nature Reviews Genetics*, *3*(11), 872–882.

Butcher, L. M., Davis, O. S. P., Craig, I. W., & Plomin, R. (2008). Genome-wide quantitative trait locus association scan of general cognitive ability using pooled DNA and 500k single nucleotide polymorphism microarrays. *Genes, Brain, and Behavior*, *7*, 435–446. http://dx.doi.org/10.1111/j.1601183X.2007.00368.x.

Butterworth, B., & Kovas, Y. (2013). Understanding neurocognitive developmental disorders can improve education for all. *Science*, *340*(6130), 300–305.

Byrne, B., Wadsworth, S., Corley, R., Samuelsson, S., Quain, P., DeFries, J. C., et al. (2005). Longitudinal twin study of early literacy development: Preschool and kindergarten phases. *Scientific Studies of Reading*, *9*(3), 219–235.

Cantlon, J. F., Brannon, E. M., Cater, E. J., & Pelphrey, K. A. (2006). Functional imaging of numerical processing in adults and 4-y-old children. *PLoS Biology*, *4*, 844–854. http://dx.doi.org/10.1371/journal.pbio.0040125.

Christopher, M. E., Hulslander, J., Byrne, B., Samuelsson, S., Keenan, J. M., Pennington, B., et al. (2015). Genetic and environmental etiologies of the longitudinal relations between prereading skills and reading. *Child Development*, *86*, 342–361.

Cirino, P. T., Morris, M. K., & Morris, R. D. (2007). Semantic, executive, and visuospatial abilities in mathematical reasoning of referred college students. *Assessment*, *14*(1), 94–104.

Davies, G., Tenesa, A., Payton, A., Yang, J., Harris, S. E., Liewald, D., et al. (2011). Genome-wide association studies establish that human intelligence is highly heritable and polygenic. *Molecular Psychiatry*, *16*, 996–1005.

Davis, O. S. P., Band, G., Pirinen, M., Haworth, C. M. A., Meaburn, E. L., Kovas, Y., et al. (2014). The correlation between reading and mathematics ability at age twelve has a substantial genetic component. *Nature Communications*, http://dx.doi.org/10.1038/ncomms5204.

Davis, O. S. P., Haworth, C. M. A., & Plomin, R. (2009). Dramatic increase in heritability of cognitive development from early to middle childhood an 8-year longitudinal study of 8,700 pairs of twins. *Psychological Science*, *20*(10), 1301–1308.

Davis, O. S. P., Kovas, Y., Harlaar, N., Busfield, P., McMillan, A., Frances, J., et al. (2008). Generalist genes and the Internet generation: Etiology of learning abilities by web testing at age 10. *Genes, Brain, and Behavior*, *7*, 455–462.

DeFries, J. C., & Fulker, D. W. (1985). Multiple regression analysis of twin data. *Behavior Genetics*, *15*, 467–473. http://dx.doi.org/10.1007/BF01066239.

DeFries, J. C., & Fulker, D. W. (1988). Multiple regression analysis of twin data: Etiology of deviant scores versus individual differences. *Acta Geneticae Medicae et Gemellologiae*, *37*, 205–216.

Dehaene, S., Piazza, M., Pinel, P., & Cohen, L. (2003). Three parietal circuits for number processing. *Cognitive Neuropsychology*, *20*, 487–506. http://dx.doi.org/10.1080/02643290244000239.

Docherty, S. J., Davis, O. S. P., Kovas, Y., Meaburn, E. L., Dale, P. S., Petrill, S. A., et al. (2010). A genome-wide association study identifies multiple loci associated with mathematics ability and disability. *Genes, Brain, and Behavior*, *9*(2), 234–247.

Docherty, S. J., Kovas, Y., & Plomin, R. (2011). Gene-environment interaction in the etiology of mathematical ability using SNP sets. *Behavior Genetics*, *41*, 141–154.

Feigenson, L., Carey, S., & Spelke, E. S. (2002). Infants' discrimination of number vs. continuous extent. *Cognitive Psychology*, *44*, 33–66.

Fiez, J. A., & Petersen, S. E. (1998). Neuroimaging studies of word reading. *Proceedings of the National Academy of Sciences*, *95*(3), 914–921.

Fuchs, L. S., Geary, D. C., Compton, D. L., Fuchs, D., & Hamlett, C. L. (2014). Sources of individual differences in emerging competence with numeration understanding versus multidigit calculation skill. *Journal of Educational Psychology*, *106*, 482–498.

Geary, D. C. (2011). Consequences, characteristics, and causes of mathematical learning disabilities and persistent low achievement in mathematics. *Journal of Developmental and Behavioral Pediatrics: JDBP*, *32*(3), 250.

Geary, D. C., Bailey, D. H., & Hoard, M. K. (2009). Predicting mathematical achievement and mathematical learning disability with a simple screening tool: The number sets test. *Journal of Psychoeducational Assessment, 27*, 265.

Geary, D. C., Berch, D. B., & Mann Koepke, K. (2015). *Evolutionary origins and early development of number processing*. San Diego, CA: Academic Press.

Georgiewa, P., Rzanny, R., Gaser, C., Gerhard, U. J., Vieweg, U., Freesmeyer, D., et al. (2002). Phonological processing in dyslexic children: A study combining functional imaging and event related potentials. *Neuroscience Letters, 318*(1), 5–8.

Greven, C. U., Kovas, Y., Willcutt, E. G., Petrill, S. A., & Plomin, R. (2014). Evidence for shared genetic risk between ADHD symptoms and reduced mathematics ability: A twin study. *Journal of Child Psychology and Psychiatry, 55*, 39–48.

Halberda, J., Mazzocco, M. M. M., & Feigenson, L. (2008). Individual differences in nonverbal estimation ability predict maths achievement. *Nature, 455*, 665–668.

Harlaar, N., Dale, P. S., & Plomin, R. (2007). From learning to read to reading to learn: Substantial and stable genetic influence. *Child Development, 78*(1), 116–131.

Hart, S. A., Petrill, S. A., & Kamp Dush, C. M. (2010). Genetic influences on language, reading, and mathematics skills in a national sample: An analysis using the National Longitudinal Survey of Youth. *Language, Speech, and Hearing Services in Schools, 41*, 118–128.

Hart, S. A., Petrill, S. A., & Thompson, L. A. (2010). A factorial analysis of timed and untimed measures of mathematics and reading abilities in school aged twins. *Learning and Individual Differences, 20*, 63–69.

Hart, S. A., Petrill, S. A., Thompson, L. A., & Plomin, R. (2009). The ABCs of math: A genetic analysis of mathematics and its links with reading ability and general cognitive ability. *Journal of Educational Psychology, 101*(2), 388–402.

Haworth, C. M. A., Kovas, Y., Dale, P. S., & Plomin, R. (2008). Science in elementary school: Generalist genes and school environments. *Intelligence, 36*(6), 694–701.

Haworth, C. M. A., Kovas, Y., Harlaar, N., Hayiou-Thomas, E. M., Petrill, S. A., Dale, P. S., et al. (2009). Generalist genes and learning disabilities: A multivariate genetic analysis of low performance in reading, mathematics, language and general cognitive ability in a sample of 8000 12-year-old twins. *Journal of Child Psychology and Psychiatry, 50*, 1318–1325. http://dx.doi.org/10.1111/j.14697610.2009.02114.x.

Haworth, C. M. A., Kovas, Y., Petrill, S. A., & Plomin, R. (2007). Developmental origins of low mathematics performance and normal variation in twins from 7 to 9 years. *Twin Research and Human Genetics, 10*, 106–117.

Haworth, C. M. A., & Plomin, R. (2011). Genetics and education: Toward a genetically sensitive classroom. In K. R. Harris, S. Graham, & T. Urdan (Eds.), *APA educational psychology handbook, Vol. 1: Theories, constructs, and critical issues*. Washington, DC: American Psychological Association. http://dx.doi.org/10.1037/13273-018.

Haworth, C. M. A., Wright, M. J., Luciano, M., Martin, N. G., de Geus, E. J. C., van Beijsterveldt, C. E. M., et al. (2010). The heritability of general cognitive ability increases linearly from childhood to young adulthood. *Molecular Psychiatry, 15*, 1112–1120. http://dx.doi.org/10.1038/mp.2009.55.

Houde, O., Rossi, S., Lubin, A., & Joliot, M. (2010). Mapping numerical processing, reading, and executive functions in the developing brain: An fMRI meta-analysis of 52 studies including 842 children. *Developmental Science, 13*, 876–885.

Joint Council for Qualifications. (2013). http://www.jcq.org.uk/ Accessed 12.29.14.

Kere, J. (2014). The molecular genetics and neurobiology of developmental dyslexia as model of a complex phenotype. *Biochemical and Biophysical Research Communications, 452*(2), 236–243.

Knopik, V. S., Alarcón, M., & DeFries, J. C. (1997). Comorbidity of mathematics and reading deficits: Evidence for a genetic etiology. *Behavior Genetics*, *27*, 447–453.

Koten, J. W., Wood, G., Hagoort, P., Goebel, R., Propping, P., Willmes, K., et al. (2009). Genetic contribution to variation in cognitive function: An FMRI study in twins. *Science*, *323*(5922), 1737–1740.

Kovas, Y., Garon-Carrier, G., Boivin, M., Petrill, S. A., Malykh, S. B., Spinath, F., et al. (2015). Why children differ in motivation to learn: Insights from over 13,000 twins from 6 countries. *Personality and Individual Differences*, *80*, 51–63.

Kovas, Y., Harlaar, N., Petrill, S. A., & Plomin, R. (2005). "Generalist genes" and mathematics in 7-year-old twins. *Intelligence*, *33*, 473–489.

Kovas, Y., Haworth, C. M. A., Dale, P. S., Petrill, S. A., & Plomin, R. (2007). Mathematical ability of 10-year-old boys and girls: Genetic and environmental etiology of typical and low performance. *Journal of Learning Disabilities*, *40*(6), 554–567.

Kovas, Y., Haworth, C. M. A., Dale, P. S., & Plomin, R. (2007). The genetic and environmental origins of learning abilities and disabilities in the early school years. *Monographs of the Society for Research in Child Development*, *72*(3) vii, 1–160.

Kovas, Y., Haworth, C. M. A., Harlaar, N., Petrill, S. A., Dale, P. S., & Plomin, R. (2007). Overlap and specificity of genetic and environmental influences on mathematics and reading disability in 10-year-old twins. *Journal of Child Psychology and Psychiatry*, *48*, 914–922.

Kovas, Y., Petrill, S. A., & Plomin, R. (2007). The origins of diverse domains of mathematics: Generalist genes but specialist environments. *Journal of Educational Psychology*, *99*(1), 128–139.

Kovas, Y., & Plomin, R. (2012). Genetics and genomics: Good, bad and ugly. In S. Della Salla & M. Anderson (Eds.), *Neuroscience in education: The good, the bad and the ugly* (pp. 155–173). Oxford: Oxford University Press.

Krapohl, E., Rimfeld, K., Shakeshaft, N. G., Trzaskowski, M., McMillan, A., Pingault, J.-B., et al. (2014). The high heritability of educational achievement reflects many genetically influenced traits, not just intelligence. *Proceedings of the National Academy of Sciences of the United States of America*, *111*(42), 15273–15278. www.pnas.org/cgi/doi/10.1073/pnas.1408777111.

Libertus, M. E., Feigenson, L., Halberda, J., & Landau, B. (2014). Understanding the mapping between numerical approximation and number words: Evidence from Williams syndrome and typical development. *Developmental Science*, *17*(6), 905–919.

Loehlin, J. C., & Nichols, J. (1976). *Heredity, environment and personality*. Austin: University of Texas.

Ludwig, K. U., Sämann, P., Alexander, M., Becker, J., Bruder, J., Moll, K., et al. (2013). A common variant in myosin-18B contributes to mathematical abilities in children with dyslexia and intraparietal sulcus variability in adults. *Translational Psychiatry*, *3*(2). e229.

Luo, Y. L., Kovas, Y., Haworth, C. M., & Plomin, R. (2011). The etiology of mathematical self-evaluation and mathematics achievement: Understanding the relationship using a cross-lagged twin study from age 9 to 12. *Learning and Individual Differences*, *21*(6), 710.

Marsh, H. W., Byrne, B. M., & Yeung, A. S. (1999). Causal ordering of academic self-concept and achievement: Reanalysis of a pioneering study and revised recommendations. *Educational Psychologist*, *34*(3), 155–167.

Marsh, H. W., & Yeung, A. S. (1997). Causal effects of academic self-concept on academic achievement: Structural equation models of longitudinal data. *Journal of Educational Psychology*, *89*(1), 41–54.

Martens, M. A., Wilson, S. J., & Reutens, D. C. (2008). Research review: Williams syndrome: A critical review of the cognitive, behavioral, and neuroanatomical phenotype. *Journal of Child Psychology & Psychiatry*, *49*(6), 576–608.

Martin, N., Boomsma, D., & Machin, G. (1997). A twin-pronged attack on complex traits. *Nature Genetics*, *17*(4), 387–392.

Martin, N. G., Jardine, R., & Eaves, L. J. (1984). Is there only one set of genes for different abilities? A reanalysis of the National Merit Scholarship Qualifying Test (NMSQT) data. *Behavior Genetics*, *14*(4), 355–370.

Mazzocco, M. M. M., Feigenson, L., & Halberda, J. (2011). Impaired acuity of the approximate number system underlies mathematical learning disability (dyscalculia). *Child Development*, *82*(4), 1224–1237.

Meaburn, E. L., Harlaar, N., Craig, I. W., Schalkwyk, L. C., & Plomin, R. (2008). Quantitative trait locus association scan of early reading disability and ability using pooled DNA and 100K SNP microarrays in a sample of 5760 children. *Molecular Psychiatry*, *13*(7), 729–740.

Meyler, A., Keller, T. A., Cherkassky, V. L., Lee, D., Hoeft, F., Whitfield-Gabrieli, S., et al. (2007). Brain activation during sentence comprehension among good and poor readers. *Cerebral Cortex*, *17*(12), 2780–2787.

Moll, K., Fussenegger, B., Willburger, E., & Landerl, K. (2009). RAN is not a measure of orthographic processing. Evidence from the asymmetric German orthography. *Scientific Studies of Reading*, *13*(1), 1–25.

National Assessment of Educational Progress. (2014). http://www.nationsreportcard.gov/reading_math_g12_2013/files/Results_Appendix_Math_G12.pdf Accessed 12.12.14.

Neale, M. C., Boker, S. M., Xie, G., & Maes, H. H. (2006). *MX: Statistical modeling* (7th ed.). Richmond, VA: Department of Psychiatry.

Oliver, B., Harlaar, N., Hayiou-Thomas, M. E., Kovas, Y., Walker, S. O., Petrill, S. A., et al. (2004). A twin study of teacher-reported mathematics performance and low performance in 7-year-olds. *Journal of Educational Psychology*, *96*, 504–517.

Oliver, B. R., & Plomin, R. (2007). Twins' Early Development Study (TEDS): A multivariate, longitudinal genetic investigation of language, cognition and behavior problems from childhood through adolescence. *Twin Research and Human Genetics*, *10*(01), 96–105.

Petrill, S. A., Deater-Deckard, K., Thompson, L. A., Schatschneider, C., DeThorne, L. S., & Vandenbergh, D. J. (2007). Longitudinal genetic analysis of early reading: The Western Reserve reading project. *Reading and Writing*, *20*(1–2), 127–146.

Pettigrew, K. A., Fajutrao Valles, S. F., Moll, K., Northstone, K., Ring, S., Pennell, C., et al. (2015). Lack of replication for the myosin-18B association with mathematical ability in independent cohorts. *Genes, Brain, and Behavior*, *14*(4), 369–376.

Pinel, P., & Dehaene, S. (2013). Genetic and environmental contributions to brain activation during calculation. *NeuroImage*, *81*, 306–316.

Plomin, R., & Deary, I. J. (2014). Genetics and intelligence differences: Five special findings. *Molecular Psychiatry*, *20*(1), 98–108. http://dx.doi.org/10.1038/mp.2014.105.

Plomin, R., DeFries, J. C., Knopik, V. S., & Neiderhiser, J. M. (2013). *Behavioral genetics* (6th ed.). New York: Worth Publishers.

Plomin, R., Haworth, C. M. A., & Davis, O. S. P. (2009). Common disorders are quantitative traits. *Nature Reviews Genetics*, *10*, 872–878. http://dx.doi.org/10.1038/nrg2670.

Plomin, R., & Kovas, Y. (2005). Generalist genes and learning disabilities. *Psychological Bulletin*, *131*(4), 592.

Plomin, R., Owen, M. J., & McGuffin, P. (1994). The genetic basis of complex human behaviors. *Science*, *264*(5166), 1733–1739.

Price, G. R., Holloway, I., Räsänen, P., Vesterinen, M., & Ansari, D. (2007). Impaired parietal magnitude processing in developmental dyscalculia. *Current Biology*, *17*(24), R1042–R1043.

Pugh, K. R., Mencl, W. E., Jenner, A. R., Katz, L., Frost, S. J., Lee, J. R., et al. (2000). Functional neuroimaging studies of reading and reading disability (developmental dyslexia). *Mental Retardation and Developmental Disabilities Research Reviews*, *6*(3), 207–213.

Rhea, S. A., Bricker, J. B., Wadsworth, S. J., & Corley, R. P. (2013). The Colorado adoption project. *Twin Research and Human Genetics*, *16*(01), 358–365.

Richlan, F., Kronbichler, M., & Wimmer, H. (2009). Functional abnormalities in the dyslexic brain: A quantitative meta-analysis of neuroimaging studies. *Human Brain Mapping*, *30*(10), 3299–3308.

Rimfeld, K., Kovas, Y., Dale, P. S., & Plomin, R. (in press). From mathematics to art—Pleiotropy across academic subjects at the end of compulsory education. *Scientific Reports*.

Rimrodt, S. L., Clements-Stephens, A. M., Pugh, K. R., Courtney, S. M., Gaur, P., Pekar, J. J., et al. (2009). Functional MRI of sentence comprehension in children with dyslexia: Beyond word recognition. *Cerebral Cortex*, *19*(2), 402–413.

Rodic, M., Gaysina, D., Docherty, S., Malykh, S., Rimfeld, K., Plomin, R., et al. (in press). Studying rare genetic syndromes as a method of investigating aetiology of normal variation in educationally relevant traits. In Y. Kovas, S. B. Malykh, & D. Gaysina (Eds.), *Behavioural genetics for education*. Palgrave Macmillan.

Rosselli, M., Matute, E., Pinto, N., & Ardila, A. (2006). Memory abilities in children with subtypes of dyscalculia. *Developmental Neuropsychology*, *30*(3), 801–818.

Science, Technology, Engineering and Math: Education for Global Leadership. U.S. Department of Education. (2014). http://www.ed.gov/stem Accessed 12.08.14.

Siegler, R. S., & Opfer, J. (2003). The development of numerical estimation: Evidence for multiple representations of numerical quantity. *Psychological Science*, *14*, 237–243.

Simos, P. G., Fletcher, J. M., Bergman, E., Breier, J. I., Foorman, B. R., Castillo, E. M., et al. (2002). Dyslexia-specific brain activation profile becomes normal following successful remedial training. *Neurology*, *58*(8), 1203–1213.

Spinath, F. M., Spinath, B., & Plomin, R. (2008). The nature and nurture of intelligence and motivation in the origins of sex differences in elementary school achievement. *European Journal of Personality*, *22*(3), 211–229.

Stanescu-Cosson, R., Pinel, P., van de Moortele, P. F., Le Bihan, D., Cohen, L., & Dehaene, S. (2000). Understanding dissociations in dyscalculia. A brain imaging study of the impact of number size on the cerebral networks for exact and approximate calculation. *Brain*, *123*, 2240–2255. http://dx.doi.org/10.1093/brain/123.11.2240.

Swanson, H. L., & Jerman, O. (2007). The influence of working memory on reading growth in subgroups of children with reading disabilities. *Journal of Experimental Child Psychology*, *96*, 249–283.

The Psychiatric GWAS Consortium. (2009). Genomewide association studies: History, rationale and prospects for psychiatric disorders. *The American Journal of Psychiatry*, *166*(5), 540–556.

Thompson, L. A., Detterman, D. K., & Plomin, R. (1991). Associations between cognitive abilities and scholastic achievement: Genetic overlap but environmental differences. *Psychological Science*, *2*, 158–165.

Tosto, M. G., Petrill, S. A., Halberda, J., Trzaskowski, M., Tikhomirova, T. N., Bogdanova, O. Y., et al. (2014). Why do we differ in number sense? Evidence from a genetically sensitive investigation. *Intelligence*, *43*, 35–46.

Tosto, M. G., Petrill, S. A., Malykh, S., Mazzocco, M. M. M., Haworth, C. M. A., Malki, K., et al. (in preparation). Number sense and mathematics: If, which, when and how?

Tucker-Drob, E. M., & Briley, D. A. (2014). Continuity of genetic and environmental influences on cognition across the life span: A meta-analysis of longitudinal twin and adoption studies. *Psychological Bulletin*, *140*(4), 949–979.

Turkeltaub, P. E., Gareau, L., Flowers, D. L., Zeffiro, T. A., & Eden, G. F. (2003). Development of neural mechanisms for reading. *Nature Neuroscience*, *6*(7), 767–773.

van der Sluis, S., de Jong, P. F., & van der Leij, A. (2004). Inhibition and shifting in children with learning deficits in arithmetic and reading. *Journal of Experimental Child Psychology*, *87*(3), 239–266.

Visscher, P. M., Medland, S. E., Ferreira, M. A., Morley, K. I., Zhu, G., Cornes, B. K., et al. (2006). Assumption-free estimation of heritability from genome-wide identity-by-descent sharing between full siblings. *PLoS Genetics*, *2*(3) e41.

Wadsworth, S. J., Corley, R. P., Hewitt, J. K., Plomin, R., & DeFries, J. C. (2002). Parent-offspring resemblance for reading performance at 7, 12 and 16 years of age in the Colorado Adoption Project. *Journal of Child Psychology and Psychiatry*, *43*(6), 769–774.

Wai, J., Lubinski, D., & Benbow, C. P. (2009). Spatial ability for STEM domains: Aligning over 50 years of cumulative psychological knowledge solidifies its importance. *Journal of Educational Psychology*, *101*(4), 817–835.

Wang, Z., Hart, S. A., Kovas, Y., Lukowski, S., Soden, B., Thompson, L. A., et al. (2014). Who is afraid of math? Two sources of genetic variance for mathematical anxiety. *Journal of Child Psychology and Psychiatry*, *55*(9), 1056–1064.

Webb, R. M., Lubinski, D., & Benbow, C. P. (2007). Spatial ability: A neglected dimension in talent searches for intellectually precocious youth. *Journal of Educational Psychology*, *99*(2), 397–420.

Wolf, R. C., Sambataro, F., Lohr, C., Steinbrink, C., Martin, C., & Vasic, N. (2010). Functional brain network abnormalities during verbal working memory performance in adolescents and young adults with dyslexia. *Neuropsychologia*, *48*(1), 309–318.

Xu, F., & Arriaga, R. (2007). Number discrimination in 10-month-old infants. *British Journal of Developmental Psychology*, *25*, 103–108.

Yang, J., Lee, S. H., Goddard, M. E., & Visscher, P. M. (2011). GCTA: A tool for genome-wide complex trait analysis. *American Journal of Human Genetics*, *88*(1), 76–82.

Yang, J., Lee, S. H., Goddard, M. E., & Visscher, P. M. (2013). Genome-wide complex trait analysis (GCTA): Methods, data analyses, and interpretations. *Methods in Molecular Biology*, *1019*, 215–236.

Chapter 12

Genetic Syndromes as Model Pathways to Mathematical Learning Difficulties: Fragile X, Turner, and 22q Deletion Syndromes

Michèle M.M. Mazzocco[1,2], Andrea I. Quintero[3,4], Melissa M. Murphy[5], Michael McCloskey[6]

[1]Institute of Child Development, University of Minnesota, Minneapolis, Minnesota, USA, [2]Center for Early Education and Development, University of Minnesota, Minneapolis, Minnesota, USA, [3]MIND Institute, UC Davis Medical Center, University of California at Davis, Davis, California, USA, [4]Department of Psychiatry and Behavioral Sciences, University of California at Davis, Davis, California, USA, [5]School of Education, Notre Dame of Maryland University, Baltimore, Maryland, USA, [6]Department of Cognitive Science, Johns Hopkins University, Baltimore, Maryland, USA

INTRODUCTION

Children with mathematics learning disability (MLD)[1] share a lifelong struggle with at least some aspects of mathematics. However, this shared struggle does not mean that the causes, nature, and trajectory of their mathematical challenges are also shared. Do these disabilities reflect global or specific differences in cognitive abilities or processing? Are mathematics disabilities linked primarily to deficits in one or more numerical skills, such as those described by Ansari, Hyde, Mou, and Fias (Chapters 2, 3, and 8)? Alternatively, are the primary deficits underlying MLD secondary to nonnumerical competencies, such as in the memory and perceptual systems described by Menon and by Lewis and colleagues, respectively (Chapters 4 and 6)? If underlying causes of mathematics disabilities vary among children with MLD (see Chapter 7), do different

1. We use the term *mathematics learning disability* as synonymous with *dyscalculia*, to indicate persistent and significant impairment in mathematical thinking or mathematics achievement; this does not include mild or occasional mathematics difficulties.

Development of Mathematical Cognition. http://dx.doi.org/10.1016/B978-0-12-801871-2.00012-5

etiologies lead to similar behavioral outcomes, or are the differences apparent at genetic, neural, and behavioral levels?

Since MLD is considered by many to be a heterogeneous disorder (e.g., Kaufmann et al., 2013), we should attend to known and potential sources of its variation to understand its development and function. But how can we systematically study variation in a latent variable (i.e., MLD) for which there is no widely accepted diagnostic marker or working definition? One approach is to study known genetic disorders associated with poor mathematics achievement. In this chapter we exemplify this approach by summarizing research on the observable intellectual characteristics (or *cognitive phenotypes*) of individuals with fragile X, Turner, or chromosome 22q11.2 deletion syndrome. We illustrate that one benefit of syndrome research is the models of variability it provides for MLD profiles. These models reflect how variations in the development of mathematical thinking may nevertheless lead to the same *functional* outcome: poor mathematics achievement.

Why Focus on Fragile X, Turner, and 22q Deletion Syndromes to Study MLD?

There are many reasons to study these particular syndromes as models of mathematics disability. First, as we describe in subsequent sections, all three syndromes are relatively common disorders with a known genetic etiology, occurring in 1 of every 2000-9000 live births (live *female* births for Turner syndrome). Thus, research on any of these disorders is directly applicable to a significant number of individuals. Second, all three disorders have a well-described cognitive phenotype that includes mathematics difficulties. Although it is unclear whether math difficulties in children with any one of these syndromes are *primary* (i.e., caused directly by the syndrome) or *secondary* (associated with a general underlying cognitive deficit such as poor working memory), we can be confident that these math difficulties are not simply the result of overall intellectual disability. This confidence stems from how frequently research participants in studies of these cognitive phenotypes exclude individuals with intellectual disability (as has been common practice in studies of MLD in fragile X or Turner syndrome) or include IQ-matched comparison groups or other statistical means to control for effects of IQ (e.g., Quintero, Beaton, Harvey, Ross, & Simon, 2014). Third, studies of any one of these three syndromes reveal, to some extent, deficits in executive function skills (such as working memory); thus, studies of any or all of these syndromes may contribute to our understanding of how executive function skills contribute to mathematical learning and to the development of MLD. Finally, there is much variability (both within and between these three populations) in terms of (a) the degree to which MLD (and other cognitive or behavioral difficulties) is evident, (b) when the disability is manifested, and (c) the skills with which performance difficulties are correlated. This variability may further help explain variation in mathematics ability in the general population in addition to persons with one of these syndromes.

Contributions of Syndrome Research to Understanding MLD

To assess learning disability phenotypes associated with developmental disorders, it is helpful to draw from the "three-level framework" that considers the roles of biology, cognition, and behavior in development (Frith, 2001). When studying known genetic disorders, specific *biological* contributions are more clearly defined than is the presence or absence of MLD, at least for syndromes for which an identifiable biological cause is known. This is the case for fragile X, Turner, and 22q11.2 deletion syndromes. (In contrast, MLD is linked to performance outcomes that vary across studies.) *Cognition*, the internal processes influenced by both biology (e.g., genes, brain function) and environment (e.g., instruction, intervention, classroom influences), is further influenced by *behavior* (e.g., attention in the classroom, arousal levels during test taking). The distinction between *cognition* and *behavior* is relevant to work on syndrome phenotypes, because initial phenotype research was often based on standardized testing of broadly defined *behavioral* outcomes, such as mathematics achievement—for which associated or underlying cognitive deficits were inferred but not always specified (e.g., specific aspects of spatial, working memory, attention, or number sense deficits). One goal of the syndrome research summarized in this chapter is to specify these underlying cognitive contributions to poor mathematics achievement, thus devoting more attention to cognition than the global behaviors of poor mathematics achievement.

Why study *multiple* phenotypes to understand MLD? Broadly speaking, many syndrome phenotypes are associated with difficulties in mathematics-related behaviors, suggesting that studying any one of them may specify cognitive underpinnings of MLD. For instance, fragile X, Turner, and 22q11.2 deletion syndromes are all associated with poor mathematics achievement, poor working memory, and deficient spatial skills, despite relatively spared verbal expression and reading abilities. Can one disorder, perhaps the most feasible to study, serve as the model of MLD? The most direct answer is no, because differences in underlying skills such as attentional impairments or short-term verbal memory strengths contradict the misleading notion of "similar" phenotypes as they pertain to mathematics abilities. Indeed, even the nature of the broad behavioral deficits may differ. These variations reflect the widespread heterogeneity of MLD in the general population. Poor mathematics achievement may manifest as slow but accurate performance in some cases and as consistent computational errors in others. This inconsistency poses challenges for understanding (and defining) MLD, if participants with discrepant profiles comprise one MLD group in research. But classifying children with MLD into subgroups can be challenging because the methods used to do so may be somewhat arbitrary (e.g., using cut-off scores) or based on limited available data (i.e., using scores at one point in time). Accordingly, studying multiple syndromes allows us to group participants who are at risk for MLD in a nonarbitrary manner that may model distinct pathways and unique MLD subtypes.

SYNDROMES AS MODELS OF MLD

Before summarizing mathematics-related phenotypes of these syndromes, we very briefly describe each disorder. The reader is referred elsewhere for more in-depth summaries of fragile X syndrome (Cornish, Levitas, & Sudhalter, 2007; Murphy, 2009), Turner syndrome (Davenport, Hooper, & Zegar, 2007; Mazzocco, 2009), and 22q11.2 deletion syndrome (De Smedt, Reynvoet, et al., 2009; De Smedt, Swillen, et al., 2009; Simon, Burg, & Gothelf, 2007).

Fragile X Syndrome

Fragile X syndrome is the leading known familial cause of intellectual disability, with an incidence of 1:4000 to 1:9000 live births (Crawford, Acuna, & Sherman, 2001). Fragile X results from a mutation of a single gene located on the X chromosome (Verkerk et al., 1991), and is referred to as FMR1 (i.e., the fragile X mental retardation gene mutation). The highly variable physical characteristics of fragile X syndrome are insufficient indicators of risk, particularly because some of those features (e.g., a long face and protruding ears) are not observed in all persons with the syndrome and are observed among many persons in the general population. Likewise, the severity of intellectual impairment varies across individuals with fragile X, ranging from moderate intellectual disability to normal intelligence.

As an X-linked disorder, fragile X syndrome affects males (who have only one X chromosome) more severely than females (who have two X chromosomes). Approximately 50% of females with fragile X have intellectual disability (Rousseau et al., 1994) whereas intellectual disability characterizes nearly all males with fragile X (Bailey, Hatton, & Skinner, 1998). The variability seen in females is partially caused by random X inactivation, a process by which one of the two X chromosomes typically present in females is inactivated (Lyon, 1991), or essentially turned off. The inactivation process results in two populations of cells, whereby either the maternally or paternally inherited X chromosome is inactive. In females, the percentage of *active* chromosomes that have the mutation can vary, since the mutation is on only one of the two X chromosomes. This variability is associated with syndrome severity (Abrams et al., 1994), as is expression of FMR protein (FMRP), a protein coded by the FMR gene that influences synaptic maturation and neuronal pruning (Greenough et al., 2001). Females with fragile X generally produce more FMRP than males; still, within females, FMRP levels are highly variable.

Research on MLD in children with the fragile X mutation has typically involved only females without intellectual disability (persons with IQ scores > 70). From this point forward in the chapter, references to math performance in persons with fragile X pertain to only these relatively high functioning females with fragile X unless otherwise indicated.

Turner Syndrome

Turner syndrome is a sporadic monosomy disorder (i.e., a disorder characterized by one chromosome missing from a pair) that occurs only in females, with an incidence of 1:2000 to 1:5000 live female births (Davenport et al., 2007). It results from either complete or partial loss of the second X chromosome that is typically present in a female (Lippe, 1991). Like fragile X syndrome, the physical characteristics of Turner syndrome vary in frequency and severity; however, the most common feature—short stature—is evident in most persons with this disorder (Sävendahl & Davenport, 2000, p. 457). Although overall intellectual ability in girls with Turner syndrome is typically in or above the average range, the hallmark feature of the cognitive phenotype is a discrepancy between verbal and nonverbal abilities, typically manifested as intact reading, vocabulary, and verbal comprehension, despite slower lexical retrieval (i.e., word finding), processing speed, and spatial reasoning. The discrepancy between these higher "verbal" versus lower "performance" competencies varies from insignificant (few if any points on a standardized scale) to remarkable (more than three standard deviations). The magnitude of this discrepancy (and other phenotypic features of Turner syndrome) appears linked, at least in part, to whether there is partial versus total loss of an X chromosome, with the former leading to milder outcomes (Temple & Carney, 1993). These and other variations of Turner syndrome karyotypes are discussed elsewhere in detail (Ross, Roeltgen, Kushner, Wei, & Zinn, 2000).

Chromosome 22q11.2 Deletion Syndrome

22q11.2 deletion syndrome (22q11.2DS) is the most common known microdeletion syndrome (i.e., a chromosomal deletion spanning several genes that is too small to be detected by conventional cytogenetic methods), with an estimated incidence of 1:4000 to 1:6000 live births (Botto et al., 2003; Robin & Shprintzen, 2005). As with Turner syndrome, 22q11.2DS is sporadic in nature, although infrequent instances of familial transmission are reported in about 15% of cases (Swillen et al., 1998). Velocardiofacial syndrome, DiGeorge syndrome, and several other clinical syndromes are included under the classification of 22q11.2DS, as each results from a 1.5 to 3.0 Mb hemizygous microdeletion (i.e., a microdeletion occurring on only one copy of the gene instead of two) on the long (q) arm of chromosome 22 (Botto et al., 2003; Kobrynski & Sullivan, 2007; Robin & Shprintzen, 2005). Highly variable physical characteristics most frequently include malformations of the heart (reviewed by Shprintzen, 2008). Likewise, overall intellectual ability (although highly variable) is lower than average, ranging from mild intellectual impairment to low-average functioning. Variability in the level of divergence between verbal and performance IQ is small, with an average deviation of less than one half of one standard deviation, and this discrepancy is possibly due to variation in the length of an individual's microdeletion (Carvalho et al., 2014; Hwang et al., 2014).

Variability in specific skills has also been found, attributable to gender, regional brain volume differences, and catechol-O-methyltransferase (COMT) polymorphism (COMT is an enzyme that helps break down catecholamines, such as the neurotransmitters dopamine and epinephrine); male children have been reported to be more cognitively affected and have larger frontal lobe volumes than girls (Antshel et al., 2005; Kates, Antshel, et al., 2006a; Kates, Miller, et al., 2006b; Sobin, Kiley-Brabeck, Monk, Khuri, & Karayiorgou, 2009). Because of a high risk of conversion to schizophrenia in this population, phenotype research on 22q11.2DS typically focuses on this aspect of the syndrome. Thus, mathematical disability in 22q11.2DS is understudied, and when it is studied the focus is largely on visuospatial and specific quantification abilities rather than on broad measures of academic achievement.

As with the study of MLD in persons with fragile X syndrome, some researchers of MLD in children with 22q11.2DS limit participants to those without intellectual disability (persons with IQ scores >70; e.g., De Smedt et al., 2006); yet others have included a broader range of impairments and tested for effects of IQ (e.g., Quintero, Beaton, et al., 2014). Most studies of mathematics skills in 22q11.2DS have been conducted in both males and females; however, in recent studies of children with any one of these three syndromes only females were included because these studies were also focused on Turner syndrome (Quintero, 2014; Quintero, Harvey, & Simon, 2014).

MLD Frequency and Severity in Children with Fragile X, Turner, or 22q11.2 Deletion Syndromes

It is now well-established that children with any one of these syndromes are at higher risk for poor math achievement than are children in the general population. But how frequently are these mathematics deficits observed? This question has been directly assessed in studies of girls with fragile X or Turner syndrome (limited to girls without intellectual disability). In those studies, approximately 50% of girls with fragile X or Turner syndrome score below the average range in mathematics achievement by kindergarten or first grade relative to approximately 10% of children in the comparison groups (Mazzocco, 2001). (The rate is higher when females with intellectual disability are included.) This rate increases to over 75% of girls with either syndrome by the end of third grade (Murphy, Mazzocco, Gerner, & Henry, 2006), and continues through at least Grade 7 (Mazzocco, 1998; Murphy & Mazzocco, 2008a; Murphy et al., 2006) and beyond (Mazzocco, 1998). With regard to 22q11.2DS, initial reports of the phenotype made reference to learning disabilities, which by 1985 were described as difficulties with "abstract thinking" and weak numerical awareness (as reviewed by Shprintzen, 2000). Although the incidence of learning difficulties was noted to become more apparent later in primary school, studies of the incidence of mathematics difficulties per se were not pursued in light of attention to broader intellectual, psychiatric, and psychosocial disabilities (Angkustsiri et al., 2014; Karayiorgou et al., 1995; Swillen et al., 1997). Studies focused specifically on MLD were simply not as prevalent in 22q11.2DS research compared to studies of Turner syndrome and fragile X.

In all three syndromes, the severity of mathematics difficulties can be significant. During the school-age years, math achievement is, on average, two grades below age expectations for girls with Turner syndrome (Rovet, 1993; Rovet, Szekely, & Hockenberry, 1994), and also below grade and age expectations for girls with fragile X (Murphy & Mazzocco, 2008a) and children with 22q11.2DS. Since poor mathematics achievement has also been reported for adult women with fragile X (Bennetto, Pennington, Porter, Taylor, & Hagerman, 2001; Mazzocco, Pennington, & Hagerman, 1993) or Turner syndrome (Molko et al., 2003), the evidence suggests that the early deficits persist and are lifelong. In children with 22q11.2DS, mathematics difficulties appear not simply to reflect lower IQ (Hooper et al., 2013; Jacobson et al., 2010). Of the few longitudinal studies that have focused on the cognitive phenotype, there are no reported studies of continued decline or improvement in mathematics skills over time in 22q11.2DS. Although there is some indication of a slight decline in full-scale IQ scores over time, the overall pattern of IQ subtest scores appears relatively stable in this population. Thus, in all three syndromes, the mathematics difficulties that emerge constitute a developmental difference rather than a short-term developmental delay.

In summary, math difficulties are common, persistent, and significant in all three syndromes. The nature and causes of math difficulties appear to vary across these populations, as do sources of individual differences within each syndrome group.

Correlates as Indicators of Pathways to or Subtypes of MLD: Contributions and Limitations

Early research on these syndrome phenotypes was based largely on correlational studies of mathematics and associated skills. As an initial step, associations between poor math achievement and other cognitive deficits inferred from previous studies were tested as evidence for the MLD subtypes that had been proposed at that time (during the 1990s).[2] For instance, subtype models linked MLD to either poor verbal abilities (including reading and lexical retrieval skills), executive function deficits (including organizational and procedural skills and

2. More recent evidence continues to support the notion of these MLD subtypes, but also suggests subtypes specific to numerical skills or to general low functioning (Bartelet et al., 2014).

strategies), or visual-spatial deficits (e.g., Geary, 1993, 2011). These early studies were logically driven investigations given the evidence that MLD co-occurs with deficits in executive function and visual perceptual skills among females with fragile X (Bennetto et al., 2001; Cornish, Munir, & Cross, 1998; Cornish et al., 2004; Kirk, Mazzocco, & Kover, 2005; Mazzocco et al., 1993) or Turner syndrome (Buchanan, Pavlovic, & Rovet, 1998; Haberecht et al., 2001; Kirk et al., 2005; Rovet & Netley, 1982; Tamm, Menon, & Reiss, 2003; Temple & Carney, 1995; Temple et al., 2006; Waber, 1979), and deficits in visuospatial memory and working memory specifically in children with 22q11.2DS (Bearden et al., 2001; Sobin, Kiley-Brabeck, & Karayiorgou, 2005). Likewise, reports of significant visual-spatial difficulties in girls with Turner syndrome led some researchers to propose that Turner syndrome is a good model of visual-spatial MLD.

But tests of these hypotheses are not straightforward. Co-occurrence does not necessarily indicate causal mechanisms, and in many cases do not even reflect associations. That is, many studies of these disorders found that targeted cognitive deficits were *not* significantly correlated with poor mathematics achievement: Rovet et al. (1994) determined, for instance, that visual-spatial and mathematics deficits were independent of each other in Turner syndrome, and De Smedt and colleagues (2008) similarly reported that working memory deficits observed in 22q11.2DS did not account for MLD in this population. With some exceptions, similar attempts to link such specific cognitive deficits to MLD in the general population have also failed (Price & Ansari, 2012). Since the relation between visuospatial abilities and mathematics can vary with age and the types of items on the math task (e.g., Li & Geary, 2013; Mazzocco & Myers, 2003), firm conclusions must await more detailed studies of spatial deficits as well as strengths and weaknesses across different mathematical domains.

Although it is widely understood that correlations do not necessarily indicate causation, the absence of a given correlation need not negate potential causal influences. That is, associations between math and domain-general skills may be masked when measures of either the outcome (e.g., "math achievement scores") or the targeted correlated skills (executive function) are too broad to specify a distinct pathway to MLD, or when the developmental time course is not considered (e.g., Murphy & Mazzocco, 2008b). Attending to greater specificity of both the mathematics outcome skills of interest and the cognitive skills believed to influence those outcomes have thus advanced our efforts to better understand pathways to (and manifestations of) MLD. This shift in approach from general to specific, and the resulting findings, are summarized next.

Correlates as Indicators of MLD Specificity in Fragile X, Turner, and 22q11.2DS

Characterizing MLD in Girls with Fragile X Syndrome

Early studies documented the existence of mathematics difficulties in fragile X, and later studies searched for causal pathways or, initially, cognitive correlates.

Such attempts by the first author and her colleagues centered on fragile X as a model of a procedural MLD subtype, in view of reports of poor executive function skills in both children (Kirk et al., 2005) and adults (Mazzocco, Hagerman, Cronister-Silverman, & Pennington, 1992) with the syndrome. The study failed to provide support for this hypothesis by not finding higher rates of procedural errors in school-age girls with fragile X relative to age-matched peers or girls with Turner syndrome (Mazzocco, 1998). (In fact, the Turner syndrome group showed more such errors than girls with fragile X, to be discussed subsequently.) Still, although procedural errors are considered a hallmark of the proposed procedural MLD subtype, the retrospective analysis of written calculation problems may have been insufficient to detect them. That is, procedural errors are not limited to overt written calculation errors, and may have been more apparent if strategy use had been evaluated in the original study. Moreover, procedural errors are only one potential indicator of an association between executive functions and mathematics. Finally, procedural errors are often due to developmental delays (Geary, 1993) and thus might only be detectable during the early stages of learning the procedure.

Later studies succeeded in identifying more specific associations between aspects of mathematics and components or indicators of executive function. In a landmark functional imaging study, Rivera, Menon, White, Glaser, and Reiss (2002) presented girls and women with two- and three-operand arithmetic statements, and asked these participants to report if the statements were true or false. Of interest was how the corresponding increases in working memory demands between the two-operand ($2+3=4$) and three-operand ($2+3+1=5$) statements affected brain activation in typically developing individuals (who showed increased activation in the prefrontal and parietal cortices during the three- vs. two-operand problems), but not females with fragile X (who, as a group, showed less activation on both sets of problems compared to the females without fragile X). Notably, the degree to which females with fragile X *did* show this increase in activation was correlated with FMRP expression, suggesting an important biological pathway to the working memory and mathematics impairment.

Overt behavioral group differences were also observed in Rivera and colleagues' study. Females with fragile X syndrome were *less* accurate than their peers at evaluating whether the three-operand problem were correct or incorrect, but were *as* accurate as their peers when evaluating two-operand problems. In view of these two markers of arithmetic difficulty, Rivera and colleagues concluded that females with fragile X appear to lack the cognitive resources linked to working memory ability that are needed (and typically relied upon) to compensate for an increase in working memory demands. Likewise, in the previously described Kirk et al. (2005) study with younger (8-year-old) participants, girls with fragile X and an IQ-matched comparison group both showed increased difficulty on an executive function task (the Contingency Naming Task, Anderson, Anderson, Northam, & Taylor, 2000) when working memory demands were increased. Still, when working memory demands were only moderate, girls with fragile X made more errors than did girls in the comparison group, despite taking the same amount of time to complete the task (Kirk et al., 2005). These findings suggest that working memory limitations in females with fragile X cannot be attributed solely to low IQ (Kirk et al., 2005), and that lower thresholds for experiencing working memory overload may contribute to the mathematics difficulties in this group (Murphy & Mazzocco, 2009).

When working memory demands interfere with otherwise effortless tasks, it may be necessary to rely on supporting or compensatory mechanisms. But generating a strategy and successfully using it to solve a taxing arithmetic task both require at least a rudimentary understanding of numbers and arithmetic operations involved. Is there evidence that girls with fragile X have at least a basic knowledge of numbers to support, for instance, dealing with the demands of three-operand arithmetic problems? Their intact performance on two-operand problems in the Rivera et al. study suggests so. But is this evidence sufficient? Participants in that study were 10- to 22-year-olds (mean age 16 years), and had a mean full-scale IQ score of 84 points. Single-digit two-operand addition problems are fairly overlearned by 10 years of age (about Grade 5), and success on these problems at ages 10-22 years may simply reflect sound memory rather than arithmetic or numerical understanding.

Indeed, strong verbal memory is a phenotypic characteristic of fragile X syndrome, and good rote numerical skills have been reported in girls with fragile X up to Grade 7 (Murphy & Mazzocco, 2008b). For instance, first- and second-grade girls with fragile X syndrome are far more accurate than their same-age MLD peers at oral number tasks such as reading numbers, counting aloud from 1, counting backwards, or skip counting (e.g., counting by 10s; Murphy et al., 2006). In fact, girls with fragile X perform nearly as well if not better than their peers *without MLD* on all of these tasks. Furthermore, they seem skilled at memorizing arithmetic facts, particularly at grades when such facts are over-rehearsed. However, they are as impaired as their MLD peers on more conceptual numerical tasks such as verbal magnitude comparison (reporting which of two numbers is larger, identifying a specific ordinal position (e.g., identifying the 4th person in line), and at using one-to-one correspondence when counting (despite facility at counting aloud forward or backward and skip counting; Murphy et al., 2006). Even kindergarten-age girls with fragile X have lower scores than same-age peers on test items that measure counting *principles*, despite good rote counting (Mazzocco, 2001). A similar pattern emerges at Grades 6 and 7 (Murphy & Mazzocco, 2008a), when girls with fragile X are indistinguishable from their non-MLD peers at reading names of decimals (0.20, 0.05) outperform their MLD peers on this task, but fail the conceptually based task of rank-ordering a combination of fractions and decimals (such as 1/2, 1/4, 0.20, and 0.40). In fact, in this particular study, all girls with fragile X failed this latter task despite stronger than average performance naming decimal values.

These findings suggest three points about inferring causes of mathematics difficulties. First, accuracy on basic numerical tasks, such as counting (or, in the

Rivera et al. study, two-operand math problems) does not necessarily determine mastery of number (or arithmetic) knowledge. Second, we cannot assume that the working memory demands of a mathematics task are responsible for mathematics difficulties simply because a less taxing version of the task poses no evident challenge. Third, working memory (or other skills) on which typically developing children can rely to solve taxing mathematics tasks may not be sufficient for children who have weak numerical knowledge despite strong rote number skills. In this latter case, if the rote skill is not accompanied by knowledge, it may not serve a child well on problem solving.

To what can we attribute the weak numerical principles seen in girls with fragile X? Early studies reported significant correlations between counting principles and other select skills in girls with fragile X but not in girls from the general population (or even girls with Turner syndrome) at kindergarten through later school-age years (Mazzocco, 1998). For example, among girls with fragile X, the ability to distinguish individual shapes within a design (figure-ground discrimination), and the ability to recall the correct location of items within an array (local vs. global visual short-term memory), were both positively correlated with evaluating correct versus incorrect counting procedures (Mazzocco, Bhatia, & Lesniak-Karpiak, 2006). These visual perception and discrimination task scores were also positively correlated with paper-and-pencil math calculation skills. IQ scores did not account for these correlations, and the correlations failed to emerge from either a same-age peer group matched on IQ or from girls with Turner syndrome also matched to the participants with fragile X on age and IQ. These findings do not indicate that a relation between math and spatial skills is unique to fragile X, because other researchers have shown that spatial and mathematics skills are related in children from the general population, albeit using different measures of spatial ability than those we describe above, including standardized IQ subtests (e.g., Geary & Burlingham-Dubree, 1989) or number line tasks (e.g., Gunderson, Ramirez, Beilock, & Levine, 2012). However, our findings suggest that girls with fragile X may be processing numbers differently from their peers, without explaining how or why this is so.

Considered together, the research summarized to this point shows that girls with fragile X have math difficulties that differ from girls with MLD in the general population and from typically developing children; that those difficulties occur despite (and may be masked by) strong rote skills or knowledge; and that both executive function and spatial skills may account for some of these difficulties. Rivera et al. (2002) that girls with fragile X fail to engage more prefrontal and parietal brain activation during three- versus two-operand arithmetic, coupled with decreased accuracy on three- versus two-operand problems, may signal disengaging from a task that is too difficult rather than attempting but failing a task due to limited cognitive resources. If their threshold for engaging working memory resources in effortful contexts is below average (Murphy & Mazzocco, 2009), girls with fragile X may rely on strong verbal memory skills that are nevertheless insufficient to support more complex (or more abstract) mathematics, including three-operand arithmetic. This reliance on verbal memory may promote rote skills that may (incorrectly) implicate stronger number knowledge than exists. Correlational analyses cannot confirm these proposed pathways, but do implicate multiple associations that may interact in ways that differ across syndromes.

Characterizing MLD in Girls with Turner Syndrome

Girls with Turner syndrome present with a different constellation of mathematics difficulties than observed in fragile X syndrome. Relative to the latter (and to typically developing peers), calculation skills in this group are, on average, accurate but slow. This may be interpreted as evidence of poor retrieval skills—a common feature observed in children with MLD from the general population—except that in the latter case the poor retrieval is typically inaccurate (and not always slow). Moreover, poor retrieval is common in MLD that co-occurs with reading disability, and girls with Turner syndrome have reading and vocabulary skills within or above the average range (Temple & Carney, 1996), with intact phonological decoding and other reading-related skills at age 5 years (Mazzocco, 2001; Temple & Shephard, 2012), middle school (Murphy & Mazzocco, 2008a) and adulthood (Temple & Shephard, 2012). The implication that slow calculation reflects global processing-speed deficits is countered by observations that dysfluency in Turner syndrome appears on select tasks only (e.g., oral fluency [Temple, 2002]), and not on typical processing-speed tasks such as rapid automatized naming (Murphy & Mazzocco, 2008a).

If slow calculation performance in Turner syndrome does not model math fact retrieval or global processing-speed deficits, perhaps the slower response times reflect reliance on atypical arithmetic strategies, strategies that are reinforced over time in view of the fact that they lead to successful (accurate) performance. Calculation that is as accurate as that of one's peers need not reflect comparable arithmetic processing. To explore this notion, Kesler, Menon, and Reiss (2006) engaged 7- to 24-year-olds with Turner syndrome in a two- and three-operand arithmetic verification task. Recall that girls with fragile X were as accurate as their peers on two-operand but not three-operand addition, and failed to demonstrate the increased engagement of executive function skills seen in their peers (i.e., did not show the same pattern of increased prefrontal activation) on three- versus two-operand items (Rivera et al., 2002). In contrast, Kesler and colleagues found that girls with Turner syndrome were as accurate as their peers on both tasks. Compared with two-operand problems, three-operand problem verification was associated with increased reliance on executive function skills in the control group (as was also the case in Rivera's study), but with increased reliance on verbal skills in girls with Turner syndrome—an area in which girls with Turner syndrome excel. It is unclear what those strategies may be as verbal skills may range from explicit counting to verbal mediation. Nevertheless, the type of strategies used during accurate performance may differentiate girls with versus without Turner syndrome.

A similar result emerged from a number combination task with variable working memory demands. On this timed task, sixth and seventh graders were asked to select (by circling) from an array of number pairs only those pairs that summed to a visible target. Once again, girls with Turner syndrome were as accurate, but slower, than their age-matched peers on this task. When working memory demands were increased by requiring participants to simultaneously cross out number pairs that summed to a different target (>23), children in both groups slowed down (completed fewer items), but accuracy *decreased* in the comparison group and *increased* in the Turner syndrome group (Mazzocco & Hanich, 2010). It may be that as a group, girls with Turner syndrome are more experienced with strategies that require slowing down and thus adopted a more effective strategy on this task (performance was slower, but more accurate compared to when no working memory demand was imposed) than their peers (whose performance was slower and less accurate compared to when no working memory load was imposed). Regardless, the strategies used across groups may have differed.

Calculation is not consistently accurate in girls with Turner syndrome, however. Weak calculation skills compared to peers occur on complex (multidigit) problems, typically in the form of procedural errors (e.g., Mazzocco, 1998; Rovet et al., 1994; Temple & Marriott, 1998). These errors, and the co-occurring visual-spatial deficits reported in this population, motivated early explorations of a causal pathway to poor math in this group (Rovet et al., 1994). These early and later studies failed to find significant correlations between measures of mathematics and spatial skills used in the study, leading Rovet and colleagues to conclude that visual-spatial and mathematics skills are independent of each other in Turner syndrome. For instance, standardized scores of math achievement, composite math ability scores, and more specific measures such as counting skills were found not to correlate with visual-spatial reasoning scores in young (Mazzocco et al., 2006; Mazzocco & McCloskey, 2005) or older girls with Turner syndrome (Mazzocco, 1998). Likewise, item analyses failed to reveal any selective impairment in girls with Turner syndrome on math items that have a strong visual-spatial component, such as rank-ordering items according to height or volume (in primary school-age years) or geometry items on the Key Math standardized test in older girls with Turner syndrome. Still, these null findings were not definitive, and conclusions about the independence of spatial and math skills in Turner syndrome would later prove to be premature.

Although "spatial" at a surface level, the procedural errors observed in girls with Turner syndrome may instead reflect weak executive function skills, such as poor organization or inadequately keeping track of steps needed during complex problem solving (with complex defined as involving more than one step), or neglecting to apply or complete a correct "regrouping" procedure (Rovet et al., 1994) relative to age, grade, and verbal IQ-matched peers. Likewise, the selective nature of slowed or otherwise less efficient verbal performance reported for the Turner syndrome phenotype appears linked to measures that tap executive function skills, despite robust vocabulary. Consider the aforementioned oral fluency findings, that adolescent girls with Turner syndrome are slower naming as many words as possible on demand (Temple, 2002). Maximum efficiency during oral fluency tasks requires an inherent organization in how words are retrieved. For example, if asked to name as many words as possible that begin with the letter "B" within 1 min, one could retrieve names randomly, or adopt a more deliberate approach such as retrieving words alphabetically (Ba__, Be___, Bi___, etc.) in which case cuing could facilitate word retrieval. Likewise, on verbal fluency tasks that require little executive function, such as sentence completion tasks, women with Turner syndrome perform as well as their peers, but when demands require inhibiting a prepotent response (completing a sentence so that it does not make sense), performance is less efficient and less accurate despite significantly slower performance in women with Turner syndrome. This response profile is similar to the slower, less accurate, and therefore less efficient performance by girls with Turner syndrome on tasks that involve heightened working memory demands (Kirk et al., 2005), and with failure to inhibit responses during an ocular motor antisaccades task (Lasker, Mazzocco, & Zee, 2007). It is unknown how, or whether, these associations interact with spatial or numerical components of mathematics in this population.

Girls with Turner syndrome, as a group, do not exhibit global number deficiencies (Bruandet, Molko, Cohen, & Dehaene, 2004; Murphy et al., 2006; Rovet et al., 1994; Temple & Marriott, 1998) based on age-appropriate performance on simple arithmetic, number comprehension, number production, rote or applied counting skills (Murphy et al., 2006), magnitude judgment accuracy, and estimation accuracy in childhood (Mazzocco, 2001; Murphy et al., 2006); but estimation errors in adulthood (Bruandet et al., 2004) and other findings implicate other specific numerical deficits. For instance, lower arithmetic scores in this group may result at least in part from fewer problems attempted (vs. more errors) within a given time frame regardless of whether the arithmetic problems include larger versus smaller addends (Molko et al., 2003; Temple & Marriott, 1998). This slower response time is observed even for magnitude judgments of very small item sets that are typically judged very quickly (i.e., subitized rather than counted) (e.g., Kaufman, 1949). There is evidence that women with Turner syndrome actually count small sets of items whereas most adults rapidly recognize the magnitude of small sets (<4 items; Molko et al., 2003).There is also evidence for a similar reliance on counting versus subitizing from studies of MLD in the general population (e.g., Koontz & Berch, 1996). Counting instead of subitizing may explain, rather than exemplify, the slow response times.

Characterizing MLD in Children with 22q11.2DS

Compared to the studies described previously for the fragile X and Turner syndrome phenotypes, there are predictable similarities and striking differences in studies of MLD in children with 22q11.2DS. Like the syndrome studies described previously, the study of MLD in girls with 22q11.2DS was motivated by an observed weakness in broader mathematics achievement, compared to a relative strength in reading achievement (Moss et al., 1999). But unlike the early studies on MLD in fragile X and Turner syndrome, even early research on MLD in 22q11.2DS paid very particular attention to specific numerical skills rather than broad math achievement scores. For instance, in a series of experiments, De Smedt et al. (2006, 2007), De Smedt, Reynvoet, et al. (2009), and De Smedt, Swillen, et al. (2009) presented a battery of numerical tasks to 6- to 12-year-old boys and girls with 22q11.2DS, enrolling only those children who did not have intellectual disability (IQs were >70). The tasks selected were designed to test the notion of a specific cognitive system specialized for processing information about quantity that relies on using analog representations (for a review see Geary, Berch, & Mann Koepke, 2015). This system is one of three proposed neuroanatomical subsystems for processing numbers described by Dehaene, Piazza, Pinel, and Cohen (2003). The two remaining, untested subsystems (which we do not address further) are the verbal and visual subsystems (described elsewhere in this volume). According to Dehaene's "triple code" theory, over the course of development the brain learns to map symbolic information about number to quantitative representations in a one-to-one fashion. It is in this way that numerical representations are developmentally linked to object-based attention and visuospatial competence (Ansari et al., 2003; Brannon & Van de Walle, 2001; Simon, 1999).

De Smedt and colleagues found select numerical abilities and difficulties in children with 22q11.2DS, and that the pattern of these skills varied with age. Among 6- to 10-year-olds, children with 22q11.2DS recognized numbers, differentiated which of two numbers was larger, counted, added, subtracted, and multiplied, for both single and multidigits, at a similar speed and accuracy level as their peers. They solved word problems with lower accuracy despite similar response times on these problems across groups. In a second, older cohort of 10- to 12-year-olds with 22q11.2DS, no group difference was seen in numeral recognition. Despite no significant differences in accuracy at age 6-10, performance was now slower than the typically developing children on tasks that involved comparing magnitudes (e.g., "Which is more?"), counting, single-digit adding and subtracting, or regrouping that required carrying a number. As with the younger children in the study, performance on word problems was less accurate, but this older cohort responded more slowly than their typical counterparts. It seems that typically developing children gain more in terms of overall speed of processing with age than children with 22q11.2DS.

This profile of impaired quantification in children with 22q11.2DS but preserved number processing when the verbal system was tapped was interpreted to represent a distinct subtype of MLD. Results also suggest a bifurcation in development of math skills in children with and without 22q11.2DS, with the latter possibly continuing to increase their reaction time. It must be noted that this profile does not represent the full range of performance for children with 22q11.2 due to exclusion of children with low IQ scores (<70). For example, a set of studies by Simon and colleagues report differences in response times while counting, for 7- to 14-year-old boys and girls with 22q11.2DS, relative to typically developing peers (Simon, Bearden, Mc-Ginn, & Zackai, 2005; Simon et al., 2008), whereas De Smedt et al. (2006, 2007) reported group differences in counting speed among the 6- to 10-year-olds but not the 10- to 12-year-olds. In the Simon studies, the mean IQ of the 22q11.2DS cohort was in the mid 70s, and it extended past the cut-off used by De Smedt. Although Simon et al. (2008) did not find a correlation between relevant IQ subscores (i.e., the Perceptual Reasoning Index or Organizational factor scores) and counting response time, it is possible that the lack of overlapping IQ scores across studies and the restricted range of IQ scores represented in some studies limited meaningful testing for correlations. Additionally, these were cross-sectional studies; so without longitudinal data, it is possible that the results implicating age differences are due to cohort effects.

Although there is some variation across studies in the age at which numerical impairments are observed, the impairment noted—counting and magnitude comparison (Quintero, 2014; Quintero, Harvey, et al., 2014; Simon et al., 2005, 2008)—points to visuospatial impairment specific to 22q11.2DS. The presence of numerical and visuospatial deficits is also observed in Turner syndrome, leading these researchers (Simon, Quintero) to compare the two phenotypes. Building upon previous studies of children with 22q11.2DS or Turner syndrome, Simon's (2008) comparisons revealed that enumeration (counting and subitizing) performance did not differ between the two disorders. Quintero and colleagues (Quintero, 2014; Quintero, Harvey, et al., 2014) directly tested the "spatiotemporal hypergranularity hypothesis" previously tested in 22q11.2DS (Simon, 2008), which predicts that a reduction in the resolution of mental representations of spatial information influences the type of performance errors made by children with 22q11.2DS. This hypothesis specifically pertains to the mental representation of quantity.

According to this hypothesis, objects and their features are relatively easy to perceive when spread out across a large space, and they become more difficult to perceive as they get closer together because the competition between objects or features increases. This interference increases the probability of perceiving two neighboring objects as one. Simon proposed that the effect of this so-called spatiotemporal hypergranularity on attentional processing is largely determined by the "grain size" of attentional resolution (Simon, 2008; Simon & Luck, 2011). Using the analogy of a digital camera, reduced attentional resolution would be similar to a lower resolution camera that yields fewer but larger pixels (or grains) to represent visual information, with a commensurate loss in resolution of mental representations of spatial information. Specifically, reduced resolution of spatial representations would produce a bias toward underestimation during counting, reflected as undercounting the represented set and a smaller subitizing range. Any test of this hypothesis should therefore include tests of subitizing and counting performance. Since the hypothesis pertains to the mental representation of quantities, care should be taken to avoid overgeneralizing these principles to the perception of symbolic arithmetic.

When linear regression models were used to test whether diagnosis and chronological age predicted differences in subitizing, counting, or error rates in girls with 22q11.2DS, relative to typically developing girls aged 7-15 years, girls with 22q11.2DS showed impaired performance in counting (the enumeration of items with a set size larger than 3), but not subitizing (the enumeration of items with a set size smaller than 4). This difference is not accounted for by processing speed, given the similar verbal response times despite, on average, lower IQ scores in the 22q11.2DS group (Simon et al., 2005, 2008). Furthermore, confirming a prediction of the spatiotemporal hypergranularity hypothesis, girls with 22q11.2DS produced more undercount than over-count errors.

The spatiotemporal hypergranularity hypothesis is one of many plausible interpretations of the behavioral results. An alternative core impairment may be attentional search or, perhaps, executive control, as seen previously (Quintero, Beaton, et al., 2014). If the former is the case, children with 22q11.2DS may underestimate because they are not searching the entire visual space; if the latter alternative is the case, overestimating may occur when items to be counted are revisited. The results do not rule out either possibility, but impaired attentional search and unimpaired executive function would lead to the group differences observed in under- and over-count rates.

Arithmetic processing has also been tested in male and female adolescents with 22q11.2DS (Eliez, Barnea-Goraly, Schmitt, Liu, & Reiss, 2002). Using two operand ($2+3=4$) and three operand ($2+3+1=5$) verification problems like those used by Rivera et al. (2002), adolescents with 22q11.2DS were less accurate than their peers on both types of problems, despite similar response times (unlike lower accuracy on only the three-operand problems in girls with fragile X). This inverted performance profile relative to girls with fragile X or Turner syndrome was accompanied by typical prefrontal activation in adolescents with 22q11.2DS (i.e., this group, overall, did not show the same pattern of decreased prefrontal activation as seen in females with fragile X; Rivera et al., 2002) and increased activity in the left supramarginal gyrus.

In typically developing individuals, mathematical, numerical, and spatial processing are subserved by parietal circuitry, including the inferior parietal lobe, superior parietal lobe, and supramarginal gyrus. This increased recruitment during more difficult arithmetic problems may reflect a known reliance on retrieval strategy over the use of calculation (De Smedt et al., 2006, 2007), which unlike the fragile X results, does not point to different working memory thresholds per se. Indeed, in studies of working memory in children with 22q11.2DS relative to typically developing children, less accurate performance is observed on visuospatial but not verbal working memory tasks (Wong, Riggins, Harvey, Cabaral, & Simon, 2014). The spatiotemporal errors seen in this group occur despite longer delays than typically developing children, which could be due to increased susceptibility to visual memory degradation, or inefficiency of visual memory retrieval, as opposed to errors encoding visual information. But these weaknesses are not apparent on verbal working memory tasks such as standardized list learning (i.e., the California Verbal Learning Test), consistent with the successful retrieval strategy implicated on three-operand addition. Regardless of the precise pathways leading to these group differences, it is interesting to note that, as a group, each syndrome shows atypical arithmetic performance, but the manifestation of that impairment differs among the three groups.

Models of Relations Between Math, Executive Function, and Attention

To this point, we have presented group differences in biology, aspects of cognition, and behavior in the form of math performance levels. We highlighted broad similarities among three phenotypes (deficits in math, executive functions including working memory), similarities and differences in aspects of executive function, and the importance of attending to numerical thought (not just mathematics procedures or achievement) to understand these relationships. (See Table 1 for a summary of group-level behavioral and MLD characteristics in persons with fragile X, Turner, or 22q DS reviewed in this chapter.) Further gains in distinguishing these contributions to mathematics abilities and to MLD phenotype profiles come from studies of mathematical thinking that involve multiple syndromes.

Studies of the lower-order cognitive phenomena that give rise to numerical thinking are paramount for understanding how impairments in this domain emerge. Evidence from studies of these three genetic disorders and of typical development suggests that impairments in visuospatial representations, spatial attention, and processing of number and space can cascade into problems acquiring basic numerical and arithmetical abilities (Walter, Mazaika, & Reiss, 2009). That numerical understanding (e.g., understanding the concept of quantity, principles of counting, and relations among numbers) is learned based on the functions of visuospatial and attentional processes is a hypothesis that has been put forward by many researchers (Dehaene, 1999; Göbel, Walsh, & Rushworth, 2001; Simon & Vaishnavi, 1996). For example, reduced attentional capacity may negatively impact the ability to correctly determine the larger of two items (Anobile, Cicchini, & Burr, 2012). Given that the visuospatial representation of objects depends on visual attentional processes, such a reduction in resolution and the consequent loss of information in the resulting representation may influence the development of numerical ability. While associations between the two have been found, a causal developmental link is difficult to establish.

Because numerical and mathematical impairments are well-established in a subset of neurodevelopmental disorders (including the three genetic disorders on which we focus in this chapter), the study of children with these disorders allows us to test whether these "higher-order" cognitive impairments (e.g., multidigit arithmetic) can be recapitulated in "lower-order" cognitive skills (e.g., efficient working memory, numerical processing). Interestingly, individuals with

TABLE 1 Summary of Group-Level Behavioral and MLD Characteristics in Persons with Fragile X, Turner, or 22q DS[a]

Characteristic	Features Shared Across Syndromes	Key Features That Vary Across Syndromes (Relative to Comparison Ss)		
		Fragile X Syndrome	**Turner Syndrome**	**22q11.2 DS**
Global math profiles reported at group level	Poor math achievement, not linked solely to IQ. Poor math performance persists, is a developmental difference, not a delay	– Intact early number knowledge, but low math ability or achievement skills that emerge in, and continue well beyond, primary school – Strong rote skills in early and later school years that are misaligned with related conceptual tasks (e.g., correctly name decimal values but rank order them incorrectly)	– Slow and often deliberate problem solving, accurate when tasks are untimed even when working memory demands are imposed – Inconsistent profile of numerical skills. Early number knowledge is strong, distant effect is evident, but no differentiation in brain activation in adults' exact and approximate arithmetic performance – Accurate calculation, but inaccurate estimation	Preserved number processing on a range of tasks (e.g., magnitude comparison, counting, word problems) for which accuracy levels persist with age, but response times do not decrease with age, suggesting lack of shifting toward automaticity
Relevant related skills	Verbal abilities are spared	Adequate to average reading; verbal memory strengths appear to affect math performance profile	Above average vocabulary, reading, and comprehension at all ages may support compensatory skills in math problem solving	Reading > numerical and math skills, not linked to IQ in higher functioning individuals although more significant among those with global intellectual impairment
	Select atypical spatial skills	Visual discrimination and spatial location skills correlated with math calculations and counting	Few correlations between reported spatial and math difficulties	Impaired attentional search, linked to spatiotemporal function or executive control, may underlie observed under- versus over-counting errors

Continued

TABLE 1 Summary of Group-Level Behavioral and MLD Characteristics in Persons with Fragile X, Turner, or 22q DS–Cont'd

Characteristic	Features Shared Across Syndromes	Key Features That Vary Across Syndromes (Relative to Comparison Ss)		
		Fragile X Syndrome	**Turner Syndrome**	**22q11.2 DS**
Executive function	Aspects of EF, including working memory, differ relative to general population	Comparable response times (RTs) but lower accuracy on relatively easy EF tasks, and different brain activation patterns during tasks with high working memory loads; the latter suggests lower working memory threshold	Slower RTs across a wide range of tasks suggest processing-speed differences that are nevertheless not global. Fluency difficulties across tasks suggest disorganized retrieval strategies	Poor working memory observed but not found to account for profile of math achievement or numerical abilities
How is this syndrome a model of MLD?	Considered together, these syndromes illustrate that MLDs are heterogeneous	For some children with MLD, accuracy on rote skills may not reflect mastery of principles. Rather than relying only on verbal memory strengths as a strategy to overcome MLD, numerical and other foundational skills need support.	Slow but accurate performance. While fluency and other timed math test scores are low for age, the profile is unlike the typical fact retrieval errors' profile associated with retrieval difficulties. Does not seem linked to general processing-speed deficits. Potential model of atypical strategies used to compensate for atypical (not automatic) number processing. Verbal mediation may be such a strategy	The observed profile of impaired quantification (i.e., estimation) but preserved number processing may represent a distinct subtype of MLD

[a] *Summary is limited to studies reviewed in this chapter.*

fragile X, Turner, or 22q11.2DS demonstrate neither the exact same nor unique impairments in visuospatial and attention tasks. Each disorder is characterized by some impairment in attention, which suggests a lack of specificity in the effects that these impairments have on numerical processing.

Behavioral and structural neuroimaging research has demonstrated that disruption in the development of visuospatial processing may cascade into problems acquiring basic numerical and arithmetical abilities (Ansari & Karmiloff-Smith, 2002; Ashkenazi, Rosenberg-Lee, Metcalfe, Swigart, & Menon, 2013; Verdine, Irwin, Golinkoff, & Hirsh-Pasek, 2014). If the neural systems that underpin basic spatial representations are not well-established early in development, then the construction of more complex numerical cognition skills that depend on well-developed visuospatial cognitive competence, such as arithmetic, are likely to be compromised. Given that many cognitive processes interact when numerical processes are being learned and later used, it is no surprise that one or many impairments can manifest as impairments in higher-order skills subsumed under the domain of mathematical thinking and behavior. These difficulties with higher-order skills may reflect impairments in lower-order cognition.

What is unclear is not only the extent to which impairment in components of spatial attention and their interactions with visuospatial processing contribute to numerical processing deficits in fragile X, Turner, or 22q11.2, but also whether such impairments are uniform across these syndromes. Using an attention networks task, the common impairment in attention is differentiated both in comparison to typically developing age-matched children, but also across these three genetic disorders (Quintero, Beaton, et al., 2014). The Attention Network Test, developed by Posner and colleagues, tests the cognitive processes that comprise attention; these can be divided into three major subsystems, alerting, orienting, and executive control (Fan, McCandliss, Fossella, Flombaum, & Posner, 2005; Fan, McCandliss, Sommer, Raz, & Posner, 2002; Posner & Petersen, 1990). Where alerting and executive indices were similar among groups, performance on the orienting index was different across groups (only girls with fragile X syndrome struggle to appropriately orient their attention when presented valid spatial cues). Girls in all three syndrome groups did manifest impaired attention, but the exact manifestations were unique for each disorder.

Syndrome Models Contribute to Understanding Neuro-Correlates of MLD

In addition to providing a model of MLD, syndrome research can inform us of possible biological mechanisms that account for some of the variation seen in mathematics ability in the general population. This approach complements traditional studies of the genetics of MLD discussed by Petrill and Kovas (Chapter 11). For instance, fragile X and Turner syndromes inform us of potential roles of X chromosome genes on the development of mathematical thinking and outcomes, as indicated by associations between impairment severity and FMRP levels, for instance. These genetic contributions may explain some variation in mathematics, but of course, individuals within the populations studied will still differ from each other on mathematics and other cognitive behaviors. Not all individual profiles conform to a "typical" syndrome-specific phenotype. In fact, degree of such variability appears to differ between groups: girls with Turner syndrome have more specific effects (specific rather than global numerical impairment is reported) compared to girls with fragile X, who have more widespread difficulties with numerical and mathematics skills.

Likewise, similarities across these groups reveal important pathways to math achievement or difficulties. Functional neuroimaging studies point to remarkable overlap of brain regions associated with the processing of numbers, quantitative comparisons, and attention mechanisms, particularly in parietal regions (Göbel et al., 2001; Holloway & Ansari, 2010; Kadosh, Bien, & Sack, 2012; Shuman & Kanwisher, 2004). Research on each disorder suggests structural and functional brain abnormalities that are common in these populations; however, few studies have involved comparisons across genetic disorders using the same tasks to determine whether children with different disorders perform similarly in affected domains of function. Comparisons of structural MRI studies illustrate the complex interaction between genes and brain structure. Across studies of adults and children, many of the brain regions that differ between typically developing individuals and individuals with a specific disorder are unique to the particular disorder.

Regions with structural abnormalities across all three genetic disorder groups are the caudate nucleus, superior temporal gyrus, and cerebellar subregions (Bish et al., 2006; Chow et al., 2011; Gothelf, Penniman, Gu, Eliez, & Reiss, 2007; Gothelf et al., 2008; Kesler et al., 2003; Marzelli, Hoeft, Hong, & Reiss, 2011; Molko et al., 2004; Peng et al., 2013; Reiss et al., 1993, Reiss, Mazzocco, Greenlaw, Freund, & Ross, 1995; Sugama et al., 2000). The corpus callosum and parietal lobe are abnormal in 22q11.2DS and Turner syndrome, although different segments are affected in each group (Brown et al., 2004; Eliez, Schmitt, White, & Reiss, 2000; Fryer, Kwon, Eliez, & Reiss, 2003; Machado et al., 2007); fragile X and 22q11.2DS both differ from typically developing peers in the amygdala and enlarged lateral ventricles (Bearden et al., 2009; Kates, Miller, et al., 2006b; Lee et al., 2007). Data from functional neuroimaging in typically developing adults implicate subregions of the parietal lobe in numerical processing (Ansari, 2008) and of the anterior cingulate in executive control (Marrocco & Davidson, 1998). Though these regions have been associated with only two of the three genetic disorders, behavioral impairments have been noted in all three. Whether these similarities are a result of the genetic influence of each disorder or a result of similar life experiences, which can clearly drive maturation, is still unclear. (For a summary of biological and group-based neuro-functional characteristics of persons with fragile X, Turner, or 22q DS reviewed in this chapter, see Table 2.)

TABLE 2 Summary of Biological and Group-Based Neuro-Functional Characteristics of Persons with Fragile X, Turner, or 22q DS[a]

Characteristic	Features Shared Across Syndromes	Key Features That Vary Across Syndromes		
		Fragile X Syndrome	Turner Syndrome	22q11.2 DS
Biological underpinning	Known genetic diagnosis	X-linked single gene mutation (FMR1 gene)	Total or partial loss of one X chromosome	De novo microdeletion on the long (q) arm of chromosome 22
Gender differences in biological underpinnings	Within-diagnosis variability	Occurs in males and females. Males more significantly affected.	Diagnosis occurs only in females	Diagnosis occurs in males and females with equal incidence
Marker of variation	None	X inactivation, FMR1 protein expression	Karyotype variation	1.5- to 3-megabase deletion
Atypical brain structure relative to typical development[b]	Caudate nucleus (22q ↓, TS ↑, FXS ↑), superior temporal gyrus (22q ↓, TS ↑, FXS ↓), and cerebellar subregions (22q ↓, TS ↓, FXS ↓)	Lateral ventricles ↑ and amygdala ↓	Corpus callosum ↓and parietal lobe ↓	Lateral ventricles ↑, amygdala ↑, corpus callosum ↑and parietal lobe ↓

↓ indicates relative reduction in volume. ↑ indicates relative increase in volume.
[a] Summary is limited to studies reviewed in this chapter.
[b] For a direct comparison of white matter structure see Villalon-Reina, J., Jahanshad, N., Beaton, E., Toga, A. W., Thompson, P. M., & Simon, T. J. (2013). White matter microstructural abnormalities in girls with chromosome 22q11.2 deletion syndrome, Fragile X or Turner syndrome as evidenced by diffusion tensor imaging. NeuroImage, 81, 1-14.

The comparison of behavior and brain structure, between persons with or without known genetic disorders, is fundamentally limited because children with a neurodevelopmental disorder never experience the same developmental trajectory as their typical counterparts. This applies to genetic disorders or other neurodevelopmental disorders of unknown origin or that are not a consequence of a genetic abnormalities. They may start on a different path that then follows a developmental trajectory unique to that disorder. Thus, the functional associations between behavioral performance and brain structure observed in typically developing individuals might be quite different in individuals with a neurodevelopmental disorder. Accordingly, structural comparisons between individuals with neurodevelopmental disorders and typically developing individuals need to be tested for functional relevance (Johnson, Halit, Grice, & Karmiloff-Smith, 2002; Karmiloff-Smith, 2010). This concept of unique brain-behavior associations suggests that examination of the interaction between structural differences and functional differences in neurodevelopmental disorders, regardless of whether structural changes alone are observed, may be the most informative approach. These findings also argue for a developmental approach to studying syndromes or MLD.

CONCLUSIONS

Returning to the questions posed at the beginning of the chapter, we reflect upon the answers indicated by studies of genetic disorders in general, including studies of fragile X, Turner, and 22q11.2 deletion syndromes.

"How specific are difficulties in mathematics?"

Although math difficulties may sometimes be part of a constellation of mild or moderate cognitive impairment (so-called garden variety MLD described by Bartelet, Ansari, Vaessen, & Blomert, 2014; reviewed by Geary, 2011), many children experience specific mathematics disabilities despite otherwise intact intellectual ability. And even though the notion that math difficulties may be specific *or* part of global cognitive impairments is not particularly controversial, models of a range of manifestations are difficult to obtain from heterogeneous groups of children with MLD. What genetic syndrome research contributes to this dialogue is a set of models of potential underlying causes of MLD. When evaluated in the aggregate, studies of different syndromes illustrate variation in specificity, such that not all MLD profiles reflect difficulties with the same types of mathematics domains. This variation likely applies to children with MLD from the general population. Thus, genetic syndromes help demonstrate that IQ alone is neither the cause of MLD nor the sole source or variability in math performance—whether in syndrome groups or in the general population.

"Are mathematics disabilities linked primarily to deficits in numerical skills, such as those described by ansari, hyde & mou, and fias (chapters 2, 3, and 8)?

> *Alternatively, are the deficits underlying mld secondary to non-numerical competencies, such as in the memory and perceptual systems?"*

The evidence from research on these three common syndromes is a clear indication that the answer to *both* questions is yes. This means that the combined evidence from these bodies of research has important implications for the likelihood—or lack thereof—of a single core deficit for MLD. This evidence can be interpreted in the following ways: On the one hand, the combined evidence may differentiate individuals with MLD, such that numerical competencies drive the development of MLD for some individuals and nonnumerical competencies drive the development of MLD in another subset of individuals. On the other hand, both sources of variation may operate on the development of MLD in all individuals with this learning disability, albeit to different degrees. Thus, numerical and nonnumerical influences on mathematical thinking and on MLD may account for both within- and across-group variation reported for children with MLD and those other mathematics difficulties in the general population (e.g., Geary, Hoard, Nugent, & Bailey, 2012; Murphy et al., 2007).

Although the findings we have summarized to this point are not exhaustive overviews of the three syndrome phenotypes that we target (and they further omit additional informative phenotypes), they demonstrate the benefits of attention to cascading or otherwise dynamic relations among biological, cognitive, and behavioral contributions to the development of MLD. These interactions give rise to the possibility—indeed, the likelihood—that affirms both sources of influences on MLD, and emphasize developmental as well as individual differences.

> *"If underlying causes of mathematics disabilities vary among children with mld, do different etiologies lead to similar behavioral outcomes, or are the differences apparent at genetic, neural, and behavioral levels?"*

Despite hypothesized similarity in phenotypes of numerical impairment that stem from overlapping impairment in these cognitive domains, different profiles of mathematical strengths and weakness emerge, and these may reflect different cognitive profiles and pathways that lead to poor mathematics achievement.

As models of MLD, fragile X syndrome, Turner syndrome, and 22q11.2DS demonstrate the kind and degree of variability that can be expected if unique impairments can converge and result in a shared phenotype of numerical impairment. In research including all three populations, children with a genetic disorder exhibited impairments in the numerical and arithmetic domains, relative to their typically developing counterparts, but within the domains children with any one of these disorders were distinguishable from children with one of the other disorders. In light of these differences, the theory that a single cognitive impairment is common to each disorder is not supported, and adds to the growing body of evidence that MLD is heterogeneous. Syndrome research reveals why any attempt to identify a single core deficit for MLD will always be inconclusive.

ACKNOWLEDGMENTS

This chapter is an adaptation of earlier work by Mazzocco, M.M.M., Murphy, M.M., & McCloskey, M (2007), "The Contribution of Syndrome Research to Understanding Mathematical Learning Disability: The Case of Fragile X and Turner Syndromes," in D.B. Berch & M.M.M. Mazzocco (Eds), *Why Is Math So Hard For Some Children? The Nature and Origins of Mathematical Learning Difficulties and Disabilities* (pp. 173–193), Baltimore, MD., Paul H. Brookes Publishing Co., Inc. Adapted with permission.

REFERENCES

Abrams, M. T., Reiss, A. L., Freund, L. S., Baumgardner, T. L., Chase, G. A., & Denckla, M. B. (1994). Molecular-neurobehavioral associations in females with the fragile X full mutation. *American Journal of Medical Genetics*, *51*, 317–327.

Anderson, P., Anderson, V., Northam, E., & Taylor, H. G. (2000). Standardization of the Contingency Naming Test (CNT) for school-aged children: A measure of reactive flexibility. *Clinical Neuropsychological Assessment*, *1*, 247–273.

Angkustsiri, K., Goodin-Jones, B., Deprey, L., Brahmbhatt, K., Harris, S., & Simon, T. (2014). Social impairments in chromosome 22q11.2 deletion syndrome (22q11.2DS): Autism spectrum disorder or a different endophenotype? *Journal of Autism and Developmental Disorders*, *44*, 739–746.

Anobile, G., Cicchini, G. M., & Burr, D. C. (2012). Linear mapping of numbers onto space requires attention. *Cognition*, *122*(3), 454–459.

Ansari, D. (2008). Effects of development and enculturation on number representation in the brain. *Nature Reviews Neuroscience*, *9*(4), 278–291.

Ansari, D., Donlan, C., Thomas, M. S., Ewing, S. A., Peen, T., & Karmiloff-Smith, A. (2003). What makes counting count? Verbal and visuo-spatial contributions to typical and atypical number development. *Journal of Experimental Child Psychology*, *85*(1), 50–62.

Ansari, D., & Karmiloff-Smith, A. (2002). Atypical trajectories of number development: A neuroconstructivist perspective. *Trends in Cognitive Sciences*, *6*(12), 511–516.

Antshel, K. M., Conchelos, J., Lanzetta, G., Fremont, W., & Kates, W. R. (2005). Behavior and corpus callosum morphology relationships in velocardiofacial syndrome (22q11.2 deletion syndrome). *Psychiatry Research: Neuroimaging*, *138*(3), 235–245.

Ashkenazi, S., Rosenberg-Lee, M., Metcalfe, A. W., Swigart, A. G., & Menon, V. (2013). Visuo-spatial working memory is an important source of domain-general vulnerability in the development of arithmetic cognition. *Neuropsychologia*, *51*(11), 2305–2317.

Bailey, D. B., Jr., Hatton, D. D., & Skinner, M. (1998). Early developmental trajectories of males with fragile X syndrome. *American Journal of Mental Retardation*, *103*, 29–39.

Bartelet, D., Ansari, D., Vaessen, A., & Blomert, L. (2014). Cognitive subtypes of mathematics learning difficulties in primary education. *Research in Developmental Disabilities*, *35*(3), 657–670.

Bearden, C. E., van Erp, T. G., Dutton, R. A., Lee, A. D., Simon, T. J., Cannon, T. D., et al. (2009). Alterations in midline cortical thickness and gyrification patterns mapped in children with 22q11.2 deletions. *Cerebral Cortex*, *19*(1), 115–126.

Bearden, C. E., Woodin, M. F., Wang, P. P., Moss, E., McDonald-McGinn, D., Zackai, E., et al. (2001). The neurocognitive phenotype of the 22q11.2 deletion syndrome: Selective deficit in visual-spatial memory. *Journal of Clinical and Experimental Neuropsychology*, *23*(4), 447–464.

Bennetto, L., Pennington, B. F., Porter, D., Taylor, A. K., & Hagerman, R. J. (2001). Profile of cognitive functioning in women with the fragile X mutation. *Neuropsychology*, *15*, 290–299.

Bish, J. P., Pendyal, A., Ding, L., Ferrante, H., Nguyen, V., McDonald-McGinn, D., et al. (2006). Specific cerebellar reductions in children with chromosome 22q11.2 deletion syndrome. *Neuroscience Letters*, *399*(3), 245–248.

Botto, L. D., May, K., Fernhoff, P. M., Correa, A., Coleman, K., Rasmussen, S. A., et al. (2003). A population-based study of the 22q11.2 deletion: Phenotype, incidence, and contribution to major birth defects in the population. *Pediatrics*, *112*(1), 101–107.

Brannon, E. M., & Van de Walle, G. A. (2001). The development of ordinal numerical competence in young children. *Cognitive Psychology*, *43*(1), 53–81.

Brown, W. E., Kesler, S. R., Eliez, S., Warsofsky, I. S., Haberecht, M., & Reiss, A. L. (2004). A volumetric study of parietal lobe subregions in Turner syndrome. *Developmental Medicine & Child Neurology*, *46*(09), 607–609.

Bruandet, M., Molko, N., Cohen, L., & Dehaene, S. (2004). A cognitive characterization of dyscalculia in Turner syndrome. *Neuropsychologia*, *42*, 288–298.

Buchanan, L., Pavlovic, J., & Rovet, J. (1998). A reexamination of the visuospatial deficit in Turner syndrome: Contributions of working memory. *Developmental Neuropsychology*, *14*, 341–367.

Carvalho, M. R. S., Vianna, G., Oliveira, L.D.F.S., Costa, A. J., Pinheiro-Chagas, P., Sturzenecker, R., et al. (2014). Are 22q11.2 distal deletions associated with math difficulties? *American Journal of Medical Genetics Part A*, *164*(9), 2256–2262.

Chow, E. W., Ho, A., We, I. C., Voormolen, E. H., Crawley, A. P., & Bassett, A. S. (2011). Association of schizophrenia in 22q11.2 deletion syndrome and gray matter volumetric deficits in the superior temporal gyrus. *American Journal of Psychiatry*, *168*(5), 522–529.

Cornish, K. M., Levitas, A., & Sudhalter, V. (2007). Fragile X syndrome: The journey from genes to behavior. In M. M. M. Mazzocco & J. L. Ross (Eds.), *Neurogenetic developmental disorders: Manifestations and identification in childhood*. Cambridge, MA: MIT Press.

Cornish, K. M., Munir, F., & Cross, G. (1998). The nature of the spatial deficit in young females with Fragile-X syndrome: A neuropsychological and molecular perspective. *Neuropsychologia*, *36*, 1239–1246.

Cornish, K., Swainson, R., Cunnington, R., Wilding, J., Morris, P., & Jackson, G. (2004). Do women with fragile X syndrome have problems in switching attention: Preliminary findings from ERP and fMRI. *Brain and Cognition*, *54*, 235–239.

Crawford, D. C., Acuna, J. M., & Sherman, S. L. (2001). FMR1 and the fragile X syndrome: Human genome epidemiology review. *Genetics in Medicine*, *3*, 359–371.

Davenport, M., Hooper, S., & Zegar, M. (2007). Turner syndrome throughout childhood. In M. M. M. Mazzocco & J. L. Ross (Eds.), *Neurogenetic developmental disorders: Manifestations and identification in childhood*. Cambridge, MA: MIT Press.

De Smedt, B., Reynvoet, B., Swillen, A., Verschaffel, L., Boets, B., & Ghesquière, P. (2009). Basic number processing and difficulties in single-digit arithmetic: Evidence from Velo-Cardio-Facial Syndrome. *Cortex*, *45*(2), 177–188.

De Smedt, B., Swillen, A., Devriendt, K., Fryns, J. P., Verschaffel, L., Boets, B., et al. (2008). Cognitive correlates of mathematical disabilities in children with velo-cardio-facial syndrome. *Genetic Counseling*, *19*(1), 71–94.

De Smedt, B., Swillen, A., Devriendt, K., Fryns, J. P., Verschaffel, L., & Ghesquiere, P. (2006). Mathematical disabilities in young primary school children with velo-cardio-facial syndrome. *Genetic Counseling*, *17*(3), 259–280.

De Smedt, B., Swillen, A., Devriendt, K., Fryns, J. P., Verschaffel, L., & Ghesquière, P. (2007). Mathematical disabilities in children with velo-cardio-facial syndrome. *Neuropsychologia*, *45*(5), 885–895.

De Smedt, B., Swillen, A., Verschaffel, L., & Ghesquiere, P. (2009). Mathematical learning disabilities in children with 22q11.2 deletion syndrome: A review. *Developmental Disabilities Research Reviews*, *15*(1), 4–10.

Dehaene, S. (1999). Thinking: Behavioral and brain-imaging evidence. *Science*, *284*, 970. http://dx.doi.org/10.1126/science.284.5416.970.

Dehaene, S., Piazza, M., Pinel, P., & Cohen, L. (2003). Three parietal circuits for number processing. *Cognitive Neuropsychology*, *20*(3), 487–506.

Eliez, S., Barnea-Goraly, N., Schmitt, J. E., Liu, Y., & Reiss, A. L. (2002). Increased basal ganglia volumes in velo-cardio-facial syndrome (deletion 22q11.2). *Biological Psychiatry*, *52*(1), 68–70.

Eliez, S., Schmitt, J. E., White, C. D., & Reiss, A. L. (2000). Children and adolescents with velocardiofacial syndrome: A volumetric study. *American Journal of Psychiatry*, *157*, 409–415.

Fan, J., McCandliss, B. D., Fossella, J., Flombaum, J. I., & Posner, M. I. (2005). The activation of attentional networks. *NeuroImage*, *26*, 471–479.

Fan, J., McCandliss, B. D., Sommer, T., Raz, A., & Posner, M. I. (2002). Testing the efficiency and independence of attentional networks. *Journal of Cognitive Neuroscience*, *14*(3), 340–347.

Frith, U. (2001). What framework should we use for understanding developmental disorders? *Developmental Neuropsychology*, *20*, 555–563.

Fryer, S. L., Kwon, H., Eliez, S., & Reiss, A. L. (2003). Corpus callosum and posterior fossa development in monozygotic females: A morphometric MRI study of Turner syndrome. *Developmental Medicine & Child Neurology*, *45*(05), 320–324.

Geary, D. C. (1993). Mathematical disabilities: Cognitive, neuropsychological, and genetic components. *Psychological Bulletin*, *114*, 345–362.

Geary, D. C. (2011). Consequences, characteristics, and causes of mathematical learning disabilities and persistent low achievement in mathematics. *Journal of Developmental and Behavioral Pediatrics*, *32*(3), 250–263.

Geary, D. C., Berch, D. B., & Mann Koepke, K. (2015). *Evolutionary origins and early development of number processing*. San Diego, CA: Academic Press.

Geary, D. C., & Burlingham-Dubree, M. (1989). External validation of the strategy choice model for addition. *Journal of Experimental Child Psychology*, *47*(2), 175–192.

Geary, D. C., Hoard, M. K., Nugent, L., & Bailey, D. H. (2012). Mathematical cognition deficits in children with learning disabilities and persistent low achievement: A five-year prospective study. *Journal of Educational Psychology*, *104*(1), 206–223.

Göbel, S., Walsh, V., & Rushworth, M. F. S. (2001). The mental number line and the human angular gyrus. *NeuroImage*, *14*(6), 1278–1289.

Gothelf, D., Furfaro, J. A., Hoeft, F., Eckert, M. A., Hall, S. S., O'Hara, R., et al. (2008). Neuroanatomy of fragile X syndrome is associated with aberrant behavior and the fragile X mental retardation protein (FMRP). *Annals of Neurology*, *63*(1), 40–51.

Gothelf, D., Penniman, L., Gu, E., Eliez, S., & Reiss, A. L. (2007). Developmental trajectories of brain structure in adolescents with 22q11.2 deletion syndrome: A longitudinal study. *Schizophrenia Research*, *96*(1), 72–81.

Greenough, W. T., Klintsova, A. Y., Irwin, S. A., Galvez, R., Bates, K. E., & Weiler, I. J. (2001). Synaptic regulation of protein synthesis and the fragile X protein. *Proceedings of the National Academy of Sciences*, *98*(13), 7101–7106.

Gunderson, E. A., Ramirez, G., Beilock, S. L., & Levine, S. C. (2012). The relation between spatial skill and early number knowledge: The role of the linear number line. *Developmental Psychology*, *48*(5), 1229–1241.

Haberecht, M. F., Menon, V., Warsofsky, I. S., White, C. D., Dyer-Friedman, J., Glover, G. H., et al. (2001). Functional neuroanatomy of visuo-spatial working memory in Turner syndrome. *Human Brain Mapping*, *14*, 96–107.

Holloway, I. D., & Ansari, D. (2010). Developmental specialization in the right intraparietal sulcus for the abstract representation of numerical magnitude. *Journal of Cognitive Neuroscience*, *22*(11), 2627–2637.

Hooper, S. R., Curtiss, K., Schoch, K., Keshavan, M. S., Allen, A., & Shashi, V. (2013). A longitudinal examination of the psychoeducational, neurocognitive, and psychiatric functioning in children with 22q11.2 deletion syndrome. *Research in Developmental Disabilities*, *34*(5), 1758–1769.

Hwang, V. J., Maar, D., Regan, J., Angkustsiri, K., Simon, T. J., & Tassone, F. (2014). Mapping the deletion endpoints in individuals with 22q11.2 deletion syndrome by droplet digital PCR. *BMC Medical Genetics*, *15*(1), 106.

Jacobson, C., Shearer, J., Habel, A., Kane, F., Tsakanikos, E., & Kravariti, E. (2010). Core neuropsychological characteristics of children and adolescents with 22q11.2 deletion. *Journal of Intellectual Disability Research*, *54*(8), 701–713.

Johnson, M. H., Halit, H., Grice, S. J., & Karmiloff-Smith, A. (2002). Neuroimaging of typical and atypical development: A perspective from multiple levels of analysis. *Development and Psychopathology*, *14*(3), 521–536.

Kadosh, R. C., Bien, N., & Sack, A. T. (2012). Automatic and intentional number processing both rely on intact right parietal cortex: A combined fMRI and neuronavigated TMS study. *Frontiers in Human Neuroscience*, *6*, 2.

Karayiorgou, M., Morris, M. A., Morrow, B., Shprintzen, R. J., Goldberg, R., Borrow, J., et al. (1995). Schizophrenia susceptibility associated with interstitial deletions of chromosome 22q11. *Proceedings of the National Academy of Sciences*, *92*(17), 7612–7616.

Karmiloff-Smith, A. (2010). Neuroimaging of the developing brain: Taking "developing" seriously. *Human Brain Mapping*, *31*(6), 934–941.

Kates, W. R., Antshel, K. M., Abdulsabur, N., Colgan, D., Funke, B., Fremont, W., et al. (2006a). A gender-moderated effect of a functional COMT polymorphism on prefrontal brain morphology and function in velo-cardio-facial syndrome (22q11.2 deletion syndrome). *American Journal of Medical Genetics Part B: Neuropsychiatric Genetics*, *141*(3), 274–280.

Kates, W. R., Miller, A. M., Abdulsabar, N., Antshel, K. M., Conchelos, J., Fremont, W., et al. (2006b). Temporal lobe anatomy and psychiatric symptoms in velocardiofacial syndrome (22q11.2 deletion syndrome). *American Academy of Child and Adolescent Psychiatry*, *45*(5), 587–595.

Kaufman, E. L., Lord, M. W., Reese, T. W., & Volkmann, J. (1949). The discrimination of visual number. *American Journal of Psychology*, *62*(4), 498–525.

Kaufmann, L., Mazzocco, M. M., Dowker, A., von Aster, M., Göbel, S. M., Grabner, R. H., et al. (2013). Dyscalculia from a developmental and differential perspective. *Frontiers in Psychology*, *4*.

Kesler, S. R., Blasey, C. M., Brown, W. E., Yankowitz, J., Zeng, S. M., Bender, B. G., et al. (2003). Effects of X-monosomy and X-linked imprinting on superior temporal gyrus morphology in Turner syndrome. *Biological Psychiatry*, *54*(6), 636–646.

Kesler, S. R., Menon, V., & Reiss, A. L. (2006). Neurofunctional differences associated with arithmetic processing in Turner syndrome. *Cerebral Cortex*, *16*(6), 849–856.

Kirk, J. W., Mazzocco, M. M., & Kover, S. T. (2005). Assessing executive dysfunction in girls with fragile X or Turner syndrome using the Contingency Naming Test (CNT). *Developmental Neuropsychology*, *28*, 755–777.

Kobrynski, L. J., & Sullivan, K. E. (2007). Velocardiofacial syndrome, DiGeorge syndrome: The chromosome 22q11.2 deletion syndromes. *Lancet*, *370*(9596), 1443–1452.

Koontz, K. L., & Berch, D. B. (1996). Identifying simple numerical stimuli: Processing inefficiencies exhibited by arithmetic learning disabled children. *Mathematical Cognition*, *2*, 1–23.

Lasker, A. G., Mazzocco, M. M. M., & Zee, D. S. (2007). Ocular motor indicators of executive dysfunction in fragile X and Turner syndromes. *Brain and Cognition*, *63*, 203–220.

Lee, A. D., Leow, A. D., Lu, A., Reiss, A. L., Hall, S., Chiang, M.-C., et al. (2007). 3D pattern of brain abnormalities in fragile X syndrome visualized using tensor-based morphometry. *NeuroImage*, *34*(3), 924–938.

Li, Y., & Geary, D. C. (2013). Developmental gains in visuospatial memory predict gains in mathematics achievement. *PLoS ONE*, *8*(7), e70160. http://dx.doi.org/10.1371/journal.pone.0070160.

Lippe, B. (1991). Turner syndrome. *Endocrinology and Metabolism Clinics of North America*, *20*, 121–152.

Lyon, M. F. (1991). The quest for the X-inactivation centre. *Trends in Genetics*, *7*, 69–70.

Machado, A. M. C., Simon, T. J., Nguyen, V., McDonald-McGinn, D. M., Zackai, E. H., & Gee, J. C. (2007). Corpus callosum morphology and ventricular size in chromosome 22q11.2 deletion syndrome. *Brain Research*, *1131*(1), 197–210.

Marrocco, R. T., & Davidson, M. C. (1998). Neurochemistry of attention. In R. Parasuraman (Ed.), *The attentive brain* (pp. 35–50). Cambridge, MA: MIT Press.

Marzelli, M. J., Hoeft, F., Hong, D. S., & Reiss, A. L. (2011). Neuroanatomical spatial patterns in Turner syndrome. *NeuroImage*, *55*(2), 439–447.

Mazzocco, M. M. M. (1998). A process approach to describing mathematics difficulties in girls with Turner syndrome. *Pediatrics*, *102*, 492–496.

Mazzocco, M. M. (2001). Math learning disability and math LD subtypes: Evidence from studies of Turner syndrome, fragile X syndrome, and neurofibromatosis type 1. *Journal of Learning Disabilities*, *34*, 520–533.

Mazzocco, M. M. (2009). Mathematical learning disability in girls with Turner syndrome: A challenge to defining MLD and its subtypes. *Developmental Disabilities Research Reviews*, *15*(1), 35–44.

Mazzocco, M. M. M., Bhatia, N. S., & Lesniak-Karpiak, K. (2006). Visuospatial skills and their association with math performance in girls with fragile X or Turner syndrome. *Child Neuropsychology*, *12*, 87–110.

Mazzocco, M. M., Hagerman, R. J., Cronister-Silverman, A., & Pennington, B. F. (1992). Specific frontal lobe deficits among women with the fragile X gene. *Journal of the American Academy of Child and Adolescent Psychiatry*, *31*, 1141–1148.

Mazzocco, M. M. M., & Hanich, L. B. (2010). Math achievement, numerical processing, and executive functions in girls with Turner syndrome: Do girls with Turner syndrome have math learning disability? *Learning and Individual Differences*, *20*(2), 70–81.

Mazzocco, M. M. M., & McCloskey, M. (2005). Math performance in girls with Turner or fragile X syndrome. In J. Campbell (Ed.), *Handbook of Mathematical Cognition* (pp. 269–297). New York: Psychology Press.

Mazzocco, M. M., & Myers, G. F. (2003). Complexities in identifying and defining mathematics learning disability in the primary school-age years. *Annals of Dyslexia*, *53*(1), 218–253.

Mazzocco, M. M. M., Pennington, B. F., & Hagerman, R. J. (1993). The neurocognitive phenotype of female carriers of fragile X: Additional evidence for specificity. *Journal of Developmental and Behavioral Pediatrics*, *14*, 328–335.

Molko, N., Cachia, A., Riviere, D., Mangin, J. F., Bruandet, M., Le Bihan, D., et al. (2003). Functional and structural alterations of the intraparietal sulcus in a developmental dyscalculia of genetic origin. *Neuron*, *40*, 847–858.

Molko, N., Cachia, A., Riviere, D., Mangin, J. F., Bruandet, M., LeBihan, D., et al. (2004). Brain anatomy in Turner syndrome: Evidence for impaired social and spatial–numerical networks. *Cerebral Cortex*, *14*(8), 840–850.

Moss, E. M., Batshaw, M. L., Solot, C. B., Gerdes, M., McDonald-McGinn, D. M., Driscoll, D. A., et al. (1999). Psychoeducational profile of the 22q11.2 microdeletion: A complex pattern. *Journal of Pediatrics*, *134*(2), 193–198.

Murphy, M. M. (2009). A review of mathematical learning disabilities in children with fragile X syndrome. *Developmental Disabilities Research Reviews*, *15*, 21–27.

Murphy, M. M., & Mazzocco, M. M. (2008a). Rote numeric skills may mask underlying mathematical disabilities in girls with fragile X syndrome. *Developmental Neuropsychology*, *33*(3), 345–364.

Murphy, M. M., & Mazzocco, M. M. M. (2008b). Mathematics learning disability in girls with Turner syndrome or fragile X syndrome during late elementary school. *Journal of Learning Disabilities*, *41*, 29–46.

Murphy, M. M., & Mazzocco, M. M. (2009). The trajectory of mathematics skills and working memory thresholds in girls with fragile X syndrome. *Cognitive Development*, *24*(4), 430–449.

Murphy, M. M., Mazzocco, M. M., Gerner, G., & Henry, A. E. (2006). Mathematics learning disability in girls with Turner syndrome or fragile X syndrome. *Brain and Cognition*, *61*, 195–210.

Murphy, M. M., Mazzocco, M. M., Hanich, L. B., & Early, M. C. (2007). Cognitive characteristics of children with mathematics learning disability (MLD) vary as a function of the cutoff criterion used to define MLD. *Journal of Learning Disabilities*, *40*(5), 458–478.

Peng, D. X., Kelley, R. G., Quintin, E.-M., Raman, M., Thompson, P. M., & Reiss, A. L. (2013). Cognitive and behavioral correlates of caudate subregion shape variation in fragile X syndrome. *Human Brain Mapping*, *35*(6), 2861–2868.

Posner, M. I., & Petersen, S. E. (1990). The attention system of the human brain. *Annual Review if Neuroscience*, *1990*(13), 25–42.

Price, G. R., & Ansari, D. (2012). Developmental dyscalculia: A case for neuroscience in education. *BJEP Monograph Series II, Number 8-Educational Neuroscience*, *1*(1), 45–62.

Quintero, A. I. (2014). *Common and specific impairments in behavior and brain circuitry: Comparison of 22q11.2 deletion, fragile X, and Turner syndromes*. Unpublished doctoral dissertation Davis, CA: University of California Davis.

Quintero, A. I., Beaton, E. A., Harvey, D. J., Ross, J. L., & Simon, T. J. (2014). Common and specific impairments in attention functioning in girls with chromosome 22q11.2 deletion, fragile X or Turner syndromes. *Journal of Neurodevelopmental Disorders*, *6*(1), 5.

Quintero, A. I., Harvey, D. L., & Simon, T. J. (2014). *Characterization of visuospatial attention and numerical processing in female children with 22q11.2 Deletion syndrome, fragile X syndrome, or Turner syndrome*. Poster presented at the second annual math cognition conference, May 19, 2014, Arlington, VA.

Reiss, A. L., Freund, L., Plotnick, L., Baumgardner, T., Green, K., Sozer, A. C., et al. (1993). The effects of X monosomy on brain development: Monozygotic twins discordant for Turner's syndrome. *Annals of Neurology*, *34*(1), 95–107.

Reiss, A. L., Mazzocco, M. M. M., Greenlaw, R., Freund, L. S., & Ross, J. L. (1995). Neurodevelopmental effects of X monosomy: A volumetric imaging study. *Annals of Neurology*, *38*(5), 731–738.

Rivera, S. M., Menon, V., White, C. D., Glaser, B., & Reiss, A. L. (2002). Functional brain activation during arithmetic processing in females with fragile X Syndrome is related to FMR1 protein expression. *Human Brain Mapping*, *16*, 206–218.

Robin, N. H., & Shprintzen, R. J. (2005). Defining the clinical spectrum of deletion 22q11.2. *Journal of Pediatrics*, *147*(1), 90–96.

Ross, J. L., Roeltgen, D., Kushner, H., Wei, F., & Zinn, A. R. (2000). The Turner syndrome-associated neurocognitive phenotype maps to distal Xp. *American Journal of Human Genetics*, *67*, 672–681.

Rousseau, F., Heitz, D., Tarleton, J., MacPherson, J., Malmgren, H., Dahl, N., et al. (1994). A multicenter study on genotype-phenotype correlations in the fragile X syndrome, using direct diagnosis with probe StB12.3: The first 2,253 cases. *American Journal of Human Genetics*, *55*, 225–237.

Rovet, J. F. (1993). The psychoeducational characteristics of children with Turner syndrome. *Journal of Learning Disabilities*, *26*, 333–341.

Rovet, J., & Netley, C. (1982). Processing deficits in Turner's syndrome. *Developmental Psychology*, *18*, 77–94.

Rovet, J. F., Szekely, C., & Hockenberry, M. N. (1994). Specific arithmetic calculation deficits in children with Turner syndrome. *Journal of Clinical and Experimental Neuropsychology*, *16*, 820–839.

Sävendahl, L., & Davenport, M. L. (2000). Delayed diagnoses of Turner's syndrome: Proposed guidelines for change. *Journal of Pediatrics*, *137*, 455–459.

Shprintzen, R. J. (2000). Velo-cardio-facial syndrome: A distinctive behavioral phenotype. *Mental Retardation and Developmental Disabilities Research Reviews*, *6*(2), 142–147.

Shprintzen, R. J. (2008). Velo-cardio-facial syndrome: 30 years of study. *Developmental Disabilities Research Reviews*, *14*(1), 3–10.

Shuman, M., & Kanwisher, N. G. (2004). Numerical magnitude in the human parietal lobe; tests of representational generality and domain specificity. *Neuron*, *44*(3), 557–569.

Simon, T. (1999). The foundations of numerical thinking in a brain without numbers. *Trends in Cognitive Sciences*, *3*(10), 363–365.

Simon, T. J. (2008). A new account of the neurocognitive foundations of impairments in space, time and number processing in children with chromosome 22q11.2 deletion syndrome. *Developmental Disabilities Research Reviews*, *14*(1), 52–58.

Simon, T. J., Bearden, C. E., Mc-Ginn, D. M., & Zackai, E. H. (2005). Visuospatial and numerical cognitive deficits in children with chromosome 22q11.2 deletion syndrome. *Cortex*, *41*(2), 145–155.

Simon, T. J., Burg, M., & Gothelf, D. (2007). Chromosome 22q11.2 deletion syndrome. In M. M. M. Mazzocco & J. L. Ross (Eds.), *Neurogenetic developmental disorders: Manifestations and identification in childhood*. Cambridge, MA: MIT Press.

Simon, T. J., & Luck, S. J. (2011). Attentional impairments in children with chromosome 22q11.2 deletion syndrome. In M. I. Posner (Ed.), *Cognitive neuroscience of attention* (pp. 441–453). New York, NY: Guilford Press.

Simon, T. J., Takarae, Y., DeBoer, T., McDonald-McGinn, D. M., Zackai, E. H., & Ross, J. L. (2008). Overlapping numerical cognition impairments in children with chromosome 22q11.2 deletion or Turner syndromes. *Neuropsychologia*, *46*(1), 82–94.

Simon, T. J., & Vaishnavi, S. (1996). Subitizing and counting depend on different attentional mechanisms: Evidence from visual enumeration in afterimages. *Perception and Psychophysics*, *58*, 915–926.

Sobin, C., Kiley-Brabeck, K., & Karayiorgou, M. (2005). Lower prepulse inhibition in children with the 22q11 deletion syndrome. *American Journal of Psychiatry*, *162*(6), 1090.

Sobin, C., Kiley-Brabeck, K., Monk, S. H., Khuri, J., & Karayiorgou, M. (2009). Sex differences in the behavior of children with the 22q11 deletion syndrome. *Psychiatry Research: Neuroimaging*, *166*(1), 24–34.

Sugama, S., Bingham, P. M., Wang, P. P., Moss, E. M., Kobayashi, H., & Eto, Y. (2000). Morphometry of the head of the caudate nucleus in patients with velocardiofacial syndrome (del 22q11.2). *Acta Paediatrica*, *89*(5), 546–549.

Swillen, A., Devriendt, K., Legius, E., Eyskens, B., Dumoulin, M., Gewillig, M., et al. (1997). Intelligence and psychosocial adjustment in velocardiofacial syndrome: A study of 37 children and adolescents with VCFS. *Journal of Medical Genetics*, *34*(6), 453–458.

Swillen, A., Devriendt, K., Vantrappen, G., Vogels, A., Rommel, N., Fryns, J. P., et al. (1998). Familial deletions of chromosome 22q11: The Leuven experience. *American Journal of Medical Genetics*, *80*(5), 531–532.

Tamm, L., Menon, V., & Reiss, A. L. (2003). Abnormal prefrontal cortex function during response inhibition in Turner syndrome: Functional magnetic resonance imaging evidence. *Biological Psychiatry*, *53*, 107–111.

Temple, C. M. (2002). Oral fluency and narrative production in children with Turner's syndrome. *Neuropsychologia*, *40*, 1419–1427.

Temple, C. M., & Carney, R. A. (1993). Intellectual functioning of children with Turner syndrome: A comparison of behavioural phenotypes. *Developmental Medicine and Child Neurology*, *35*, 691–698.

Temple, C. M., & Carney, R. A. (1995). Patterns of spatial functioning in Turner's syndrome. *Cortex*, *31*, 109–118.

Temple, C. M., & Carney, R. (1996). Reading skills in children with Turner's syndrome: An analysis of hyperplexia. *Cortex*, *32*, 335–345.

Temple, C. M., Carney, R. A., & Mullarkey, S. (1996). Frontal lobe function and executive skills in children with Turner's syndrome. *Developmental Neuropsychology*, *12*, 343–363.

Temple, C. M., & Marriott, A. J. (1998). Arithmetical ability and disability in Turner's syndrome: A cognitive neuropsychological analysis. *Developmental Neuropsychology*, *14*, 47–67.

Temple, C. M., & Shephard, E. E. (2012). Exceptional lexical skills but executive language deficits in school starters and young adults with Turners syndrome: Implications for X chromosome effects on brain function. *Brain and Language*, *120*(3), 345–359.

Verdine, B. N., Irwin, C. M., Golinkoff, R. M., & Hirsh-Pasek, K. (2014). Contributions of executive function and spatial skills to preschool mathematics achievement. *Journal of Experimental Child Psychology*, *126*, 37–51.

Verkerk, A. J., Pieretti, M., Sutcliffe, J. S., Fu, Y. H., Kuhl, D. P., Pizzuti, A., et al. (1991). Identification of a gene (FMR-1) containing a CGG repeat coincident with a breakpoint cluster region exhibiting length variation in fragile X syndrome. *Cell*, *65*, 905–914.

Waber, D. P. (1979). Neuropsychological aspects of Turner's syndrome. *Developmental Medicine and Child Neurology*, *21*, 58–70.

Walter, E., Mazaika, P. K., & Reiss, A. L. (2009). Insights into brain development from neurogenetic syndromes: Evidence from fragile X syndrome, Williams syndrome, Turner syndrome and velocardiofacial syndrome. *Neuroscience*, *164*(1), 257–271.

Wong, L. M., Riggins, T., Harvey, D., Cabaral, M., & Simon, T. J. (2014). Children with chromosome 22q11.2 deletion syndrome exhibit impaired spatial working memory. *American Journal on Intellectual and Developmental Disabilities*, *119*(2), 115–132.

Index

Note: Page numbers followed by *f* indicate figures and *t* indicate tables.

C

D

G

J

K

L

M

O

P

Q

T

FSC
www.fsc.org
MIX
Papier | Fördert
gute Waldnutzung
FSC® C083411

Kolibri 360 GmbH
Ferdinand-Jühlke-Straße 7
99095 Erfurt, Deutschland
produktsicherheit@kolibri360.de